Human Genetics

Human

Edward Novitski
UNIVERSITY OF OREGON, EUGENE

Genetics

Macmillan Publishing Co., Inc.
NEW YORK

Collier Macmillan Publishers
LONDON

To Curt Stern

Teacher, Colleague, and Friend

Macmillan Publishing Co., Inc.
866 Third Avenue, New York, New York 10022

Collier Macmillan Canada, Ltd.

Library of Congress Cataloging in Publication Data

Novitski, Edward.
  Human genetics.

  Includes bibliographies and index.
  1. Human genetics. I. Title. [DNLM: 1. Genetics, Human. QH431 N944h]
QH431.N68   573.2′1   75-43704
ISBN 0-02-388550-5

Printing: 1 2 3 4 5 6 7 8   Year: 7 8 9 0 1 2 3

Preface

There are so many excellent texts in the field of human genetics that it may seem superfluous to add still another to the growing list. Nevertheless, it seems that for one reason or another the presently existing books do not serve the purpose for which this one is intended. In writing this text I have followed an outline which I have found to be attractive to beginning students who simply want to learn a bit about the genetic make-up of the human being at the same time that they gain some understanding of the biological problems now becoming very important to modern society. For this reason subjects that demand mathematical skills or basic chemical knowledge have been minimized and other aspects, such as radiation effects, transplantation, racial differences, counseling, and genetic engineering, are covered more extensively. Insofar as it has been reasonable to do so, the human problem has been presented first, followed by the explanation for its biological basis (rather than the reverse, which is the common textbook approach).

Although the presentation of the material is elementary and not intended for professional consumption, there is probably more in the book than can be easily covered in an elementary course during a single term. For this reason an attempt has been made to arrange the chapters so that certain of the more advanced ones may be omitted without detracting from the sense of fairly complete coverage of the field that the student has when he finishes a text, but lacks when he is able to get through only the first half or three quarters during the course. The beginning chapters are essential, as are the concluding ones, but, if necessary, a number of chapters in the middle and later parts of the book may be omitted without detracting from the completeness of the presentation. The choice of omission will, of course, depend entirely on the interests and the level of the students in the course.

Similarly, there is no sanctity in the order in which subjects are taken up. The chapter on the chemistry of the gene, for example, has been put near the end of the book because students with a limited background in chemistry are often discouraged by a premature exposure to biochemical concepts. As a result, there are certain deficiencies in the treatment of other subjects (the nature of mutations, for instance). The instructor may prefer to introduce this material earlier if his audience is sufficiently sophisticated, and then to take advantage of this biochemical foundation in dealing with material in subsequent chapters.

Because the emphasis of the book is on the social implications of human genetics, and the bulk of the presentation is dedicated to that end, much material considered standard in elementary texts has been dealt with lightly or not at all. Probabilities, with permutations and combinations, are mentioned only in passing, and statistical tests not at all. The text does not start in the traditional way with the history of the field; this has been introduced only sketchily at appropriate points throughout the book. In addition, the very many contributions that have come from the study of the genetics of lower forms have been minimized on the grounds that these are more properly covered in a course in elementary general genetics. The selection of certain subjects and omission of others have been a highly subjective procedure and it is not likely that any instructor would agree entirely with my choices. However, there is no reason why the instructor should not elaborate on certain points, and introduce new ones; in fact, there is every reason for the instructor to do so. In addition, the student should be encouraged to do outside reading from time to time with some direction from the instructor. There is a very good argument for exposing the beginning student to bona fide scientific presentations at an early stage in his development, in contrast to the more general, often highly colored, and even sometimes emotional or sensationalized accounts that too often form part of the student's general reading.

In a work like this, where many of the points are controversial, there is bound to be some difference of opinion concerning the relative weight to be devoted to certain kinds of arguments (the discussion of nuclear energy is a good example). I have tried to be factual, objective, and quantitative, in an effort to keep the discussion on a scientific level. Many readers will disagree with my approach, on one side or the other, but disagreement itself should be encouraged as the basis for a continuing rational debate on these important problems. The student must always be encouraged to present his own point of view, supported by such arguments as he considers to be particularly cogent, especially when his view is a minority, or unpopular, one.

Finally, it must be acknowledged, regretfully, that this work, like all the others, must in some respects be in error. The least of these are the trivial errors attendant upon the mechanics of publication. Some important errors originate in the ignorance of the author, and these will, in time, be corrected. Probably the most important ones come from the fast pace of developments in this field—what seems self-evidently true today will become the myth of tomorrow.

Acknowledgments

The bulk of the contributions of others are acknowledged at that point in the text where they are used; I trust that those acknowledgments will be regarded as a personal note of thanks from me. The excellence of the plates used in illustrating mitosis and meiosis is evidence of the extraordinary skill of Drs. A. Bajer for mitosis and J. Kezer for meiosis, both of the Biology Department, University of Oregon. Others who have made additional substantial contributions, both materially and conceptually, include my good colleagues at the University of Oregon Medical School, Drs. F. Hecht, R. Koler, E. Louvrien, E. Magennis, B. McCaw, and H. Wyandt; Dr. J. Hall of the University of Washington Medical School; Professor C. E. Ford, University of Oxford, Sir William Dunn School of Pathology; Dr. V. Bond, Oak Ridge National Laboratory; Drs. L. Bernini, P. Meera Khan, A. T. Natarajan, K. Sankaranarayanan, and E. Vogel, University of Leiden; and Professor E. Polani, Guy's Hospital Medical School, London. Much of the work was done during a sabbatical leave at the University of Leiden in the laboratories of Professor Frits Sobels, whose help is greatly appreciated.

Special thanks are due Professors Jean W. MacCluer, Pennsylvania State University, and Allen Burdick, Purdue University, who subjected the entire manuscript to close scrutiny, primarily from the teaching viewpoint, and Peter Pearson, University of Leiden, who painstakingly went through the manuscript as a professional cytogeneticist, all of whom made large numbers of invaluable corrections and suggestions. Finally, it is not possible for me to express appropriately the contribution of the next generation, Charles, Peggy, and Paul, as well as my contemporary, Esther, in working tirelessly in an effort to maintain the accuracy of the manuscript at a high level, on the one hand, and in insisting on simplicity and comprehensibility, on the other.

vii

Contents

Detailed Contents

1

Some Common Human Syndromes

Variability in the Human Species

If all human beings were exactly alike, the science of human genetics would not exist. The variability that exists from one individual to the next, and from one group of people to the next, makes it possible to study the basic mechanisms responsible for the transmission of characteristics from parents to offspring.

The study of genetics demands the use of inherited differences, characters determined by the basic cell components, and this implies a study of as wide a variety of human types as possible. All of us carry heritable differences; perhaps sex is the best known of the differences in the normal range. All persons also carry genetic traits that would, under other circumstances, lead to clear abnormalities, but, since these are not usually expressed, most people consider themselves to be normal. In a smaller number of humans these departures from the normal do express themselves, and it is to these persons that we must go for much of our detailed knowledge of human genetics.

The contribution that a specific human type may make to our knowledge may be quite independent of its overall frequency in the population. In many cases, we shall find it profitable to consider in detail human characteristics that occur with frequencies of less than one per many thousands. In many cases a single pedigree, or even a single individual, may clarify a puzzling problem. The selection of specific human examples in the study of human genetics is determined neither by their appearance nor by their frequency, but by the extent to which they make a contribution to our knowledge of humans.

Although we must necessarily examine other human beings (and ourselves as well) for evidence of genetic factors responsible for the human make-up, we must never lose sight of the fact that these variations are minor differences superimposed on the common base of biological identity that characterizes all mankind. If the study of human genetics teaches us anything at all, it is that the line separating those within the range of "normality" and those outside that range is exceedingly fine and that the circumstances determining one or the other are entirely a matter of chance.

In many cases it will be necessary for us to look at pictures of individuals with "defects" of varying degrees of severity. It is a natural human reaction for us to feel discomfort at the sight of another human being who is less perfect than we consider ourselves to be. This uneasiness may be manifest by a refusal to admit that such cases exist, or by a preference that they be kept out of sight and not intrude on our sensibilities. Although such an attitude may appear considerate at first sight, perhaps an even more humane approach is to realize, first, that a person's genetic make-up is never his own responsibility and, second, that only by scrutinizing other people's constitutions most minutely and, where possible, relating their characteristics to identifiable biological phenomena can we approach an understanding of the biological nature of man.

Down's Syndrome

Out of every 600 births there appears one mentally defective child with a set of distinctive characteristics adding up to the diagnosis of *mongolism*. These characteristics, together referred to as a *syndrome*, include *epicanthus*, an unusual form of the upper eyelid that superficially gives the affected person an Oriental appearance (Figure 1-1). This syndrome was first recognized as a distinct medical entity by Langdon Down, who pointed out in 1866 that if two such affected but unrelated children were placed side by side it would be difficult to say that they were not of the same parents. In recognition of Down's early observations, this defect is often referred to as *Down's syndrome* in preference to the common designation "mongolian idiot," which carries unfortunate and quite inaccurate racial implications. About 10 years after Down's original description of this syndrome, it was observed that its incidence of occurrence was much higher when the mother was older than average. The differences in the eyefolds of an affected person and those of an Oriental person and of a non-Oriental are shown in Figure 1-2.

PHYSICAL CHARACTERISTICS. Some of the features of Down's syndrome are evident in Figure 1-3. Along with the typical eyefold, affected persons have *hypotonia*, a relaxed condition of the muscles. Because the mouth cavity does not grow to normal size, the tongue tends to protrude and the mouth may be held slightly open. The tongue itself is often large with a furrowed surface. In addition, the affected person's stature tends to be short, the nose is broad and somewhat depressed, and the hands are clumsy and stumpy. These individuals very often have a *simian crease*, a single fold across the palm of the hand instead of the several palmar creases found in normal individuals (Figure 1-4). They also have irregular, abnormal sets of teeth. The growth of pubic hair is sparse, as is

Figure 1-1 A young girl with Down's syndrome, with the characteristic eyefolds that superficially resemble those of the Oriental race. (Courtesy of the Medical Genetics Group, University of Oregon Medical School.)

the beard in males. The iris of the eye shows a speckling sometimes referred to as *Brushfield spots.*

Other physical characteristics include a markedly increased frequency of heart malformations and a low resistance to illness, especially respiratory infections. In the past, affected persons tended to die fairly young from infections, but with the advent of antibiotics this problem has largely been removed. The physical features characterizing this syndrome are listed in Table 1-1. No one affected individual will show all of these characteristics, but enough of them will appear in any one person with this syndrome that a diagnosis based on physical inspection alone is likely to be correct.

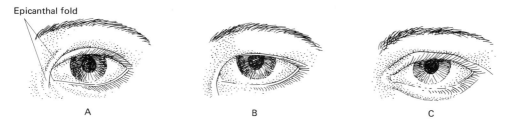

Epicanthal fold

A   B   C

Figure 1-2 A detailed drawing showing the basic difference between the eyefold of a patient with Down's syndrome (*A*) and the eyefold of an Oriental eye (*B*) and a non-Oriental eye (*C*). (O. Solnitsky, *Georgetown Med. Bull.,* **15**:276, 1962.)

Figure 1-3 Other manifestations of Down's syndrome, including the oversized tongue with partially open mouth and relaxed muscles. (Courtesy of E. W. Lovrien, University of Oregon Medical School.)

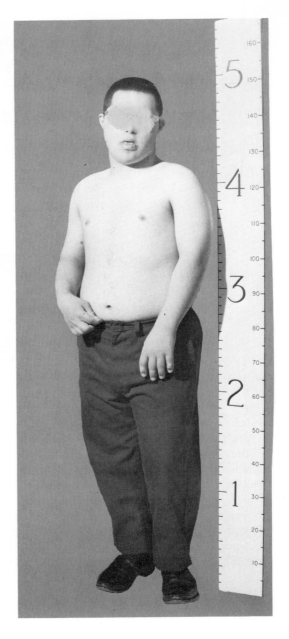

PSYCHOLOGICAL CHARACTERISTICS. Persons with Down's syndrome tend to have quite subnormal intelligence. At young ages their mental ability is generally that of a child a quarter to a half their age, that is, their intelligence quotient, or IQ, ranges from 25 to 50. Their capabilities level off early in life so that even in middle age they appear childish. The distribution of IQs in Down's patients runs from 25 to 49 in 60 to 70 percent of the cases, classifying them as "severely" retarded, and from 0 to 24 in 25 to 40 percent of the cases, putting them in the category of the "profoundly" retarded. On the other hand, in a small number of

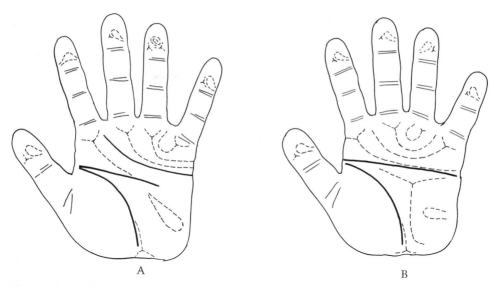

A   B

Figure 1-4 The palmar crease of a normal person (*A*) and of a person with Down's syndrome (*B*). (From R. Turpin and J. Lejeune, *Les Chromosomes Humains*, Gauthier-Villars, Paris, 1965.)

cases their intelligence may approach, or even slightly exceed normality (including one case in which an apparently affected individual became an admiral in the Royal Navy!).

About 1 in 25 can read with some understanding, and 1 in 50 can write. In general, affected children can learn simple tasks and can speak with a limited vocabulary. With proper guidance and early training, their range of capabilities can be extended considerably but will not usually extend into the normal range. They are good mimics and may have a well-developed sense of rhythm. They

Description of Symptom	Number Showing Symptom/ Number Checked	Percent Showing Symptom
Abundant neck skin	48/51	94
Mouth corners turned downward	42/50	84
General hypotonia	28/34	82
Flat face	39/49	80
At least one malformed ear	39/50	78
Epicanthus in at least one eye	38/50	76
Gap between first and second toes	35/52	67
Tongue protruding	32/51	63
Head circumference at birth not exceeding 32 cm	20/47	43
Simian crease in at least one hand	22/52	42

Table 1-1 A list of the most accurate clinical symptoms for diagnosing Down's syndrome. (J. Wahrman and K. Fried, The Jerusalem Prospective Newborn Survey of Mongolism, *Ann. N.Y. Acad. Sci.*, **171**:341–60, 1970.)

enjoy singing and dancing and are particularly partial to watching television. Although they have a reputation for being good-natured and sociable in general, some affected individuals are reported to be as unpleasant as normal children can be.

Among the inmates of institutions for the mentally retarded, Down's syndrome is found in about one case in ten and is the most frequent single diagnosis of the severely mentally retarded. In the general population it characterizes about 30 percent of all severely retarded children in the United States and Western Europe.

The Human Chromosomes

To see the basic difference between the make-up of a person with Down's syndrome and that of a normal one, we must look at the body cells of each. If white blood cells are taken from an individual, grown in culture for several days in a special medium, and then stained with a suitable dye, some of the cells will be at the proper stage of cell division to show distinct bodies called *chromosomes*. A typical set of chromosomes from a normal human cell as seen under the microscope is found in Figure 1-5.

THE KARYOTYPE. In order to examine the chromosomes more closely it is convenient to photograph the cell, cut out the chromosomes individually, and place them in order according to size and staining pattern. When this is done for normal individuals, we find that there are 46 chromosomes; of these, 44 can be assigned to seven major size-groups (labeled A to G) and in each group similarly stained chromosomes can be arranged in pairs. This standard arrangement is called a *karyotype* (Figure 1-6). Using special staining techniques it is possible to bring out variation in staining density along the length of the chromosome,

Figure 1-5 The chromosomes of a normal human female. In this case the chromosomes have been stained with an ordinary dye that simply darkens the chromosome. (Courtesy of R. E. Magenis, University of Oregon Medical School.)

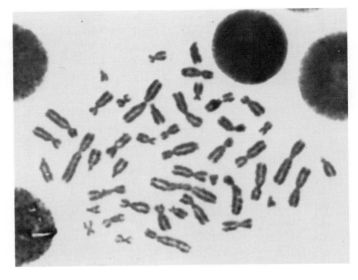

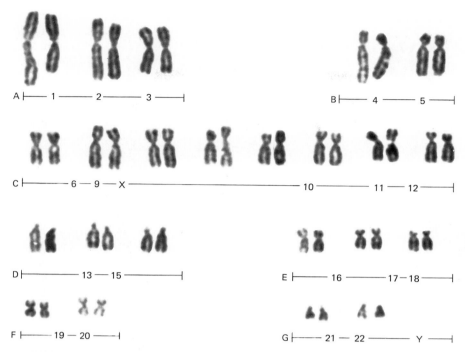

Figure 1-6 The chromosomes of Figure 1-5 arranged by pairs according to size and shape in a standard karyotype. (Courtesy of R. E. Magenis, University of Oregon Medical School.)

distinctly different for each chromosome, and when this is done, the chromosomes can be unambiguously assigned a specific number from 1 to 22.

However, 22 pairs of chromosomes make a total of 44, not 46. What about the other two chromosomes? In the cells of just about half of all humans, each chromosome has a duplicate of approximately the same size and shape, making a total of 23 pairs. Such cells come from females (Figure 1-6). In the other half of the cases, cells from males, the chromosomes do not occur as 23 similar pairs; one of the large chromosomes found in the female cell is replaced in the male by a smaller chromosome, the *Y-chromosome* (Figure 1-16, p. 17). The chromosome found twice in the female but only once in the male is the *X-chromosome*. The X- and the Y-chromosomes are referred to as the *sex chromosomes*; the other 22 pairs are *autosomes*.

When we examine the karyotypes of individuals with Down's syndrome, we discover that in 97 percent of the cases, instead of the normal 46 chromosomes (22 pairs of autosomes plus 2 sex chromosomes), the total count comes to 47, the extra chromosome being an additional autosome 21. Because there are three chromosomes 21 characteristic of Down's syndrome, it is often referred to as *trisomy-21* (Figure 1-7). In the remaining 3 percent of the Down's syndrome cases, the additional chromosome 21 material is present, but may be less obvious, being attached to one of the other chromosomes in a way to be described in a later chapter.

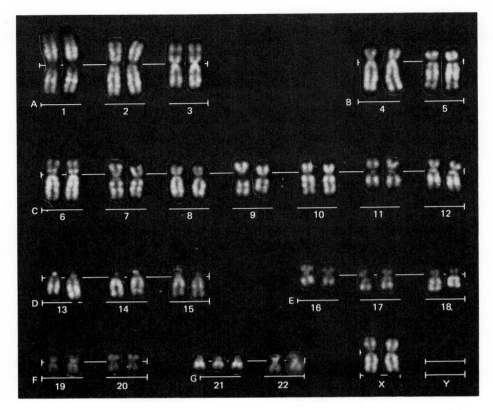

Figure 1-7 The karyotype of a person with Down's syndrome. The presence of three number 21 chromosomes (in the G group), instead of the normal two, has led to the alternative name "trisomy-21." (Courtesy of D. Hepburn, University of Oregon Medical School.)

Turner's Syndrome

On rare occasions a young girl at the onset of puberty fails to commence menstruation, and at the same time her *secondary sexual characteristics* (breasts, pubic hair) fail to develop normally. When the anxious parents take the girl to a physician, he may identify her difficulty as *Turner's syndrome*, which afflicts about one female in 2,000.

PHYSICAL CHARACTERISTICS.   Affected women of this type are almost normal in appearance. In addition to the failure of the secondary sexual characteristics to develop, such women tend to have short stature, widely spaced nipples, and a characteristic shape of chest called a "shield" chest (Figure 1-8). At birth there may be folds of skin at the neck; an extreme example of such pathological "webbing" is shown in Figure 1-9. This problem may be relieved by surgery, but older children will usually manifest a flaring of the neck as it joins the shoulders. In addition, the hairline at the nape of the neck is unusually low (Figure 1-10).

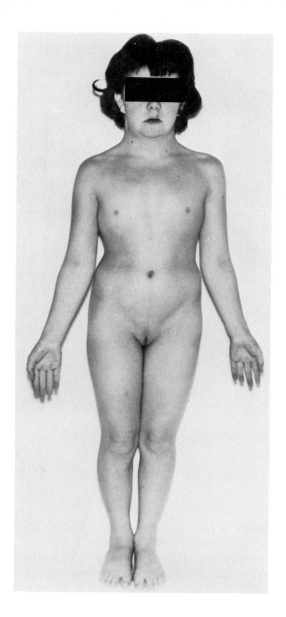

Figure 1-8 An eighteen-year-old with infantile secondary sexual characteristics and other manifestations of Turner's syndrome noted in the text. (Courtesy of P. Pearson, University of Leiden.)

In most cases the slowness of growth is noticed from birth. Administration of female hormones (*estrogens*) may cause the secondary sexual characteristics to develop and additional growth may be promoted, but the infertility of the affected individual cannot be corrected, having been irrevocably impaired prior to birth.

PSYCHOLOGICAL CHARACTERISTICS. Women affected with Turner's syndrome are quite normal in intelligence and may be rather bright; if they have any academic weakness it is likely to be in mathematics. This is probably related to

Figure 1-9 A case of pathological webbing of the neck sometimes seen as one manifestation of Turner's syndrome.

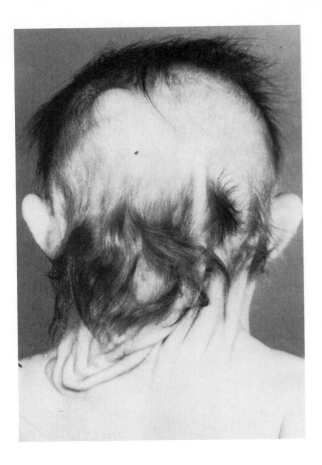

the fact that, as a group, they do appear to have one interesting psychological peculiarity—faulty spatial perception. This is illustrated in Figure 1-11.

Probably because of the diminished amount of sex hormones present, these individuals may be characterized by strong inhibitions and an absence of aggressive impulses and initiative. Interest in sex (*libido*) is usually reduced, but can be increased with estrogen treatment.

KARYOTYPE. A karyotype of the chomosomes of a woman affected with Turner's syndrome usually reveals a conspicuous deficiency: instead of 46 chromosomes, she has only 45. The missing chromosome is an X. Thus instead of having two X-chromosomes she has only one and her condition may be referred to as *XO* to distinguish it from that of the normal female, *XX*. Corresponding to the word trisomy used when there are three instead of the usual two chromosomes of a kind present, the word *monosomy* is used to indicate that only one of the normal two is present; in this case the karyotype may be called monosomy-X (Figure 1-12).

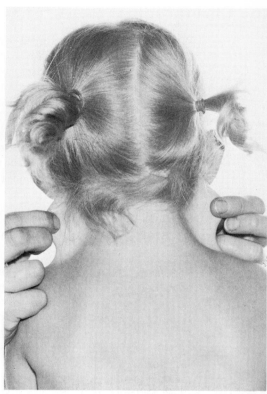

Figure 1-10 The low hairline at the base of the neck and the looseness of the skin characteristic of Turner's syndrome.

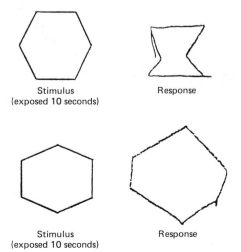

Stimulus
(exposed 10 seconds)

Response

Stimulus
(exposed 10 seconds)

Response

Figure 1-11 Two figures from a visual retention test, showing how a 10-year-old girl with Turner's syndrome reproduced the figures at the left after viewing them for 10 sec. (J. Money, in *Endocrinology and Human Behavior*, ed. by R. P. Michael, Oxford University Press, New York, 1968.)

The Chromosomal Basis for Sex Determination

We know from our examination of karyotypes of normal males and females that the female has two X-chromosomes, whereas the male has only one. But the male has a Y-chromosome which the female lacks. Is a woman female because she has two X's and a man male because he has only one X? Or is a man male because he has a Y and a woman female because she does not? The genetic constitution of the female with Turner's syndrome gives us a partial answer to this question. Since she has only one X-chromosome, like a normal male, it would appear that it is the additional presence of the Y-chromosome in the male that is ordinarily responsible for the difference between the sexes, not the number of X-chromosomes. The next syndrome to be described verifies this conclusion in another way.

Klinefelter's Syndrome

PHYSICAL CHARACTERISTICS. In about one of every 500 male births, there appears a boy who seems quite normal until the onset of puberty. At this time he may show partial breast development and other less obvious female characteristics, a slightly altered set of bodily proportions with somewhat narrow shoulders and larger hips than are typical of the normal male (Figure 1-13). Such persons may be taller than average. Generally there is a deficiency of the male hormone level, resulting in sparse and female distribution of pubic hair, small testicles, and a somewhat high-pitched voice. This person may appear to be a perfectly normal male and marry but will have no children. About 5 percent of all men who appear at fertility clinics prove to have this syndrome.

PSYCHOLOGICAL CHARACTERISTICS. In common with most cases of abnormal chromosome complements (with the exception of Turner's syndrome), these individuals tend to show a slightly depressed intelligence, with an average IQ of the order of 90. About one person in a hundred institutionalized for mental retardation has Klinefelter's syndrome.

KARYOTYPE. In Figure 1-14 we see the karyotype typical of a male with Klinefelter's syndrome. These men have two X-chromosomes as well as a Y, that is, they are *XXY*. This information confirms our earlier conclusion based on Turner's syndrome: it is the presence of the Y-chromosome that is characteristic of maleness, not the single X-chromosome. More than one X will not necessarily direct development toward femaleness, since the presence of a Y along with two X's will cause the individual to develop into a male.

The XYY Male

Figure 1-15 shows an interesting set of twins. These two boys, nonidentical twins, are grossly different in physical characteristics, one being excessively tall. Chromosome analyses show that the taller twin has two Y-chromosomes. In general, about two thirds of all affected individuals are over 6 feet tall, and about six out

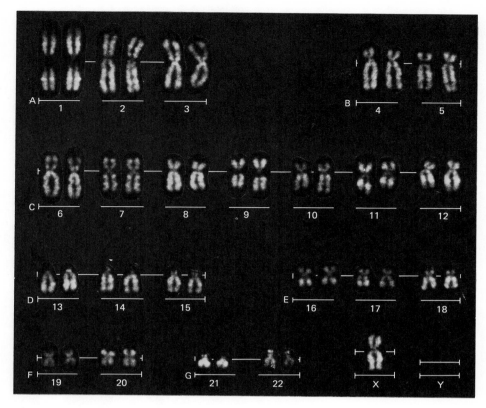

Figure 1-12 The karyotype of a Turner's syndrome patient. Note that there is only one X-chromosome and no Y. (Courtesy of G. Prescott, University of Oregon Medical School.)

of seven have an IQ of less than 100. In some instances, individuals with this chromosome anomaly appear to be quite normal in every respect.

BEHAVIORAL TRAITS.   It was pointed out by Jacobs and coworkers in 1965 that an unduly high proportion of males found in penal institutions are XYY. One such study in England showed that although only one in four hundred males in the general population was of this type, in a few prisons the frequency went as high as one in 25. These studies indicated that such individuals might be innately more aggressive than normal and therefore more likely to run afoul of the law. Others have suggested that effects of tallness and a slower mentality might combine in early childhood to promote the development of an aggressive personality that would then lead secondarily to behavioral problems. Further, because many cases are known of XYY males who are of normal height, intelligence, and behavior, the question of whether men with this chromosome constitution are necessarily characterized by innate behavioral disturbances that lead to antisocial acts is still considered a matter of debate by some. The fact remains, however, that there is a higher percentage of such men in maximum security prisons than in the population at large and the preponderance of

Figure 1-13 A male with Klinefelter's syndrome, showing partial breast development. (Courtesy of P. Pearson, University of Leiden.)

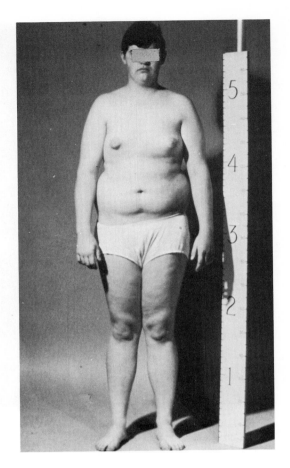

opinion among those medically trained who work with such cases is that their personality disturbances, when present, exceed any reasonable expectation based on the combination of their greater height and lesser intelligence.

This condition has been of great interest from the legal standpoint, because in a few cases (the first occurring in France and subsequent ones in Australia and Los Angeles) men accused of murder who proved to be XYY used their abnormal chromosome constitution in their defense, claiming that this abnormality led them to commit crimes over which they, as individuals, had little control. Although this consideration has not led to an acquittal of any XYY individuals, in several cases such persons have had their sentences reduced on the grounds that their chromosome complement, for which they could hardly be held responsible, was in part the cause of their antisocial behavior.

The data are fairly limited at the present time, but the evidence for a higher than expected proportion of XYY males in penal institutions comes from studies of incarcerated Whites and does not seem to be true for Blacks or the Orientals. Perhaps the basic frequency of the XYY condition is different in the different

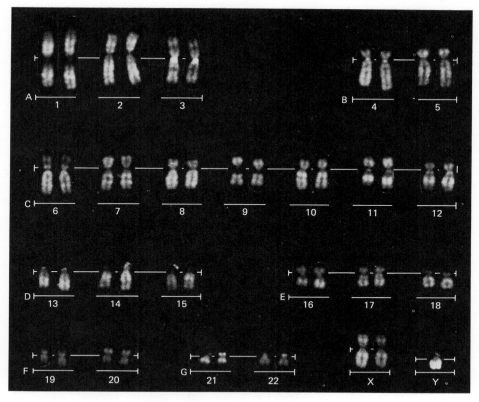

Figure 1-14 The karyotype of a male with Klinefelter's syndrome showing the presence of two X-chromosomes as well as a Y-chromosome. (Courtesy of B. McCaw, University of Oregon Medical School.)

races. Studies made to determine whether institutionalized XYY men have been predominantly in the lower socioeconomic classes show that in fact they come from a wide spectrum of middle-class backgrounds. The exact psychology of the XYY male is still a matter of controversy, and it continues to be one of the more interesting abnormalities because of the legal and ethical issues involved.

Questions Raised by These Unusual Chromosomal Types

A study of the abnormal chromosome complements associated with the syndromes described in this chapter points up the importance of the normal chromosome set in determining normal development. Abnormalities of chromosome constitution can be responsible for deviation from normality in physical, mental, and behavioral traits: this implies that an understanding of the normal human make-up depends on a greater appreciation of the role our chromosomes play.

Figure 1-15 Two nonidentical twin brothers, one XY and the other XYY, showing the increased height often associated with the presence of an extra Y-chromosome. (Courtesy of P. Pearson, University of Leiden.)

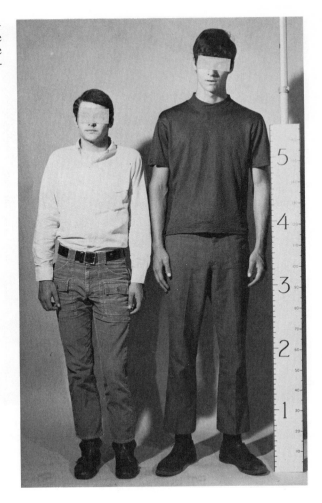

It would be possible to expand our list of genetically informative syndromes to include some caused by simple genetic factors, rather than chromosome additions or subtractions, or to describe some syndromes that appear as birth defects after exposure of the mother to harmful chemicals or to diseases. However, such discussion must be deferred; our description of these four chromosomal syndromes has raised more questions than it has answered. What about anomalies involving other chromosomes? What happens when a developing individual has, say, an extra chromosome 2 or 9? What about the autosomal monosomies, that is, those cases in which a person might have only one chromosome of the usual autosomal pair?

There are even more fundamental and more interesting questions we can ask. What is the basic mechanism by which these syndromes are produced? Is there anything that can be done to decrease the frequency with which they occur in the population? Is there any way that a family can safeguard against having such an abnormal child? Will an affected person, if fertile, produce similarly abnormal children?

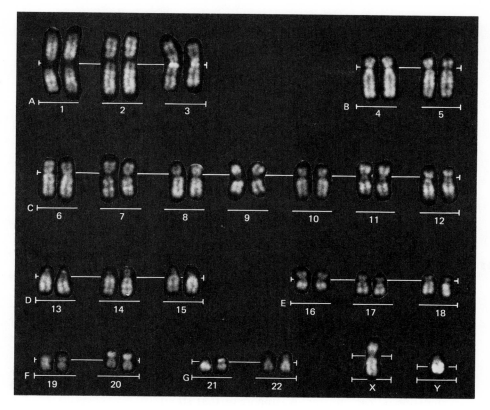

Figure 1-16 A karyotype of a normal male cell, with one Y-chromosome and one X-chromosome. In this case the chromosomes have been stained with a special dye called quinacrine that causes different parts of each chromosome to fluoresce differently when exposed to ultraviolet light. (Courtesy of F. Hecht, University of Oregon Medical School.)

As a first step in attempting to answer these questions, we will consider in some detail the mechanism by which the chromosomes are distributed during the process of cell division, or *mitosis*.

References

ALEXANDER, D., H. T. WALKER, and J. MONEY. 1964. Studies in direction sense. I. Turner's syndrome. *Arch. Gen. Psychiatry,* **10**:337–39.

APGAR, V., ed. 1970. Down's syndrome (mongolism). *Ann. N.Y. Acad. Sci.,* **171**:303–688.

BECKWITH, J., and J. KING. 1974. The XYY syndrome: a dangerous myth. *New Sci.,* **64**:474–76.

FORD, C. E., K. W. JONES, P. E. POLANI, J. C. DE ALMEIDA, and J. H. BRIGGS. 1959. A sex chromosome anomaly in a case of gonadal dysgenesis. *Lancet,* **1**:711.

GERMAN, J. 1970. Studying human chromosomes today. *Am. Sci.*, **58**:182.

HOOK, E. B. 1973. Behavioral implications of the human XYY genotype. *Science*, **179**:139–50.

JACOBS, P. A., and J. A. STRONG. 1959. A case of human intersexuality having a possible XXY sex-determining mechanism. *Nature*, **183**:302.

JACOBS, P. A., et al. 1968. Chromosome studies on men in a maximum security hospital. *Ann. Hum. Genet.*, **31**:339.

MONEY, J., and D. ALEXANDER. 1966. Turner's syndrome: further demonstration of the presence of specific cognitional deficiencies. *J. Med. Genet.*, **3**:47–8.

PENROSE, L. S., and G. F. SMITH. 1966. *Down's Anomaly.* Boston: Little, Brown.

SHAFFER, J. W. 1962. A specific cognitive deficit observed in gonadal aplasia (Turner's syndrome). *J. Clin. Psychol.*, **18**:403–6.

WAHRMAN, J., and K. FRIED. 1970. The Jerusalem prospective newborn survey of mongolism. *Ann. N.Y. Acad. Sci.*, **171**:341–60.

Questions

The purpose of the questions at the end of each chapter is to help the student master the concepts given in the chapter; for this reason they do not always refer to the specific factual material that is presented. Very often the question will appear to be satisfied with a simple yes or no answer; in such cases, a further explanation or elaboration of the answer is appropriate. Also, answers to many of the questions will involve statements of opinion or belief, and the student is encouraged to uphold his particular point of view, supported by as many factual arguments as he can muster in support of that stand.

Useful terms: syndrome, chromosome, karyotype, sex chromosome, autosome, trisomy, monosomy, mitosis.

1. Why is it necessary to make use of examples of human characteristics that depart from the average in some way in the study of human genetics?
2. Does the frequency with which an unusual human type appears bear any relation to the amount of information that it gives us about the human make-up?
3. How frequent is Down's syndrome among newborn? About how many such affected children are born per month in your city?
4. Can an obstetrician recognize a Down's syndrome infant at birth? By what characteristics? What might he do next to confirm his preliminary diagnosis?
5. What are the arguments for and against institutionalizing a Down's syndrome child? Might this depend on the degree of mental retardation?
6. What proportion of inmates of institutions for the mentally retarded have this syndrome?
7. What are the major differences in the manifestations of Turner's and Down's syndromes?
8. Humans ordinarily have 46 chromosomes. How many chromosomes would be found in a karyotype of the very rare combination, in the same person, of Turner's and Down's syndromes?

9. What does Turner's syndrome tell us about the chromosomal basis for sex? How is this confirmed by Klinefelter's syndrome?
10. If a male has an extra Y-chromosome, how is he likely to differ from the average? Why is this condition of legal interest?
11. How are the 23 pairs of chromosomes arranged in a karyotype? Is the chromosome pair numbered 2 larger or smaller than the pair numbered 9?
12. In theory, how many different monosomic types of humans could occur? How many trisomic types? Can you suggest any reason why out of all these possibilities the Down's, Turner's, and Klinefelter's syndromes were selected for the introductory discussion?
13. Discuss the relevance of the following conditions to the concepts discussed in this chapter: Down's syndrome, Turner's syndrome, Klinefelter's syndrome, XYY composition.

2

Cell Division

The Cell: The Basic Unit of Structure

All living organisms, with the exception of the minute viruses, are made up of units called cells, and all of the components of the organism are made of combinations of cells or of their products. The cells of a mature animal are specialized in their work and they have a form to carry out their function. A nerve cell that transmits information from the tip of a limb to the central nervous system may be several feet long, whereas a red blood corpuscle, which must carry oxygen through the smallest capillaries of the body to every permanently fixed cell, has a microscopic ovoid structure that allows it free passage.

All cells come from the division of a parent cell into two daughter cells. Every mature person started initially as a single cell, a fertilized egg, which, by mitosis, became two, and the two became four, and, as the total number grew, the cells gradually became different in structure in different parts of the developing organism until the mass of cells took on the appearance of a small child. The human body, as it is finally constituted, seems so complex that it is sometimes difficult for the nonbiologist to keep in mind that the entire body is made up only of minute cells and, to a small extent, their immediate products (hair, nails, bone).

THE IMPORTANCE OF THE GENETIC MATERIAL.   The internal machinery of the cell is a complicated assembly of chemical systems (Figure 2-1) for carrying on a certain number of basic functions at the same time that the cell performs some

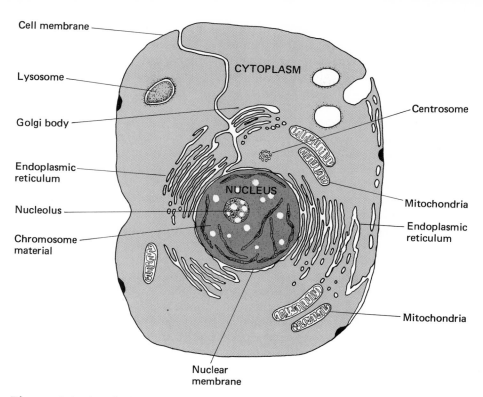

Cell membrane

CYTOPLASM

Lysosome

Centrosome

Golgi body

Endoplasmic
reticulum

Mitochondria

Nucleolus

NUCLEUS

Endoplasmic
reticulum

Chromosome
material

Mitochondria

Nuclear
membrane

Figure 2-1   An idealized representation of a "typical" cell. The synthetic activities of the cell are carried out in an intricate structural complex, including specialized bodies indicated as mitochondria, endoplasmic reticulum, Golgi bodies, and lysosomes, located in the cytoplasm. However, our attention will be focused primarily on the nucleus and on the genetic contents carried in the chromosomes that provide the instructions for the operation of the cell. (From J. Brachet, *The Living Cell*, © September 1961 by Scientific American, Inc. All rights reserved.)

special task, depending on the tissue in which it is found, whether it be muscular, nervous, dermal, or digestive. Whatever the assigned job, the directions are given by the *genes* that are located in the chromosomes of the cell. In this book we will be concerned primarily with the manner of transmission of the genes and chromosomes from one generation to the next, or, in other words, with inheritance, and secondarily with the functioning of the genes as they determine the characteristics of the final product, the fully developed organism. In many cases our attention will be directed to the chromosomes, which, as aggregates of genes, affect many human characteristics; in other cases it will be more appropriate to look at the specific gene responsible for the production of a particular characteristic.

Research over the past few decades has produced information of fundamental importance regarding chemical interactions within the human cell, but it is important to realize that present knowledge of the details of cellular mechanisms is quite incomplete. Despite newspaper reports on the great biological breakthroughs of the last several years, and the imminent possibility of "creating life," that possibility is far from being realized, and it will be many years, if ever, until man can synthesize even the most primitive kind of cell. However, this does not prevent us from making observations on the relationship between the final product, the mature individual, and the genes that are fundamentally responsible for the final form of the organism.

THE AMOUNT OF GENETIC MATERIAL IN THE CELL.   The genetic material of the chromosomes is in the form of very long thin molecules of *deoxyribonucleic acid*, called *DNA*, whose structure is of fundamental importance and will be discussed in detail later. The length of threadlike DNA in each human cell cannot be measured directly, but techniques exist for determining the total quantity of DNA present in an individual nucleus. This turns out to be approximately 20,000 times more than the amount of DNA in a virus particle. By electron microscopy the virus DNA has been shown to have a length of 60 μm (60 millionths of a meter). Since virus and human DNA are similar in basic structure, we can calculate the length of DNA in the nucleus of a human cell as 60 μm times 20,000, or 120 cm, which amount to about 4 ft.

This length must be packed into the cell nucleus, which is possible only because the thread is so thin that the total volume the thread occupies is much less than the size of the nucleus. When a cell divides to produce two new cells, the genetic material must reproduce itself exactly and partition itself precisely into the two daughter cells so that each of them receives exactly the same genetic material present in the original cell.

This is an incredibly difficult problem. Suppose that you were present at the Creation and were asked to help design a system with the following properties: that a cell of microscopic size must contain a thin thread several feet long; that this thread must be able to replicate itself exactly; and that, when this minute cell divides into two, each of the products must receive precisely one of the two long threads. This problem would baffle the most skillful and imaginative engineer. Yet this feat is performed thousands of times a second in every human, with such accuracy that only infrequently can we detect any kind of error. This problem is solved by the cell during the process of nuclear division, or mitosis.

The problem of handling the long length of genetic material within the cell nucleus is helped by its being divided into a number of individual pieces, the chromosomes. As we have seen, there are 46 of these in the body cells of humans. In this chapter we shall see how this large amount of genetic material is moved from one cell into two daughter cells and how identical sets of genetic material are delivered to the two daughter cells.

The Rate of Cell Division

INTERPHASE, THE "RESTING STAGE."   Between cycles of division, the cell goes into the stage of *interphase*, in which the chromosomes are contained within a nucleus as quite diffuse and inconspicuous bodies. The chromosomes are uncoiled into long threads and are engaged in synthetic activity. The name "resting stage" has historically been used for this period to emphasize the absence of cell division but is a complete misnomer from the standpoint of cell activity, since this is the time when the cell does its main work, whatever that may be.

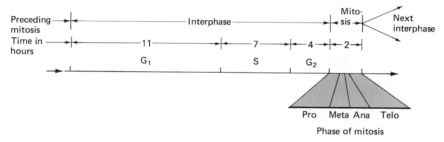

Figure 2-2 The time scale of events in hours showing the relatively long duration of the G_1, S, and G_2 stages of interphase, and the speed of the other mitotic stages, indicated as *Pro, Meta, Ana,* and *Telo* for prophase, metaphase, anaphase, and telophase respectively.

TIME SPENT IN INTERPHASE.   The cells of some tissues, such as those of the blood system and the epithelium of the intestinal tract, undergo rapid turnover with an accompanying high rate of division. Then there are other cells that do not divide at all after embryonic development, remaining in a permanent interphase condition. This is true for nerve cells. Within those cells in the process of division, an important event takes place during each interphase—the replication of the genetic material. In actively dividing human blood cells grown in culture, the whole process of cell division may average about 24 hr, of which interphase takes up 22. The chromosome replication (or synthetic) period, the S period, occupies about 7 hr, being preceded by a nonreplicating period of interphase (G_1) of 11 hr and followed by a period of completed replication (G_2) of about 4 hr. Mitosis, or nuclear division, itself actually occupies only a very small part of the total cycle, in this case, only about 2 hr (Figure 2-2).

The Details of Mitosis

SELECTION OF APPROPRIATE MATERIAL.   Since our main interest in this book is the human being, it might naturally be assumed that the best material for a demonstration of mitosis would come from humans. The fact is, however, that the phenomenon of mitosis is found in all plants and animals, with only minor variations, so that we are free to choose an organism which shows the chromosomes clearly as they go through cell division, with the understanding that the essential features of mitosis will be the same for man.

An unusually favorable species to demonstrate mitosis is a plant with very large and clear chromosomes, the blood lily from South Africa, *Haemanthus.* The cells being observed are those from tissue called the *endosperm* found in the developing seed. These cells can be removed and observed in the living state with a special optical system as they go through the complete cycle of mitosis. The first visible evidence that a cell is about to undergo mitosis and produce two identical daughter cells comes about when certain changes take place in and around the interphase nucleus. Starting with this next stage, called *prophase,* we will follow the details of mitosis.

Figure 2-3 Early prophase, showing the nucleoli within the nucleus. (Figures 2-3 through 2-14 are reproduced courtesy of Andrew S. Bajer, Department of Biology, University of Oregon.)

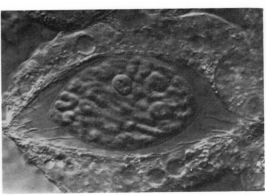

Figure 2-4 Late prophase stage with the appearance of the spindle.

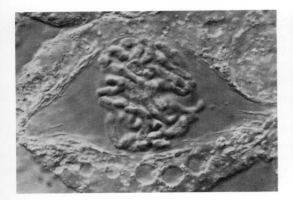

Figure 2-5 Beginning of metaphase, with the breakdown of the nuclear membrane.

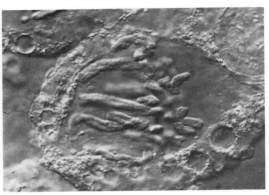

Figure 2-6 Metaphase movement of chromosomes toward the cell's equator with the duplicated structure of the chromosomes clearly visible.

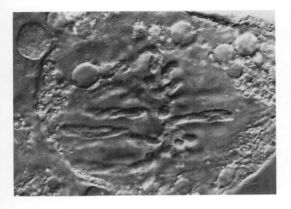

Figure 2-7 Metaphase positioning of chromosomes.

Figure 2-8 Late metaphase stage.

PROPHASE. At prophase the nucleus is clearly visible; within it are two conspicuous structures called *nucleoli* (Figure 2-3). The chromosomes, closely packed together within the nucleus, are so intimately associated that they are indistinguishable as separate entities. Previously, in interphase, the chromosomes existed as elongated, exceedingly thin strands, and these strands, in this early prophase stage, are arranged into compact coils so that they appear thicker and shorter than at interphase. Outside the nuclear envelope clear spaces are present. These indicate the formation of a special cellular mechanism, the *spindle*, which will eventually be responsible for moving the chromosomes to the daughter cells. A later prophase stage is shown in Figure 2-4. The nucleoli and the nucleus are still present but the clear zone is becoming more elongate, taking on the shape of a spindle. In *Haemanthus* endosperm, prophase lasts about 3 or 4 hr.

METAPHASE. In Figure 2-5 the nucleoli are disintegrating, the clear zone is becoming increasingly spindle-shaped, and the envelope around the nucleus no longer encases the nuclear contents. Studies with the electron microscope have shown that the nuclear envelope at this stage of mitosis, called *metaphase*, breaks into small pieces which then mostly disintegrate, a process taking only about 2 min. With the breakdown of the nuclear envelope, the chromosomes begin movements that will take them to the cell's equator. Those movements are going on in Figure 2-6. The chromosome arms are in focus only on one side of this cell—if one could focus on what appears to be a clear region on the right, he would see chromosome arms similar to those present at the left.

The fact that each chromosome consists of two strands is clearly evident in Figure 2-6. Chromosome duplication occurred back in the interphase stage preceding this mitosis, but, in living *Haemanthus* endosperm, it is only now that the two-strand structure of each chromosome can be easily seen. The two strands making up a chromosome are called *chromatids:* thus each of the chromosomes that are present in Figures 2-3, 2-4, 2-5, and 2-6 consists of two identical chromatids, as a consequence of interphase duplication.

In Figure 2-7 the chromosomes have almost achieved their final metaphase positions at the equator of the cell. In this photomicrograph, as in Figures 2-6 and 2-8, the chromosome arms in one half of the cell are out of the plane of focus. The spindle is now fully developed, its fibers radiating out from the poles at either end of the cell. It consists of long organic molecules that have formed filamentous tubular elements called microtubules which individually are beneath the resolution of the light microscope. The spindle fibers that are seen at relatively low magnifications consist of bundles of microtubules that have reached a dimension great enough for observation with light microscopy. To execute the complex movements of mitosis, each chromatid must make contact with the spindle microtubules at a particular specialized region of its axis, the *centromere*. It is the centromeres of the chromosomes that move to the equatorial position within the cell and at that position they oscillate to and fro, suggesting pulling forces that are acting first toward one pole of the spindle and then toward the opposite pole. These movements are believed to be associated with dynamic changes in the spindle microtubules.

The cell of Figure 2-8 is at late metaphase. Because we are examining a living cell, the chromosome arms in one half of the cell in Figure 2-8 are out of focus,

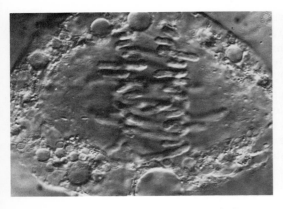

Figure 2-9 Anaphase separation of the two chromatids making up each chromosome.

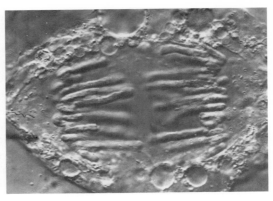

Figure 2-10 Further movements of chromosomes during anaphase.

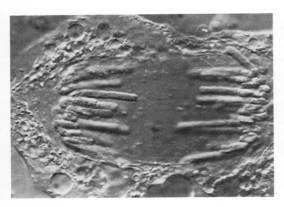

Figure 2-11 Completion of anaphase.

Figure 2-12 Beginning of telophase.

Figure 2-13 Midtelophase, with the chromosomes becoming more tightly packed.

Figure 2-14 Later telophase and the appearance of the characteristic features of the new interphase.

as a consequence of its thickness. The apparently clear area to the right is actually occupied by chromosome arms similar to those seen to the left. To visualize these metaphase chromosomes as they are in the living cell, we must think of the chromosome centromeres responding to the pulling forces of the attached microtubules so as to produce a restless oscillatory movement of the chromosomes back and forth across the cell equator.

ANAPHASE.   In Figure 2-9 the two chromatids of which each chromosome is composed have been separated from each other, producing two sets of identical chromosomes. This is the *anaphase* of mitosis. About 2 hr is required by the cell from the beginning of metaphase until the start of the anaphase movements, which are completed in 35 to 45 min. Each chromosome now consists of only a single chromatid. Figures 2-9, 2-10, and 2-11 are a sequence showing a cell advancing from the middle of anaphase in Figure 2-9 to its completion in Figure 2-11. The centromeres lead in the movements of anaphase, apparently in response to pulling forces administered by way of the spindle microtubules. It is a curious fact that such a simple and universal phenomenon as anaphase movement of the chromosomes during mitosis still defies complete explanation. Much research effort is spent in this area and many rather advanced theories have been proposed. For our purpose it is sufficient to regard the system as one like a rope and pulley, with the rope (the microtubules) pulling the chromosomes toward one end of the cell or the other.

TELOPHASE.   Anaphase has been completed in the cell illustrated by Figure 2-12 and the events of *telophase* are now beginning. The spindle fibers are disintegrating and the chromosomes are clumping together. In Figure 2-13 the telophase clumping of chromosomes has progressed to the point at which individual chromosomes no longer can be clearly distinguished. The single cell is forming two new ones. In the late telophase stage of Figure 2-14, the tightly packed gyres of the chromosome strands have relaxed, producing long, greatly extended chromosomes. The two new nuclei are now surrounded by their nuclear envelopes, the nucleoli are reappearing, and the division into two cells is complete. At this point the two daughter cells go into interphase and the mitotic cycle has been completed.

From the foregoing we can readily visualize the process of mitosis: At interphase, the long chromosomes replicate themselves exactly. At prophase, they shorten and thicken, and move to the center of the cell, where they are found at metaphase. At anaphase, the replicated chromosomes separate from each other so that each of the two products of mitosis will have precisely the same genetic make-up.

The Analysis of Abnormal Chromosome Behavior

SHORTCUTS IN CHROMOSOME REPRESENTATION.   Since many of the analyses depend on a visualization of chromosome form, or behavior, during cell division, drawings of chromosomes are indispensable in illustrating important points. It is almost never necessary to draw, laboriously, all 23 pairs of chromosomes found in the normal cell; only as many chromosomes are depicted as are necessary to illustrate the particular point being made. Very often only a few or just one pair of chromosomes are shown in a cell and the student must imagine the presence of the rest of the complement, behaving normally. Similarly, there is usually no point in drawing a chromosome very precisely; a single line will generally do quite well. A centromere may, or may not, be indicated on the chromosome, depending on its relevance to the issue under consideration. Two homologues may be included in a figure if the relationship of the two is important; otherwise only one need be shown.

ABNORMAL CHROMOSOME BEHAVIOR IN MITOSIS.   While we may marvel at the accuracy with which the chromosomes are reproduced at mitosis and partitioned to the daughter cells at each division, it is not surprising that on occasion a serious mistake occurs. The most common error in the regular distribution of chromosomes during mitosis is referred to as *mitotic nondisjunction*. Figure 2-15 shows one of several ways by which mitotic nondisjunction can occur. It is possible for a set of two chromatids, each of which would ordinarily go to each of the poles, to lag behind the rest of the chromosome complement at anaphase and remain on the metaphase plate, with their subsequent loss. The two daughter cells, then, would each have one chromosome missing.

Another possibility, shown in Figure 2-16, is that one of the two chromatids might lag at metaphase. If this accident occurs very early in development, then an embryo would be formed which, theoretically, would on the average have half normal cells and half lacking one chromosome; it would therefore be a *mosaic* with respect to chromosome constitution.

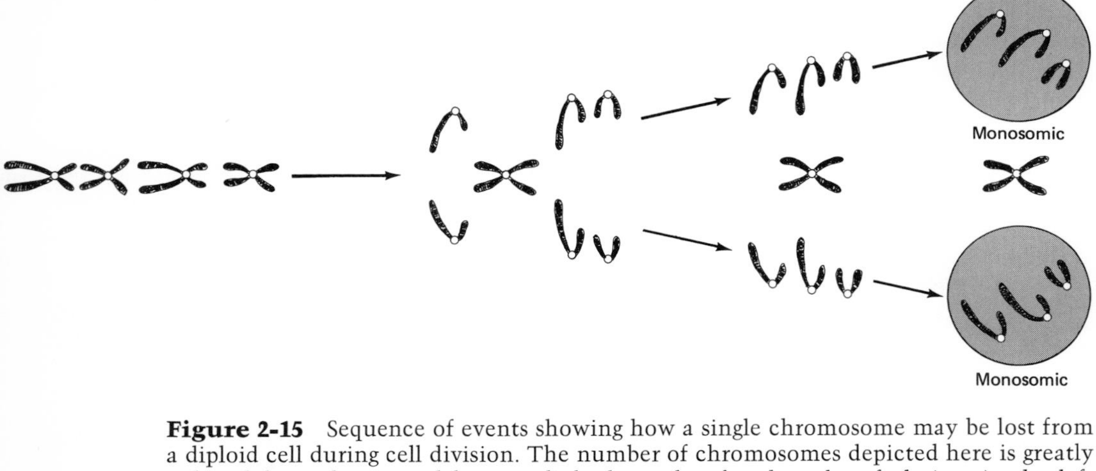

Monosomic

Monosomic

Figure 2-15 Sequence of events showing how a single chromosome may be lost from a diploid cell during cell division. The number of chromosomes depicted here is greatly reduced from the normal human diploid number for the sake of clarity. At the left are four chromosomes of the complement, each already split into two chromatids at metaphase. During early anaphase the individual chromatids separate, becoming new chromosomes. As they proceed to opposite poles one set accidentally becomes stalled on the spindle. At late anaphase the chromosomes have moved to opposite poles except that the immobile pair have remained on the spindle. After the completion of the division, when the new cell membranes form, each of the two parts of the division lacks one of the members of a pair and the two products are therefore monosomic for that particular chromosome.

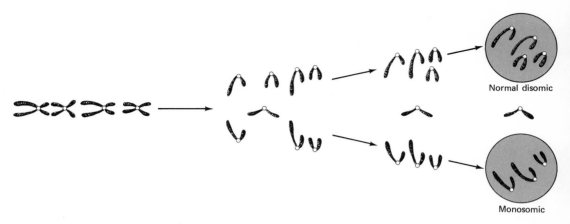

Figure 2-16 A second way in which mitotic nondisjunction can produce cells of abnormal constitution. In this instance, only one of the two chromatids has lagged, with the result that of the two products one has a normal chromosome number, whereas the second is monosomic.

Yet a third possibility is that the two chromatids of one replicated chromosome might go to the same pole (Figure 2-17). This event will give rise to two cell lines, one of which will be trisomic, the other monosomic. The fractions of the body made up of each type depend on the stage of development at which the mitotic nondisjunctional event takes place. All individuals are, without doubt, to some extent mosaic, since among the billions of somatic cells mitotic nondisjunction must have altered at least some of them. In an individual of otherwise normal constitution there will always be abnormal cells present; this frequency increases as the individual ages. Fortunately, the majority of cells

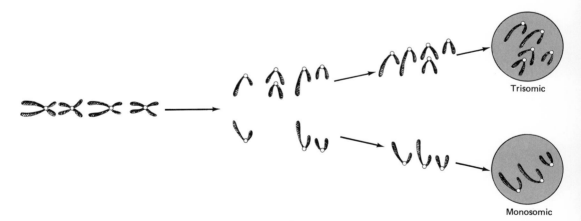

Figure 2-17 Still a third way in which mitotic nondisjunction may produce cells of altered chromosome constitution. In this case the two chromatids of a given chromosome, instead of going to opposite poles, both go to the same pole. As a result, of the two products of the division, one lacks the chromosome completely, whereas the other one has an extra copy. The result is that one cell is monosomic and the other trisomic.

with normal chromosome numbers are sufficient to carry on all of the necessary functions of life.

Individuals of both sexes produce *germ cells*, or *gametes* (sperm in the male, eggs in the female), highly specialized cells that combine to produce the fertilized egg and, eventually, the newborn. *Nondisjunction* can also occur during gamete formation and, when it does, the resulting chromosome unbalance may seriously affect development. It is, in fact, during the formation of the germ cells that the nondisjunction most often occurs which gives rise to the trisomies and monosomies that we have already discussed in Chapter 1. We shall therefore examine still other ways by which nondisjunction can produce abnormal chromosome numbers when we take up germ cell formation in Chapter 4.

The Form of the Chromosomes

THE FOUR-ARMED STRUCTURE.   In the karyotypes in Chapter 1, most of the chromosomes appeared to have four arms. This is so because these are metaphase chromosomes, with each of the two chromosome arms split into two chromatids, but with the centromere still joining them together. Chromosomes may be obtained in this condition by treating dividing cells with the drug *colchicine*, which prevents the separation of the centromere into two and thus effectively halts the mitotic process at metaphase. This treatment is almost universally used in chromosome studies because it yields a very large number of cells with the visible chromosomes of this stage; without the treatment, the number of cells at the metaphase stage would be a small fraction of the total since most cells at any given time will be in interphase, at which stage the chromosomes are virtually invisible. Chromosomes in this divided condition have several advantages: The position of the centromere is clearly marked, making it possible to distinguish chromosomes of equal total length, but with different centromere positions; the short and the long arms of the chromosomes can be identified; and, since there are two identical chromatids still attached, if one of the two chromatids for some reason is not clear in a given cell, the other presents a second opportunity.

TERMINOLOGY BASED ON CENTROMERE POSITION.   Chromosomes are given different names according to the position of their centromere (Figure 2-18). When the centromere has a central (medial) position, the chromosome is a *metacentric*; if it is slightly off center, it is *submetacentric*; if near the end of the chromosome, *acrocentric*; if at the extreme end of the chromosome (as far as can be determined), *telocentric*. Telocentric chromosomes do not normally occur in man. The acrocentric chromosomes of man may have stalklike appendages, called *satellites*, attached. The chromosomes of groups A and F are metacentric; those of B, C, and E are submetacentric; and groups D and G are acrocentric, the latter also showing satellites in favorable preparations.

THE DETAILS OF CHROMOSOME STRUCTURE.   Since so much information about a person's genetic make-up depends on a close look at the individual chromosomes, it might seem profitable to magnify them many thousands of times for greater detail. This can be done with an electron microscope (Figure 2-19); the

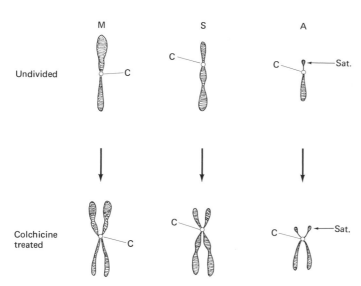

Figure 2-18 A comparison of the metaphase configurations of a metacentric (*M*), a submetacentric (*S*), and an acrocentric (*A*) chromosome. When the centromere splitting is suppressed, as it may be when chromosomes are treated by the application of colchicine, the two chromatids are held together at the centromere region, marking the location of the centromere (*C*), or primary constriction, unambiguously. The chromosome depicted in column (*A*) is like that of the D or G group, with obvious satellites (*Sat*).

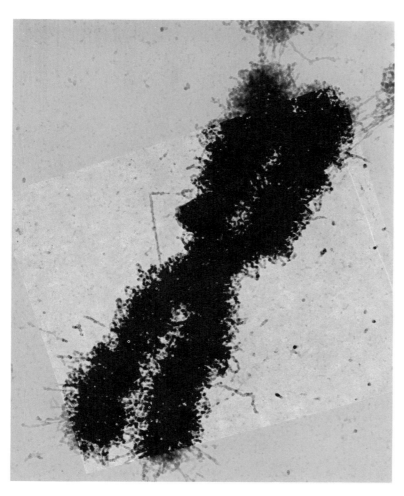

Figure 2-19 An electron microscopic photograph of a metaphase X-chromosome, showing its highly fibrous structure, and, although some differentiation along the fibers is evident, unfortunately not as much detail can be seen as might have been hoped. (G. Bahr, *Human Chromosomes and Chromatin*, Tutorial Proceedings of the International Academy of Cytology, 1972.)

Figure 2-20 One hypothetical model suggesting how extremely long strands might be packed into metaphase chromosomes. Note that at the centromere regions (shown here as double) the strands interlace, but are not attached, so that the two chromatids might separate from each other easily during cell division. (D. E. Comings, *Advances in Human Genetics*, Plenum, New York, 1971.)

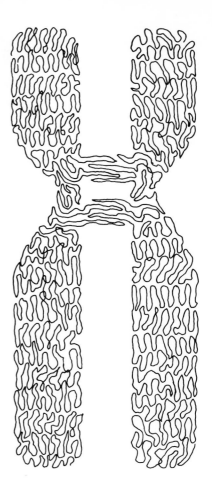

results are disappointing, at least as far as viewing genetic defects is concerned. It is not known definitely how the fibers making up the chromosome are joined together. One hypothetical model, shown in Figure 2-20, accounts for this configuration by assuming that each chromatid is a single strand folded upon itself innumerable times.

References

BARTALOS, M., and T. A. BARAMKI. 1967. *Medical Cytogenetics.* Baltimore: Williams & Wilkins.

HAMERTON, J. L. 1971. *Human Cytogenetics.* Vol. I: *General Cytogenetics.* New York: Academic Press.

LEVINE, H. 1971. *Clinical Cytogenetics.* Boston: Little, Brown.

LEVITAN, M., and M. F. A. MONTAGU. 1971. *Textbook of Human Genetics.* New York: Oxford University Press.

MAZIA, D. 1961. How cells divide. *Sci. Am.,* September.

SWANSON, C. P. 1969. *The Cell.* Englewood Cliffs, N.J.: Prentice-Hall.

THOMPSON, J. A., and M. W. THOMPSON. 1966. *Genetics in Medicine.* Philadelphia: Saunders.

WHITE, M. J. D. 1954. *Animal Cytology and Evolution.* Cambridge: Cambridge University Press.

Questions

Useful terms: gene, DNA, replication, nucleoli, endosperm, spindle, chromatid, centromere, mitotic nondisjunction, mosaic, colchicine, metacentric, submetacentric, acrocentric, telocentric, satellite.

1. How many cell divisions could produce a total of about 8 billion cells from a single cell?
2. What is the fundamental structure of the genetic material? How much of it is found in the cell? How is it possible for such a long length of material to exist in a cell of microscopic size?
3. What is the basic problem that is solved by the process of mitosis? Approximately how much time is required to accomplish this?
4. What are the stages of mitosis? Can you form an acronym using the initial letters?
5. Why is it acceptable or even desirable in a treatment of human genetics to examine the details of mitosis in a plant—in our case the blood lily?
6. What is the difference between the nucleus and the nucleolus?
7. What is nondisjunction? How does mitotic nondisjunction during early development make an individual a mosaic?
8. Why do the chromosomes in a typical karyotype appear to have four arms? In what way does this configuration change in different chromosome types (i.e., the acrocentric and metacentric types)?

3

Prenatal Development

Normal Embryonic Development

During the 9-month period between the time of fertilization of the egg by a sperm and the birth of the infant, a series of most remarkable developmental changes takes place. The total number of cells increases phenomenally from one to countless billions which then become transformed into hundreds of different types: one to become a heart muscle cell, another a nerve cell, and so on. In addition, these cells take up special positions, in some cases by migrating from one spot in the developing embryo to another, and in other cases by undergoing extra mitoses at those sites. In its final form, the newborn child is a creation of extraordinary complexity with a set of highly integrated systems that normally enable it to survive and cope with an almost infinite variety of situations for its long life-span.

To accomplish this complex development, two basic forces are at work. The first is mitosis, which allows for the precise reproduction of all the essential contents of a single cell into two daughter cells. As we have seen in the preceding chapter, our knowledge of the mechanism of mitosis is fairly complete. Also, as development proceeds, different cell lines take on new and characteristic structures and functions: this is called *differentiation*. Although differentiation has been studied in great detail from the single-cell zygote to the adult in a wide variety of animals, our knowledge of the process is, by comparison, quite primitive. We are only now beginning to understand why one of the descendants of the original fertilized egg will become a white blood cell and another descendant of that same egg will be a color-detecting cell in the retina of the eye. It will probably be many decades before we solve the most important aspects of this fundamental problem, but for now we need consider only the gross morphological changes occurring prior to birth.

FERTILIZATION.   At intervals of slightly less than a calendar month, at approximately the middle of the menstrual cycle, a fertile woman of childbearing age will experience *ovulation*, the discharge of a mature egg or *ovum* from an ovary into the corresponding *fallopian tube*. Here it may survive for as long as a day and then will degenerate unless it is fertilized by a *sperm*, which, having a longer life-span, may have been present in the female genital tract for several days. When one sperm succeeds in fertilizing the egg (Figure 3-1*A, B, C*), the egg becomes refractory and will not accept any additional sperm, with only rare exceptions.

The human egg is about 1/200 of an inch in diameter and is barely visible to the human eye; the sperm is much smaller, 1/500 of an inch, the majority of its length consisting of a long tail. Together these minute cells contain all of the instructions necessary for the production of a new human being.

THE CLEAVAGE DIVISIONS.   Some 30 hr after the union of sperm and egg, the fertilized ovum (the *zygote*) divides into two (Figure 3-2*A*) and after another 10 hr into four. These early mitoses of the fertilized egg are referred to as *cleavage divisions* (Figure 3-2*B* and *C*). As these divisions continue, the group of cells formed by mitosis takes on the appearance of a little mulberry, from which it gets its scientific name, the Latin *morula* (Figure 3-2*D*). At about 6 or 7 days after fertilization, the morula has traveled down the fallopian tube and has reached the uterus, where it may attach itself to the uterine wall in a process known as *implantation*. Implantation is one of a long series of hazards that the fertilized egg must confront in its progression toward a fully formed fetus; perhaps as many as a half of all fertilized eggs fail to implant and are lost.

The morula changes shape when a quarter or so of its cells concentrate on one side to form the embryo proper, while a cavity forms between them and other cells which will be responsible for attaching the *embryo* to the uterine wall (Figure 3-3). At the same time, the maternal tissue is reacting to the presence of the embryo by forming enlarged blood vessels at the site of implantation; later these will provide oxygen and other nutrients to the developing child. On occasion this sudden growth of blood vessels may cause bleeding, and since this occurs about 13 days after ovulation, or approximately 28 days after the last menstrual period, it can be confused with ordinary menstrual bleeding, leading to an error of a month in the estimate of the date of conception.

LATER EMBRYONIC DEVELOPMENT.   At 2 weeks the embryo has become two-layered and at 3 weeks the nervous system and main organs begin to form. At this point the heartbeat begins. During the period from 4 to 8 weeks, the change from an unrecognizable mass of cells to a new individual is apparent (Figure 3-4). At the end of the second month of development, the embryo, now less than an inch long, has progressed to the point where it is recognizable as a human being with the major features to be possessed by the newborn infant present, and its embryonic phase is at an end. From the beginning of the third month on it is referred to as a *fetus*, from the Latin meaning "a young one."

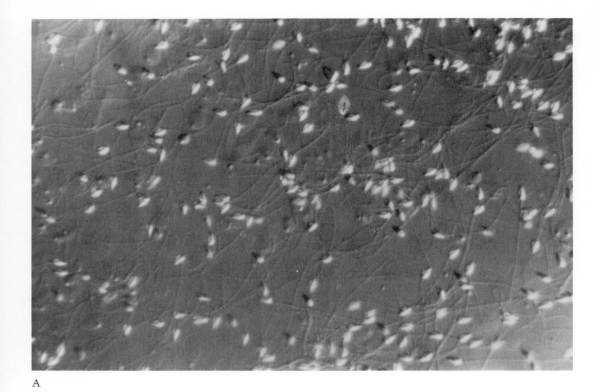

A

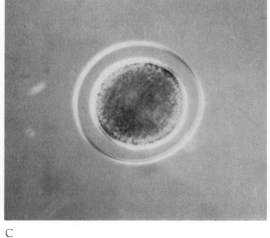

B   C

Figure 3-1 The gametes and fertilized egg of the sea urchin of the Pacific Northwest. The structure of the gametes and the appearance of the egg during early divisions are so similar in virtually all animals that good illustrative material, such as this, can be used equally well to illustrate the gametes and early development in man. (*A*) The sperm, with its head showing up clearly under special lighting conditions and high magnification, and the thinner long tail. (*B*) The egg prior to fertilization. (*C*) The egg shortly after fertilization showing how the production of a "fertilization" membrane protects the egg against entry by a second sperm. (Photographs of sea urchin development courtesy Oregon Institute of Marine Biology, University of Oregon.)

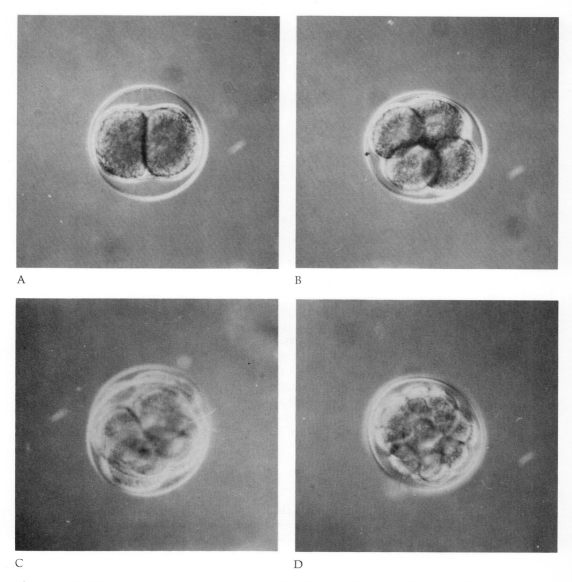

Figure 3-2 The progressive division of the fertilized egg of the sea urchin into the two-cell stage (*A*), the four-cell stage (*B*), and the more advanced stages (*C* and *D*), the last of which now resembles a morula.

FETAL DEVELOPMENT. Except for additional growth and changes in the relative dimensions of the body parts (the head, for instance, still being abnormally large), the basic elements of the complete human being have been formed. During the early fetal stages, sexual development is neutral, with primitive elements of both sexes being present. At the end of the tenth week of gestation, the structures forming the internal male organs start to form (if the fetus is male),

Figure 3-3. The transformation of the morula into a mass with a hollow center, the larger mass of cells at one end to become the embryo proper and the other cells the fetal membrane known as the placenta. (A. T. Hertig, J. Rock, and E. C. Adams, *Am. J. Anat.* **98:**435–93, 1956.)

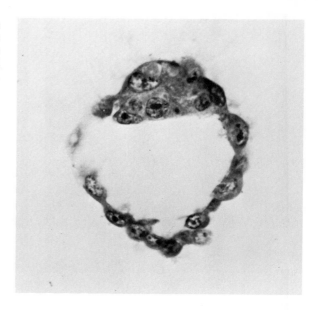

and somewhat later those of the female differentiate (if the fetus is female). By the end of the third month the external genitalia of the fetus may be distinctive, making it possible to classify the sex, in many cases, by simple inspection. The fetal heartbeat is detectable from outside the mother's abdomen, and at about 4 months fetal muscular activity may be felt by the mother. The most pronounced changes during this time are those of weight and size—at the beginning of the third month when the embryonic period has ended and the fetal period has begun, the fetus weighs only 3 gm (about a tenth of an ounce).

Abnormal Development

The series of events starting in the fertilized egg involving rapid mitosis, cell migration, and differentiation to produce the complete infant represent such an extraordinarily complex process that it may seem nothing short of miraculous that the vast majority of children are normal at birth. Nevertheless, accidents of development do occur, and such accidents, present at birth, are known as *congenital defects*. The most common are such minor anomalies as *cleft lip*, *cleft palate*, and *club foot*—usually slight defects readily repaired by minor surgery. Fortunately, most of the serious defects that occur during early development give rise to an inviable embryo which is lost in early pregnancy by *spontaneous abortion* (= *miscarriage*). Specific agents that cause maldevelopment of the embryo or fetus resulting in a miscarriage, a stillbirth, or the presence of a congenital defect are called *teratogenic agents*, or *teratogens*.

Statistics on the overall incidence of congenital defects vary widely with the country from which they come, whether they come from hospital birth records or from births at home, and the age of the child at the time the observation is made. The frequency from birth certificates seems to be less than 1 percent, and from hospital records somewhat greater. It is greatest when detailed examinations

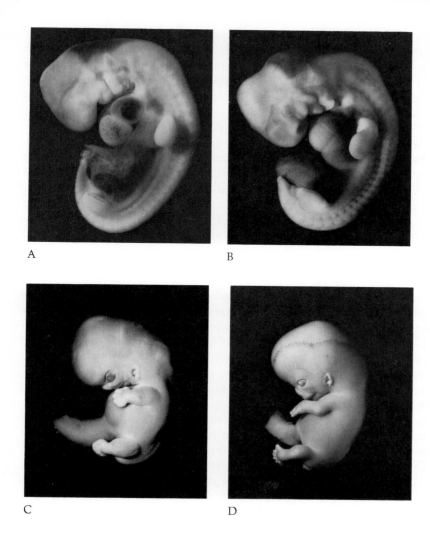

A

B

C

D

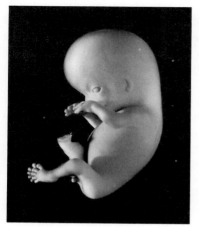

E

Figure 3-4 Successive changes in the embryo during early development. The ages, given in approximate times after ovulation of the egg, are (*A*) 29 days, (*B*) 32 days, (*C*) 37 days, (*D*) 43 days, and (*E*) 47 days. (B. Böving, Anatomy of Reproduction, in *Physiology and Conduct of Pregnancy*, Carnegie Institution of Washington, Washington, D.C., 1962.)

are made by pediatricians. Pediatric examinations in the United States have given a figure of almost 9 percent, whereas in Germany the figure is as low as 2 percent. As an overall average, it can be surmised that from 2 to 3 percent of all newborn infants show congenital malformations at birth and after 1 year this figure is twice as high because of the appearance of malformations not obvious at birth.

GERMAN MEASLES (RUBELLA).   Because the embryo and fetus are so well protected by the fetal membranes and amniotic fluid in which the developing infant rests, both from outside influences and from toxic agents in the mother's bloodstream, it had been previously assumed that the embryo was relatively immune to external damage. In 1941 an Australian eye specialist named Gregg called attention to an extraordinary increase in the frequency of eye cataracts in newborn children. Of 78 children born with opaque lenses, 68 had mothers who had contracted German measles during early pregnancy. Gregg suggested a cause-and-effect relationship between the *rubella virus* responsible for German measles and the cataracts.

Under ordinary circumstances, only four to eight pregnant women out of every 10,000 in the United States are afflicted with rubella. During the year 1964, however, this country experienced an epidemic of German measles and the incidence went as high as 220 per 10,000. There were approximately 20,000 cases of congenital malformations among newborns as a result. In addition to blindness, other defects such as heart disease, deafness, and mental retardation, as well as an extremely small size at birth, were observed among the children of infected mothers.

Stillbirths and spontaneous abortions are from two to four times more likely in pregnancies accompanied by rubella than in uncomplicated pregnancies. In addition to the congenital malformations resulting from the 1964 epidemic, 30,000 stillbirths and miscarriages were attributed to this cause.

The kind of defect caused by rubella depends on the stage of embryonic development at which it occurs: infection during the sixth week of pregnancy may cause cataracts, whereas deafness follows infection during the ninth week. Heart defects may result from infection between the fifth and tenth weeks. The overall probability of malformations in newborns is about 50 percent when the infection occurs during the first month of pregnancy, and decreases to 20 percent when the infection comes between the fifth and eighth weeks. It is less than 10 percent thereafter. These percentages would increase if inspection for defects that cannot be diagnosed immediately at birth were made somewhat later.

Fortunately, the vast majority of women are immune to the rubella virus (curiously, Japanese women show unusual resistance to it) and it is possible to vaccinate the remainder against the disease, although vaccination immediately before or during the early stages of pregnancy is not recommended because there is a slight chance that the vaccine itself might cause congenital defects.

OTHER VIRUSES.   Another virus that has been positively shown to produce congenital malformations is *cytomegalovirus*. Unfortunately, it has no visible effect on the infected mother and so will usually go unrecognized in pregnant women; in fact, the only way of knowing that this virus has this capability is by the observation that a stillborn child will sometimes carry large quantities of the virus, visible microscopically. The disease is often fatal for developing infants, and survivors can be severely mentally retarded.

Other maternal infections such as Asian flu, mumps, hepatitis, polio, and chickenpox have been reported on occasion to have preceded the births of malformed children, but the data at the present time are conflicting and suggest that, at most, they play a minor role in producing birth defects. It was thought at one time that syphilis might be a major cause of congenital defects. At present there is some doubt that it produces as many defects with as great a frequency as was once believed, although it has been repeatedly observed to lead to congenital deafness and mental retardation as well as to defects of other organs of the developing fetus.

OTHER TERATOGENIC AGENTS.　During the period from 1957 to 1961, a very effective antinauseant and sleeping pill called *thalidomide* was produced by a German pharmaceutical company and made available throughout the world. Its properties made it very valuable in combatting morning sickness and other discomforts experienced during early pregnancy. Within 4 years it became apparent that large numbers of children were being born with *phocomelia*—a condition involving flipperlike arms and legs, usually very short and sometimes entirely missing. Both in Germany and in Australia the correlation was made between the mother's taking thalidomide and the birth of a defective child. Before the drug was withdrawn from the market at least 5,000 affected babies were born in Germany and another thousand in other countries throughout the world. Because of stricter laws regulating the sale and distribution of new drugs in the United States, thalidomide was kept off the market in this country by the action of a concerned official of the Food and Drug Administration; nevertheless, some women were able to obtain the drug from outside the country with occasional unhappy results.

A study of the range of defects thalidomide causes showed that the most disastrous effects occurred if it was taken by the mother during the third and fourth weeks of the embryo's life. The timing is consistent with the observation that the limbs develop during this embryonic period. Lesser effects were noted when the drug was taken subsequent to this time. In addition to deformities of the long bones, heart abnormalities and defects of the ear and intestinal tract have been recorded.

The list of substances that have been suspected of being teratogenic includes quinine, certain antibiotics, antihistamines, tranquilizers, alcohol (in excess), and some hormones. Cigarette smoking and LSD have been suggested as teratogens; even aspirin is on the list of chemicals suspected of producing occasional congenital malformations if taken in quantity during early pregnancy.

There are several difficulties in proving that any particular chemical is teratogenic. Congenital defects are relatively infrequent after the ingestion of known teratogens; the manifestation of defects may not only be erratic but also be variable with respect to the specific organs affected. Further, it is obviously not possible to test potential teratogens deliberately by feeding them to pregnant women. Experimental work with animals usually involves feeding excessive amounts of the material, and very often different species of animals differ in their responses to these chemicals. Even different strains or races within a given species may manifest different reactions to a given teratogen. Simple prudence would suggest that no mother expose herself to any unusual chemical during her child's prenatal life, especially during the very sensitive period of early embryonic development. Unfortunately, women are not always aware that they are pregnant during this critical early period.

RADIATION.   Of pregnant Japanese women who were exposed to heavy doses of radiation at the time of the atomic bomb explosions at the end of World War II, 25 percent produced children who died during their first year of life and another 25 percent gave birth to infants with severe congenital anomalies.

Embryos developing in women who must be given pelvic radiation treatments in early pregnancy are particularly at risk because of the high doses of radiation used in such medical treatments—as many as 50 percent of the offspring of such treated women may be abnormal. These abnormalities include *microcephaly* (an unusually small head) accompanied by severe mental retardation, skull defects, blindness, cleft palate, and defects of the limbs. As with other teratogenic effects, the radiation-induced defects are most common and most severe when the mother is x-rayed during the first 2 months of pregnancy, decreasing markedly thereafter. Some possible effects of radiation such as the production of changes in the chromosomes and genes will be discussed in a later chapter of this book.

Abnormalities of Chromosome Number

The condition of duplicate chromosomes, or *diploidy*, is characteristic of the vast majority of existing species on this planet. One might suspect, therefore, that the presence of two of each kind of chromosome has some advantage in the survival of most forms of life. The argument that such an advantage exists is very simple: diploidy provides a back-up system in the event that some of the essential genetic material in a cell is destroyed. During the long life-span of a multicellular organism, it can be anticipated that some of the genetic information carried by the chromosomes will be lost or destroyed by normal molecular agitation, by factors from the external environment such as ultraviolet or cosmic radiation, or by poisonous chemicals. In that event, the unaffected chromosome of the pair is usually able to take over the function of the damaged one. It is easy to understand that this would be much more important to a long-lived organism than to a short-lived one; most of the organisms that are *haploid* (i.e., with only one set of chromosomes) such as bacteria and molds have relatively short life-spans.

Because the *diploid* organism has become adjusted to the presence of two chromosome sets over countless generations of natural selection, the addition of an extra chromosome throws the cellular mechanism off balance—the genetic factors no longer interact in the proper proportions. When this occurs in a germ cell, an abnormal individual is produced. As a simple analogy, let us suppose that an expert mechanic has designed an excellent two-cylinder gasoline engine. If one were to try to add to this machine one additional spark plug or cylinder or piston, the net result would be chaotic; the engine, after such a piecemeal addition, could never work as well as the original optimum design. Quite likely, it would not work at all.

TRISOMY-18 AND TRISOMY-13. We have already seen how trisomy-21 results in Down's syndrome (Chapter 1). Other autosomal trisomies are even more disastrous to the developing embryo. Trisomy-18 is one of the more common; generally embryos with such trisomies succumb prior to birth; when affected infants do survive to term, they usually do not live beyond a few months (although a few rare cases have been recorded of survival beyond several years). Some surveys have suggested frequencies of incidence as high as, or even slightly higher than, 1 per 10,000 and four-fifths of the newborn so affected are females.

Like trisomy-18, trisomy-13 is usually lethal; affected fetuses are most often aborted spontaneously, or, when viable as a newborn, the infant rarely survives beyond the age of 3 months. The incidence of this abnormality among newborns is about 1 per 10,000 births. A list of physical characteristics in these two syndromes is given in Table 3-1. Although we can expect occasional medical reports of viable trisomies other than 13, 18, or 21 (and, indeed, trisomy-22 and trisomy-8 have also been reported on several occasions as specific syndromes), it is a general rule that the other autosomal trisomies are lethal during early prenatal development (Figure 3-5).

SPONTANEOUS ABORTION. Because the occurrence of a spontaneous abortion has medical and legal implications to society, it is generally defined by national and state laws and therefore its definition is subject to considerable variation. For our purpose we shall define spontaneous abortion as the natural termination of pregnancy before the fetus has reached an age of 22 weeks or a weight of 500 gm (slightly over 1 lb). The spontaneous loss of the fetus after this period is a *stillbirth*.

The estimate of the frequency of spontaneous abortions in the population depends on the source of the data. Estimates based on personal interviews with women suggest a figure close to 15 percent, but this is undoubtedly an underestimate since most women may not be aware of early embryonic losses. Studies of the early stages of pregnancy in women normal in fertility show that approximately 30 percent of all fertilized eggs are clearly physically abnormal and incapable of further development. This additional 30 percent represents fertilized

	Trisomy-13 (percent)	Trisomy-18 (percent)
Mental retardation	51	85
Low-set ears	51	82
Malformed ears	77	75
Congenital heart disease	83	82
Flexed fingers	40	75
Overlapping fingers	14	50
Rocker-bottom feet	—	40
Microcephaly	54	—
Receding chin	—	87
Defects of eyeball	88	—
Polydactyly	63	—
Cleft lip or palate	71	—

Table 3-1 Frequencies of developmental anomalies associated with trisomy-13 and trisomy-18. (Condensed from J. Nusbacher and K. Hirschhorn, *Adv. Teratol.*, **3**:11–53, 1968.)

Figure 3-5 A grossly abnormal early embryo, identified as having trisomy-16. (D. H. Carr, Chromosomes and Abortion, in *Advances in Human Genetics*, copyright © Plenum, New York, 1971.)

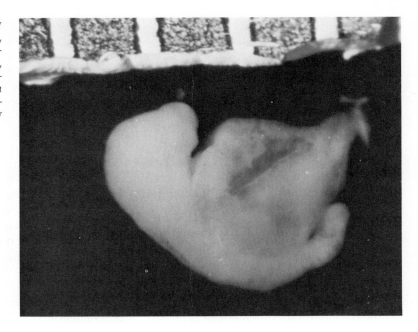

eggs that would have been lost prior to 3 weeks of pregnancy, i.e., probably before the women themselves would have been aware of their existence. This gives a total of 45 percent for spontaneous loss of very early embryos.

Nearly half of the embryos and fetuses spontaneously aborted are morphologically abnormal. It seems quite likely that an additional percentage are abnormal to some degree not obvious to ordinary inspection. From this point of view, then, it would appear that in most cases early spontaneous abortions have a therapeutic effect in preventing grossly abnormal fetuses from coming to term.

Chromosome Anomalies in Spontaneous Abortions

In a large series of spontaneous abortions examined for abnormal chromosome numbers, about 50 percent proved to have some deviation from diploidy, or *aneuploidy*. Of the aneuploid specimens, about 50 percent had simple trisomy. The frequency of chromosomally abnormal abortuses depends on the mother's age; both very young and very old females show higher incidences than women of age 20 to 30 years (Figure 3-6). Furthermore, the actual frequency may depend on the age of the embryo or fetus at the time it was aborted. Those that are aborted at less than $2\frac{1}{2}$ months of age are abnormal almost twice as often as those aborted between $2\frac{1}{2}$ and $3\frac{1}{2}$ months, and those aborted after $3\frac{1}{2}$ months may have as few as 5 percent abnormalities. Table 3-2 shows the relative frequencies of trisomy for the different chromosome groups out of a total of 679 trisomic spontaneous abortions. Even those trisomies that appear at times in liveborn infants—13, 18, and 21—result in abortion far more often than they do in live births.

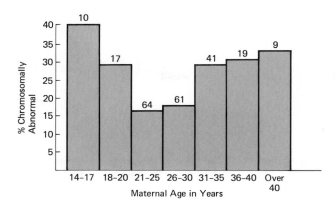

Figure 3-6 The incidence of chromo- somally abnormal abortuses as a function of the age of the mother. The numbers above each bar on the graph represent the number of cases investigated. (D. H. Carr, Chromosomes and Abortion, in *Advances in Human Genetics*, copyright © Plenum, New York, 1971.)

MONOSOMY. From our earlier discussions of nondisjunction, we have seen that trisomy and monosomy may be complementary products of one event (Figure 2-17), but that monosomy may arise in ways that do not also produce a trisomic product. We would therefore predict that the frequency of embryos carrying a monosomy for one of the chromosomes would be greater than the frequency of trisomies for that chromosome. In fact, the frequency of observed monosomies is very much less, the reason being that monosomies seem to be inviable and are presumably lost prior to the time of birth—in fact, even before they could be detected as miscarriages. The rare occasional cases of monosomy reported in the medical literature probably represent mosaic individuals with some normal cells present that allow them to develop more normally than a pure monosomic would or, more likely, individuals with complicated chromosome anomalies who appear to be monosomics but in fact are not.

An exception to this rule is monosomy for the X-chromosome. X0 individuals may survive to term and in fact are relatively normal except for disorders of fertility (Turner's syndrome, Chapter 1). The curious fact is that although X0 newborns are relatively uncommon, one in about 2,500 newborn females, they comprise about one fifth of all abortuses with abnormal chromosome numbers.

Table 3-2 The frequencies of trisomies for the seven chromosome groups in a series of abortions. The "expected" percentage is based simply on the number of chromosomes belonging to each group. A comparison of the observed percentages with the expected suggests that, except for the F group, the smaller chromosomes are more likely (and the larger ones less likely) to be found in an inviable trisomy recoverable from a spontaneous abortion. (Courtesy of C. E. Ford, Oxford University.)

	Chromosome Group							Total Trisomies
	A	**B**	**C**	**D**	**E**	**F**	**G**	
Number	20	9	112	150	252	22	125	679
Percentage observed	4	1	16	22	27	2	18	
Percentage expected	14	9	31	14	14	9	9	

Figure 3-7 Triploidy in a human abortus. Note that there are three chromosomes of each type instead of two, except for chromosome 16 and the sex chromosomes. (D. H. Carr, Chromosomes and Abortion, in *Advances in Human Genetics*, copyright © Plenum, New York, 1971.)

It can be calculated that less than 5 percent of all fertilized eggs that start out as X0 in fact come to term.

POLYPLOIDY.   In about 20 percent of the abortions with chromosome abnormalities, three sets of chromosomes (*triploidy*, Figure 3-7) are found instead of two (69 chromosomes instead of 46), and about 5 percent have *tetraploidy*, or four sets instead of two (92 instead of 46) (Table 3-3). The possession of more

Table 3-3 Types of chromosome abnormalities identified in human spontaneous abortions and their relative frequencies. (Courtesy of C. E. Ford, Oxford University.)

Type	Number of Observations	Percent
Monosomics	262	19
Trisomics	703	51
Triploids	239	18
Tetraploids	76	6
Mosaics and rearrangements	79	6
Total	1,359	100

than two chromosome sets, *polyploidy,* is lethal in humans. Why this is so is not known; not only is this condition quite viable in plants and in some animals other than mammals, but also in many cases polyploids are difficult to distinguish from the diploids. A few humans have been found in which a sizable fraction of their cells, but not all, were triploid; these individuals were themselves somewhat abnormal.

References

AUSTIN, C. R. 1961. *The Mammalian Egg.* Springfield, Ill.: Thomas.

AUSTIN, C. R. 1965. *Fertilization.* Englewood Cliffs, N.J.: Prentice-Hall.

BOUÉ, J., and A. BOUÉ. 1973. Chromosomal anomalies in spontaneous abortions. Chromosomal Errors in Relation to Reproduction Failure. Institut National de la Santé et de la Recherche Médicale, Paris.

BÖVING, B. G. Physiology and conduct of pregnancy. In *Anatomy of Reproduction.* Washington, D.C.: Carnegie Institution of Washington.

CARR, D. H. 1971. Chromosomes and abortion. In H. Harris and K. Hirschhorn, eds., *Advances in Human Genetics,* **2**:201–49. New York: Plenum.

JACOBS, P. A., M. MELVILLE, and S. RATCLIFFE. 1974. A cytogenetic survey of 11,680 newborn infants. *Ann. Hum. Genet.,* **37**:359–76.

JANERICH, J. T., J. M. PIPER, and D. M. GLEBATIS. 1974. Oral contraceptives and congenital limb reduction defects. *N. Engl. J. Med.,* **291**:697.

JONES, K. L., et al. 1973. Pattern of malformation in offspring of chronic alcholic mothers. *Lancet,* 1267.

LANGMAN, J. 1969. *Medical Embryology.* Baltimore: Williams & Wilkins.

NUSBACHER, J., and K. HIRSCHHORN. 1968. Advances in teratology. In D. H. M. Woollam, ed., *Autosomal Anomalies in Man.* New York: Academic Press.

PATTEN, B. M. 1968. *Human Embryology.* New York: McGraw-Hill.

SIEGEL, M. 1973. Congenital malformations following chickenpox, measles, mumps and hepatitis. *JAMA,* **31**:1521–24.

Questions

Useful terms: differentiate, ovum, sperm, cleavage division, implantation, embryo, fetus, congenital defect, miscarriage, spontaneous abortion, teratogenic agent, diploidy, haploidy, stillbirth, aneuploidy, triploidy, tetraploidy, polyploidy.

1. In rough outline, give the sequence of events from the fertilization of the egg to the birth of an infant.
2. What is the distinction between an embryo and a fetus?
3. What is your personal opinion about the time at which human life begins during prenatal existence?
4. What are the most common congenital defects?
5. At what stage of development are teratogenic agents likely to be most injurious?

6. Is there a relationship between the type of congenital defect produced and the teratogenic agent involved?

7. What distinction, if any, is there between a miscarriage, a spontaneous abortion, and a stillbirth?

8. How often do spontaneously aborted fetuses prove to carry an abnormal chromosome number? How does this vary with (a) the age of the mother and (b) the age of the embryo or fetus?

9. More trisomies than monosomies are found in spontaneous abortions. Does this imply that the trisomic condition is more harmful to the developing embryo than monosomies are?

10. What is the outstanding exception to the general rule that monosomy has a disastrous effect on development?

11. Discuss the relevance of the following conditions to the concepts discussed in this chapter: cleft lip, cleft palate, clubfoot; German measles (rubella); phocomelia; microcephaly; trisomy-18; trisomy-13; Turner's syndrome.

4

The Formation of the Germ Cells

August Weismann was one of the many brilliant biologists of the last half of the last century. His name is commonly associated with experiments designed to demonstrate the lack of inheritance of acquired characters by the amputation of rats' tails for a number of generations to show that newborn rats with amputated ancestors still had tails of normal length. Among his more substantial contributions was his deduction that there must exist, as a matter of simple logic, a stage in the formation of the germ cells when the genetic material is halved. The reasoning went as follows:

It is a matter of common observation that two individuals of different races produce children intermediate in their racial characteristics. This must mean that both parents make a contribution to the next generation, and, since the children are pretty much the same regardless of which parent is of which race—that is, the same in *reciprocal crosses*—the contribution of each parent must be about equal. However, this simple conclusion leads us into a logical absurdity, for if each parent contributes his characteristic genetic material to his progeny they will have the material of both parents, or twice as much as either one alone. This would mean that the genetic material would double generation after generation until the human body would be too small to hold all the genetic material it was receiving from its ancestors.

To resolve this dilemma, Weismann postulated that the genetic material must be precisely halved at some point in the formation of the gametes, so that when two gametes join to form the zygote, each progeny will have the same amount of

49

genetic material as each parent. This halving of the genetic material is accomplished during two distinctive cell divisions that occur only in those cells that are the immediate ancestors of the gametes. These divisions are designated as *meiosis* and consist of two divisions, the first and second meiotic divisions. Since the cellular events of meiosis determine what the parents transmit to their offspring, an understanding of the behavior of the chromosomes during the two meiotic divisions is basic to an understanding of genetics.

The Occurrence of Chromosomes in Homologous Pairs

Since the zygote from which all the cells of the mature individual are derived is formed from the fusion of the egg and the sperm, it must contain two sets of chromosomes, one set derived from the male parent and the other set from the female parent. For each chromosome brought in by the sperm, there will be a "partner" chromosome in the set contributed by the egg. In man there are 46 chromosomes in the zygote, 23 brought in by the sperm and 23 by the egg, resulting in 23 pairs of chromosomes.

Gametes are formed by certain specialized cells called *gonial* cells found in the testes of the male and the ovaries of the female. These divide by mitosis, producing millions of other cells like themselves. After a certain number of mitotic divisions in the male testis, a *spermatogonium* becomes a *spermatocyte*, ready for the sequence of changes that will transform it into a mature *sperm*. Similarly, in the female ovary, the *oogonial* cells produced by mitosis differentiate into *oocytes* and, after the two meiotic divisions, differentiate into mature eggs, or ova. Although a number of events of genetic importance occur during meiosis, the basic accomplishment of the meiotic divisions is that the homologous chromosomes are separated from each other so that the resulting gametes carry a haploid set of 23 chromosomes.

Outline of Meiosis

Before considering the two meiotic divisions in detail, we shall take a quick overview of the process. In the upper left of Figure 4-1, a simplified cell with six chromosomes is shown. Note that the six chromosomes occur in pairs: chromosomes A and A' are homologues, chromosomes B and B' are homologues, as are C and C'.

In the interphase prior to this stage, duplication of the chromosomes occurred, so that each chromosome entering the first meiotic prophase consists of two identical strands. Note that the two strands (sister strands or chromatids) of each chromosome are still held together at the centromere region.

PROPHASE OF THE FIRST MEIOTIC DIVISION. Early in the prophase of the first meiotic division, each chromosome moves to a position alongside its homologue in a process known as *synapsis*. This is a most remarkable phenomenon: At this stage the chromosomes are still long threads, with 46 individual chromosomes scattered more or less at random throughout the nucleus in a fashion resembling

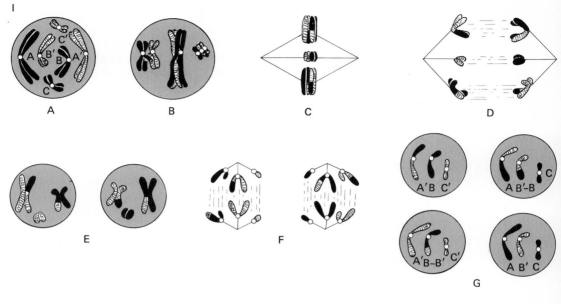

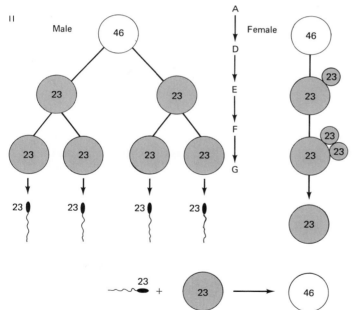

Figure 4-1 *I.* A schematic diagram illustrating the important aspects of chromosome behavior during the two meiotic divisions. For simplicity, only three pairs of homologues are shown. The detailed explanation of this is given in the text. *II.* The sequence of events in meiosis showing the reduction of the chromosome number and the formation of four sperm in the male, or an egg in the female, from the final products. The letters *A* to *G* indicate where the corresponding stages of chromosome behavior in Figure 4-1*I* are found.

a tangled web of string. Yet somehow the homologues of each of the 23 pairs manage to synapse without becoming hopelessly entangled with the others.

Each of the chromosomes making up a synapsed pair consists of two identical strands, making a total of four strands. This group of four is called a *bivalent*. By synapsis, the six chromosomes become three bivalents (Figure 4-1, *IB*).

METAPHASE OF THE FIRST MEIOTIC DIVISION.   The bivalents, which were more or less randomly placed in the nucleus of the cell, now line up along one plane

(Figure 4-1, *IC*). The nuclear membrane disappears and fibers can be seen connecting the centromere regions of the chromosomes to two opposite poles.

ANAPHASE OF THE FIRST MEIOTIC DIVISION.   At anaphase of the first meiotic division, the spindle fibers appear to pull the two homologous centromeres of each bivalent to opposite poles (Figure 4-1, *ID*). Although the homologous chromosomes are effectively separated at this division, each chromosome still consists of two chromatids held together at the centromere.

THE SECOND MEIOTIC DIVISION.   At the conclusion of the first meiotic division, the chromosome number is halved, not as a random selection of chromosomes, but with precisely one of each of the homologous pairs—no more and no less (Figure 4-1, *IE*).

The interphase between the first and second divisions is atypical, usually of short duration without the return of the chromosome to its active extended state. The chromosomes do not replicate between the first and second meiotic divisions. At metaphase of the second division, they simply line up on the division plane, and the centromere which had held the two strands together during the preceding division now divides. At second-division anaphase (Figure 4-1, *IF*), the individual strands, the chromatids, go to opposite poles. The important result of meiosis is that the two divisions produce four products, each of which has precisely an accurate haploid set of chromosomes from the diploid gonial cell (Figure 4-1, *IG*).

These haploid products now go on to differentiate into the mature gametes. In the male, each of the four products differentiates into a mature sperm. In the female, however, the four products of meiosis, although equal in chromosome distribution, are grossly unequal in size. At the first division, one of the two products retains the mass of the original cell, while the other is pinched off as a tiny *polar body*. At the second division, the process is repeated for the larger cell, and the polar body may (or may not) divide again, with the result that one large cell, the functional egg, and two or three smaller nonfunctional polar bodies are produced (Figure 4-1, *II*).

Production of New Gene Combinations

SHUFFLING OF WHOLE CHROMOSOMES.   Meiosis separates the homologues so that the gametes have only one member of each of the homologous pairs. Just which member of a given pair of homologues gets into a particular gamete depends on the orientation of the bivalent relative to the poles of the meiotic apparatus at the first meiotic metaphases (Figure 4-1, *IC*). This orientation is entirely random from one pair of homologues to the next so that the various gametes may contain any possible combination of the homologous chromosomes. If we take the simplified case in which a zygote is formed from the combination of three chromosomes, ABC from one parent and a homologous (but not genetically identical) set A'B'C' from the other parent, the diploid condition may be represented by AA'BB'CC'. A gamete from such a person could, by chance, happen to have the three chromosomes from one parent (ABC) or from the other (A'B'C'). With equal frequency, however, it could have one of the other combinations ABC', AB'C, A'BC, AB'C', A'BC', and A'B'C. With three sets of homologues, there are 2^3 or 8 different possible products. In the case of man, with 23 pairs of homologous chromosomes, there would be 2^{23} (or more than 8 million) possible different combinations of maternal and paternal chromosomes in the gametes.

REDISTRIBUTION OF CHROMOSOME SEGMENTS.   In addition to the random distribution into the gamete of either the maternally or paternally derived member of each homologous set, another process shuffles the genetic material even further. In the prophase of the first meiotic division, homologous strands within a particular bivalent exchange segments with each other. This process, called *crossing over*, is shown in the second column of the page of diagrams. In Figure 4-1, *IIB*, two strands have crossed over with a resulting exchange of material between them. The effect of this crossover on the composition of the chromosomes of the four nuclei that result from the meiosis is shown in Figure 4-1, *IIH*. Since the homologues usually differ in the genes they carry, crossing over produces strands that contain new combinations of genetic material.

EVOLUTIONARY SIGNIFICANCE OF RECOMBINATION.   By these mechanisms, genes found in certain combinations in the parents are *recombined* during meiosis of each of the gametes, and an additional randomization occurs when two gametes fuse to form a zygote. In this way, each new birth represents a unique combination of genes found in a person unlike any ever to have appeared earlier in the history of man, or ever to appear again.

When two or more advantageous genes appear in different lines, *recombination* provides the opportunity for them to be combined in a single individual, with obvious beneficial results. For this reason, recombination is considered one of the fundamental processes of evolution.

The Details of Meiosis

Meiosis, like mitosis, is a universal phenomenon, found in all higher organisms. It has been the subject of research for almost a century, with certain organisms such as grasshoppers, salamanders, and corn among the preferred objects of study because of the clear pictures provided by their large, and relatively few, chromosomes. It is only within the past decade that reasonably good preparations of meiosis have been made in man; these preparations do not show anything that had not been seen previously in other organisms. Since we have a choice in the selection of organisms for the demonstration of fundamental properties of the cell, in this case we choose to make use of a species of the salamander in which meiosis in the male is unusually clear. The species used here is one of a large family of lungless salamanders, consisting of more than 200 species, found in the United States and Latin America.

The First Meiotic Division

PROPHASE.   The prophase of the first meiotic division is long and complicated, differing in a number of important respects from the prophase of mitosis. Figure 4-2 shows an unfixed and unstained cell in a tissue culture chamber just entering

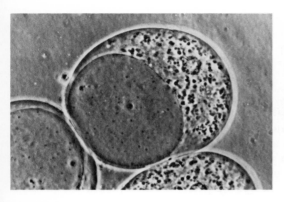

Figure 4-2 Cell just entering first meiotic division. (Figures 4-2 through 4-21 are reproduced courtesy J. Kezer, Department of Biology, University of Oregon.)

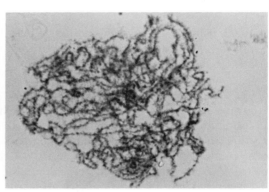

Figure 4-3 Chromosomes just entering prophase of first meiotic division.

Figure 4-4 Beginning of synapsis.

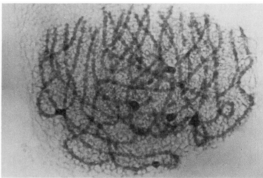

Figure 4-5 Spermatocyte nucleus in which synapsis has been completed.

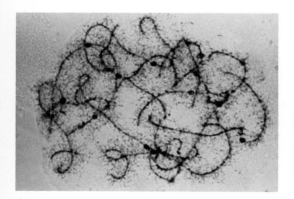

Figure 4-6 Squashed nucleus showing structures of bivalents.

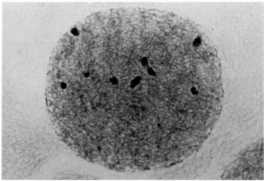

Figure 4-7 DNA spun out into elongated loops.

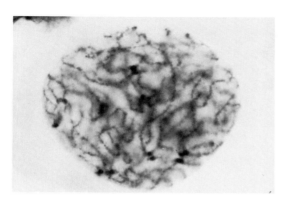

Figure 4-8   Nuclei just emerging from the diffuse stage.

Figure 4-9   Bivalents becoming visible in prophase, showing chiasmata.

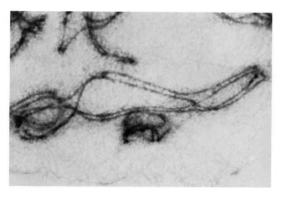

Figure 4-10   Detailed structure of a bivalent.

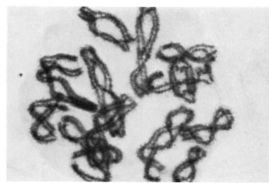

Figure 4-11   Twists and turns in strands of bivalents when chiasmata are present.

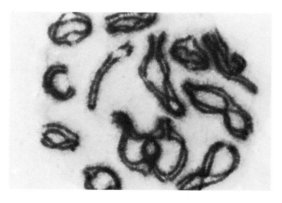

Figure 4-12   Beginning of first meiotic metaphase.

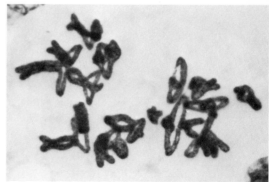

Figure 4-13   Metaphase of first meiotic division.

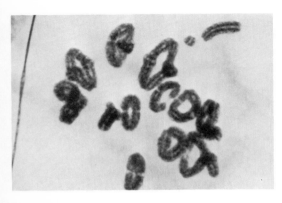

Figure 4-14 First meiotic metaphase bivalents.

Figure 4-15 Early anaphase of first meiotic division.

Figure 4-16. Completion of first anaphase.

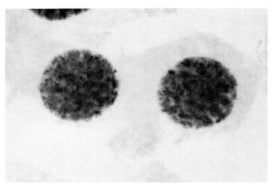

Figure 4-17 Two nuclei in interphase between first and second meiotic divisions.

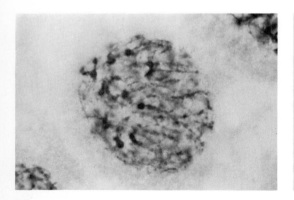

Figure 4-18 Nucleus in early prophase of second meiotic division.

Figure 4-19 Chromatids still attached at centromere in prophase.

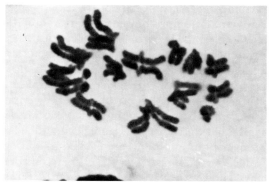
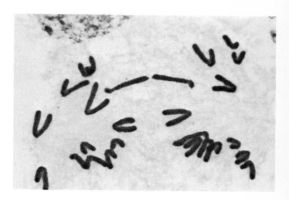

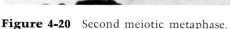

Figure 4-20   Second meiotic metaphase.

Figure 4-21   Second meiotic anaphase.

the first meiotic division. The large oval nucleus contains an almost centrally located nucleolus. Although the chromosome strands cannot be seen, the parts of the chromosomes associated with the centromeres are visible as small dark granules scattered throughout the nucleus. The cytoplasm contains many large, dense granules and rodlike structures called *mitochondria*. The conspicuous spherical structure in the cytoplasm is a *centrosome*. A pair of *centrioles*, not visible in the photomicrograph, is embedded in the centrosome. During the first meiotic division, this centrosome divides, giving rise to the poles from which the spindle fibers radiate.

Figure 4-3 illustrates the appearance of chromosomes that are just entering the prophase of the first meiotic division. At this early stage of meiosis, the chromosomes are resolved by the microscope as elongate single strands, all tangled together within the nucleus. Although the microscope cannot reveal it, the chromosomes consist of two identical strands because of the duplication that took place during the immediately preceding interphase. In Figure 4-3 it is possible to see that these elongate strands do not have an even surface contour but rather seem to be constructed of small granules and lumps strung together by thinner regions. The small granules are called *chromomeres* and this chromomeric structure is characteristic of meiotic chromosomes. It is also apparent that the chromosome strands have a "fuzzy" appearance. In favorable preparations, it can be seen that this "fuzz" consists of pairs of fine, threadlike loops that originate from the chromomeres.

As has been noted earlier, the genetic material is deoxyribonucleic acid, abbreviated DNA. A chromosome consists of DNA associated with protein. How these two components are arranged relative to each other to produce the chromosomes of the higher organisms is not known, although current research is providing important information regarding this complex problem. Some of the protein is thought to comprise part of the actual structure of the chromosome, and other proteins may determine which parts of the DNA will be active at any given time during differentiation. Chromomeres can be visualized as consisting of DNA that has become packed and folded along the chromosome axis to produce the characteristic granular appearance.

In these cells, it can be seen that synapsis begins simultaneously at both ends of a pair of homologues and proceeds inward, bringing homologous chrommomeres together. In Figure 4-4 there is a nucleus in which synapsis is just beginning. At the periphery of the nucleus are stretches of more deeply staining material, indicating that the pairing of the ends of the homologous strands has begun.

Figure 4-5 illustrates a spermatocyte nucleus in which synapsis has been completed. The chromosomal structures present in this nucleus are now complete four-strand bivalents but the two members of a homologous pair have become so closely associated by synapsis that they appear as a single strand. The pairs of fine loops of DNA coming from the chromomeres are packed in among the bivalent strands. The relatively large, deeply staining lumps of material that are conspicuous in this nucleus mark the location of the centromeres. This centromere-associated material consists of condensed DNA and is called *centromeric heterochromatin*.

The structure of the bivalents at this stage of meiosis is more clearly shown in the severely squashed nucleus of Figure 4-6. In this nucleus, the bivalents have been separated and pushed out of their normal position. The centromeres of the bivalents are marked by deeply staining centromeric heterochromatin. The chromomeres and their associated loops are clearly visible.

In the next stage, Figure 4-7, the DNA of the chromomeres of the preceding stage spins out into such greatly elongated loops that the bivalent axes almost disappear. The bivalent axes are only faintly visible and the entire nucleus is packed with the greatly extended filamentous threads of the loops. The material associated with the bivalent centromeres, the centromeric heterochromatin, remains condensed and is conspicuous as deeply staining granulelike particles. At the earliest stage of the first meiotic prophase, threadlike loops of DNA springing from the chromomeres were present. It is at the stage illustrated in Figure 4-7 that they reach their maximum extension, producing a diffuse nucleus lacking clearly visible bivalent strands. It seems that at this diffuse stage of meiosis the DNA that had been packed in the chromomeres is spun out so as to place it in intimate contact with the metabolic pool of the nucleus.

To produce the nuclei of the next stage in the prophase of the first meiotic division, the greatly extended loops of genetic material of the diffuse stage must again become folded into the chromomeres of the bivalent axes. The appearance of nuclei just emerging from the diffuse stage is illustrated in Figure 4-8. As the bivalents become more clearly visible (Figure 4-9), it can be seen that the component halves, so closely associated earlier, are separated except at certain positions where they are engaged in intimate contact. These places of intimate association of the strands are known as *chiasmata* (singular *chiasma*). It is at these locations that the breaks and exchanges of crossing over are believed to occur. The detailed structure of a bivalent at this stage is illustrated in Figure 4-10. Chiasmata are present and there is a terminal contact of strands at each end of the bivalent. The centromeres appear as more deeply staining short regions on each of the four strands.

Figure 4-11 shows the bivalents at this stage of meiosis with their four strands involved in the intricate twists and turns caused by the occurrence of chiasmata between the homologous members of the bivalents.

METAPHASE.   In Figure 4-12 we see the chromosomes as they approach the first meiotic metaphase. The bivalent axes possess the same kinds of filamentous loops that were present in the earlier meiotic stages.

Metaphase of the first meiotic division is illustrated in Figure 4-13. The bivalents have moved to the equatorial position within the cell, but in this illustration they have been pushed out of their normal equatorial arrangement during preparation. The homologous centromeres of the bivalent halves have become more widely separated, suggesting the pulling forces of the spindle microtubules that are operating at this stage. The arms on either side of the centromeres of the bivalent halves remain locked in chiasmata, but the chiasmata have moved toward the chromosome ends during the late prophase stages in a process known as "terminalization." The first meiotic metaphase bivalents are shown in Figure 4-14, the equatorial arrangement again having been disturbed by the squashing.

ANAPHASE.   Early anaphase of the first meiotic division is illustrated in Figure 4-15. Homologous centromeres of the bivalent halves are moving in opposite directions so that the bivalent halves are separated to give two sets of two-stranded structures. In Figure 4-16, the first anaphase has been completed and the resulting chromosome sets have been separated.

The Second Meiotic Division

INTERPHASE AND PROPHASE.   The two nuclei presented in Figure 4-17 are in the interphase between the first and second meiotic divisions. Chromosome duplication does not occur during this interphase. The interphase nuclei enlarge and enter prophase of the second meiotic division, indicated by the gradual appearance of the half bivalents. In the early prophase nucleus, the location of the centromeres is marked by the more deeply staining centromeric heterochromatin from which the faintly visible strands extend (Figure 4-18). Somewhat later, the chromosomes become completely distinguishable, with the two chromatids of each chromosome still attached at the centromere regions (Figure 4-19).

METAPHASE AND ANAPHASE.   The second meiotic metaphase is shown in Figure 4-20 with the chromosomes somewhat squashed out of their normal equatorial position. The centromeres of a centrally located chromosome are stretched so that they have the shape of a V, suggesting the pulling forces of the spindle fibers. Anaphase of the second meiotic division (Figure 4-21) separates the sister centromeres of the metaphase half bivalents. Thus the two meiotic divisions are completed with the production of four products, each with a precisely halved—haploid—set of chromosomes.

To summarize the major features: (1) Chromosomal duplication takes place during the interphase prior to the prophase of the first meiotic division. This doubleness of the chromosome strands does not become visible until later. (2) Synapsis occurs early in the first meiotic prophase, bringing the homologous strands into intimate contact. (3) Exceedingly fine, threadlike pairs of loops spin out from the chromomeres, so greatly depleting the substance of the chromomeres that the bivalent axes almost disappear from view. (4) The loops return in large part to their respective chromomeres, producing bivalents in which the halves are separated except at the chiasmata, which are the positions at which the breaks and exchanges of crossing over are believed to occur. (5) The four-stranded bivalents are now separated by the two anaphases: the first anaphase separates the homologous centromeres and the second anaphase separates the sister centromeres. (6) The four cells that result from the two meiotic divisions contain only half the number of chromosomes of the cell from which they have been derived, since there was no duplication of the genetic material in the interphase between the first and second divisions. (7) The two members of each homologous pair of chromosomes, genetically modified by crossing over, are separated from each other by meiosis and distributed into different nuclei so that each of the four products of a meiosis will have only one member of a given homologous pair. These four haploid products of meiosis differentiate into four sperm in the male and one egg and three nonfunctional polar bodies in the female. The diploid number is then restored for the next generation when the haploid egg and haploid sperm unite to produce a diploid zygote.

Continuity of the Germ Line

During the very early stages of embryonic development, a small number of cells are set aside to become the germ cells for the forthcoming generation. All other cells are destined to go through mitosis, to differentiate, and to form specific tissues of the developing embryo. These somatic cells will be represented by their mitotic descendants in the mature animal and will cease to exist when the animal dies. The germ cells, however, will form gametes and it is the combination of two gametes that will form a new individual. Hence the germ cells are characterized by potential immortality; these are the cells that concern us when we consider the propagation of genetic changes from one generation to the next.

The Timing of Meiosis

The course of germ cell formation in the human male parallels the natural maturation of the sexually mature male. During infancy and childhood, the potential germ cells (spermatogonia) remain relatively quiescent and do not develop into spermatocytes and mature sperm until the onset of puberty in the early teens (Figure 4-22). After this time, the male then produces sperm until the end of the reproductive lifetime, many decades later.

One might, therefore, imagine that the course of events in the human female would parallel that of the male. This is not the case. The course of oogonial proliferation is at its maximum during the middle of the fetal life, well before birth, at which time the total number of potential egg cells runs into the millions (Table 4-1). During the latter part of this period the cells enter the late prophase of the first meiotic division. The chromosomes then assume a fuzzy appearance and the cell goes into a prolonged prophase stage called *dictyotene*. At the same time, large numbers of these oocytes disintegrate so that the total number of

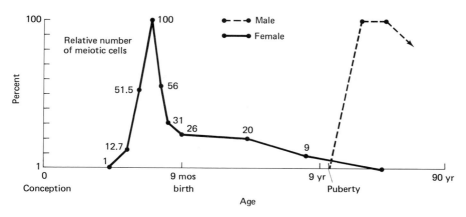

Figure 4-22 The relative number of germ cells in meiotic division in the male and female at different ages. In order to expand the prenatal period on the graph, the ages are plotted logarithmically so that the first third of the graph represents conception to 9 months (birth), the middle third represents birth to 9 years, and the last third represents 9 to 90 years. (Data for female meiosis from T. G. Baker, *Proc. R. Soc. Lond.* [*Biol.*], **158**:427, 1963.)

potential egg cells at the time of birth will have decreased to several hundred thousand. This number will continue to decrease throughout the reproductive lifetime of the woman. At the time of puberty, the egg cells are released individually (usually) at intervals of about once a month. At the time of their release from the ovary, at ovulation, the cells resume meiosis and continue as far as the metaphase of the second meiotic division, at which time the process again halts. It continues when the egg is fertilized by a sperm, after which time four meiotic products, one mature egg and three polar bodies, are present (Figure 4-23).

Age Before Birth	Total Number of Germ Cells/Ovary
2 mo	596,800
3 mo	1,421,200
4 mo	3,577,200
5 mo	6,831,600
6 mo	3,609,400
7 mo	2,277,600
At Birth	**2,023,600**
Age After Birth	
6 mo	383,000
10 mo	288,600
2 yr	402,000
7 yr	188,000

Table 4-1 Calculated populations of germ cells in the human female. Note that in the upper half of the table age is in months after conception of the female embryo. (T. G. Baker, *Proc. R. Soc. Lond.* [*Biol.*], **158**:417–33, 1963.)

Figure 4-23   Time sequence of events of meiosis in the female, showing the prolonged prophase stage of the first division, the resumption of meiosis after ovulation of the egg up to the second division, and the completion of meiosis after fertilization.

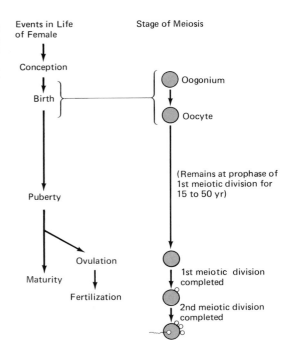

From this it can be seen that the egg which will give rise to an embryo during the middle of the reproductive life of a woman will have remained in the dictyotene stage for two or three decades.

MEIOTIC NONDISJUNCTION.   Just as chromosomes may be lost or go to the wrong pole during anaphase of mitosis, they may behave similarly during meiosis, giving rise to *meiotic nondisjunction* (Figure 4-24). It can occur at either the first or the second division of meiosis. If it occurs at the first meiotic division, and if the second meiotic division separates the sister strands normally, then the gamete may contain two homologous chromosomes which, when uniting with a gamete from the opposite sex, can produce a trisomic embryo. The complementary class will have no chromosome, and can give rise to an embryo monosomic for the chromosome involved.

Nondisjunction at the second meiotic division may give a gamete that carries two sister chromatids (Figure 4-25) and generally will be indistinguishable from first meiotic nondisjunction. There are several cases, however, in which they may be distinguished. For instance, the YY sperm that produce XYY males must come from second meiotic nondisjunction, since first meiotic nondisjunction would give only XY sperm (Figure 4-26).

TRISOMY-21.   One of the interesting characteristics of trisomy-21 is that its frequency is higher in offspring born to older mothers than in those born to younger ones (Figure 4-27). This is known to be an effect of the mother's age, not the father's age, because if the ages of both parents are checked it is found that an increased incidence of trisomy-21 is found only when the mother is

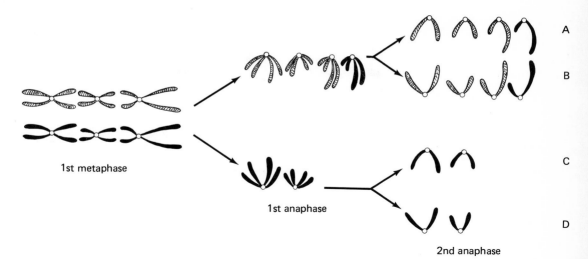

Figure 4-24 Nondisjunction of two homologous chromosomes at the first meiotic division. At the conclusion of meiosis, two of the four products (*A* and *B*) will have an extra chromosome beyond the haploid set and two (*C* and *D*) will lack one.

older. It seems reasonable to suppose that this maternal-age-dependent event is related to some alteration of normal meiosis in the older mothers.

One common and adequate explanation is that the oocytes, which have been stalled at the dictyotene stage since about the time of birth, deteriorate with age, and some 40 years later do not go through meiosis properly, but instead the chromosomes show a tendency to nondisjoin. It has been suggested that this tendency is particularly obvious for the chromosome 21 pair because they are

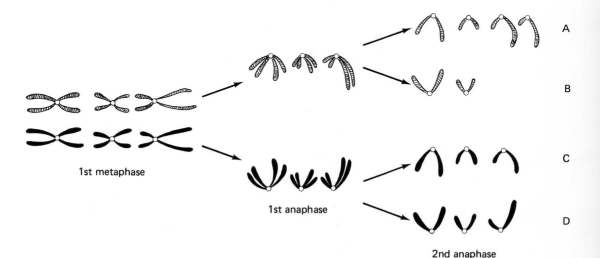

Figure 4-25 Nondisjuction of chromosomes occurring at the second anaphase division, resulting in an extra chromosome in one of the four products (*A*) and a missing one in another (*B*).

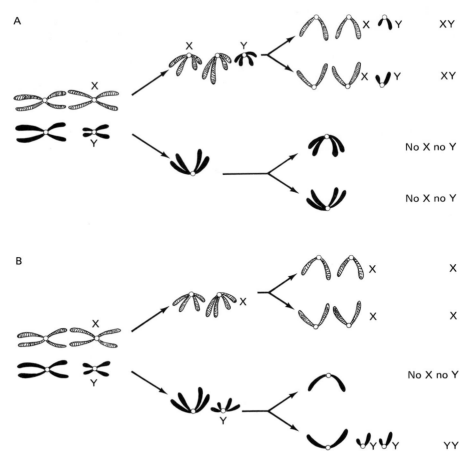

Figure 4-26   Illustration of a basic distinction between first- and second-division nondisjunction. When the homologues are different, as they are in the case of the X- and Y-chromosomes, first-division nondisjunction (*A*) can produce only XY or no-X no-Y sperm. Second-division nondisjunction (*B*), however, can produce YY sperm but not XY sperm. The unlabeled chromosome is an autosome with normal behavior.

the smallest of the complement and therefore may have fewer than average chiasmata to hold them together up to the time of the first meiotic metaphase. It should be kept in mind, however, that because chromosome 21 is small, trisomic fetuses are undoubtedly more viable and come to term more often than trisomics for the larger chromosomes.

When statistics are gathered on the frequency of occurrence of trisomy-21 as mother's age increases, not only does the risk go up markedly but so does the *recurrence risk*, the risk of a second affected child after a first has been born (Table 4-2). If an occasional spontaneous nondisjunctional event is responsible for trisomy-21, why should the occurrence of one have any effect on the chance of a second? To some extent, it may be that some women are constitutionally disposed to nondisjunction, so that the trisomies tend to occur to a small degree in sibships. However, an important consideration is that only 97 percent of all

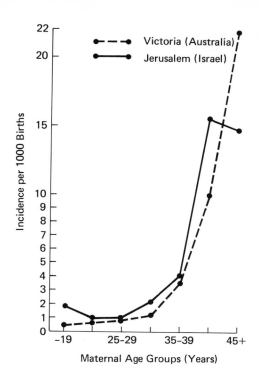

Figure 4-27 The greatly increased frequency of trisomy-21 in two widely separated populations, with increasing age of the mother. (J. Wahrman and K. Fried, *Ann. N.Y. Acad. Sci.*, **171**:341–60, 1970.)

cases of trisomy-21 result from nondisjunction of normal chromosomes; the other 3 percent are caused by a different phenomenon involving chromosome anomalies called *translocations*. These are particularly interesting because they are transmissible through a normal parent, father or mother, and may affect several offspring in a sibship, causing the appearance of trisomy-21 independent of the age of the mother. These are discussed in detail in Chapter 22.

MATERNAL AGE AND OTHER CHROMOSOME ANOMALIES.   Because of the lower frequency of birth of infants affected with other trisomies, it is more difficult to make a categorical statement about an age effect. However, what information

Table 4-2   The relationship of maternal age to trisomy-21. The chance of a trisomy-21 child for mothers of different ages is given under "risk of occurrence" when there has been no previous affected child and under "risk of recurrence" when there has been a previous affected child. (From A. Redding and K. Hirschhorn, Guide to Human Chromosome Defects. In *Birth Defects: Orig. Art. Ser.*, ed. D. Bergsma. Published by Williams & Wilkins Co., Baltimore, for The National Foundation—March of Dimes, White Plains, N.Y., Vol. IV [4], 1968.)

Maternal Age	Risk of Occurrence	Risk of Recurrence
20–30	1:1,500	1:500
30–35	1:750	1:250
35–40	1:600	1:200
40–45	1:300	1:100
45 up	1:60	1:20

is available suggests that trisomy-13 and trisomy-18 births may also appear at advanced maternal ages. This, however, is not true for the birth of infants with Turner's syndrome, suggesting that the syndrome does not arise primarily from nondisjunction in the oocyte as it proceeds through meiosis. This fits in very well with the observation that a large proportion (up to 30 percent) of all Turner's syndrome individuals are known to be mosaics, and this method of origin would require a mitotic nondisjunctional event occurring during the early cleavage divisions shortly after fertilization of the egg. Furthermore, as we will see later, many XO females carry the X-chromosome of the mother, indicating that in such cases the nondisjunctional event occurred in the father.

References

BAKER, T. G. 1963. A quantitative and cytological study of germ cells in human ovaries. *Proc. R. Soc. Lond.* [*Biol.*], Vol. 158, Oct. 22.

HAMERTON, J. L. 1971. *Human Cytogenetics.* Vol. I: *General Cytogenetics.* New York: Academic Press.

HIRSCHHORN, K., and H. L. COOPER. 1961. Chromosomal aberrations in human disease. A review of the status of cytogenetics in medicine. *Am. J. Med.,* **31**:442–70.

JAGIELLO, G. M., and P. E. POLANI. 1969. Mammalian meiosis with special reference to man (a pictorial presentation). *Guy's Hosp. Rep.,* **118**:413–31.

KEZER, J. 1969. Meiosis in salamander spermatocytes. Supplement to Chapter 6 in F. Stahl, *The Mechanics of Inheritance,* 2nd ed. Englewood Cliffs, N.J.: Prentice-Hall.

LEVINE, H. 1971. *Clinical Cytogenetics.* Boston: Little, Brown.

REDDING, A., and K. HIRSCHHORN. 1968. Guide to human chromosome defects. In D. Bergsma, ed., *Birth Defects: Orig. Art. Ser.,* Vol. IV. Baltimore: Williams & Wilkins.

RHOADES, M. M. 1961. Meiosis. In J. Brachet and A. E. Mirsky, eds., *The Cell,* Vol. 3. New York: Academic Press.

WAHRMAN, J., and K. FRIED. 1970. The Jerusalem prospective newborn survey of mongolism. *Ann. N.Y. Acad. Sci.,* **171**:341–60.

WHITE, M. J. D. 1974. *Animal Cytology and Evolution.* Cambridge: Cambridge University Press.

Questions

Useful terms: reciprocal cross, meiosis, gonial, synapsis, spermatogonium, spermatocyte, oogonia, oocytes, bivalent, polar body, recombination, crossing over, centromeric heterochromatin, chiasma, dictyotene, meiotic nondisjunction, translocation, recurrence risks.

1. What line of argument might a nonbiologist (a philosopher, say) use to predict the existence of a meiotic process?
2. List the most important ways by which the two meiotic divisions differ from two successive mitotic divisions.

3. By what two major mechanisms are genes recombined during meiosis? What is the evolutionary significance of recombination?

4. At what stages of meiosis do the following become evident: chromatids, synapsis, bivalents, polar bodies, crossing over, centrosomes, chromomeres, centromeric heterochromatin, chiasmata, dictyotene?

5. Describe the difference between meiosis in the male and in the female, with respect to (a) the different kinds of cells produced during the meiotic divisions and (b) the timing of the different meiotic stages during the lifetime of the two sexes.

6. How does meiotic nondisjunction differ from mitotic nondisjunction? Can nondisjunction at the second meiotic division produce gametes different from those that come from nondisjunction at the first division?

7. Women do not stand the same frequency of producing a trisomy-21 offspring at all ages. How does this frequency change with age? Does this change have a very simple explanation?

8. If a couple have a child with Down's syndrome, which of the parents is responsible for the anomaly?

9. The ages of married couples must be highly correlated. How do we know, then, that the frequency of trisomy-21 depends on the mother's age, and not the father's age?

10. Large sums of money are expended for the prevention of birth defects. To what extent are such programs likely to be successful in preventing the occurrence of an abnormal embryo in utero caused by teratogenic agents compared to those caused by nondisjunction?

11. Discuss the relevance of the following to the concepts discussed in this chapter: trisomy-21, age effect; zygote.

5

Simple Inheritance

From even the most casual observations of parents and children it is clear that there exist many characteristic differences inherited from one generation to the next. A more detailed examination of families by physicians, human geneticists, biochemists, and similar inquisitive people yields a total of several thousand inheritable characteristics, but the vast majority of them are of interest primarily to specialists in the field.

Two common characteristics with a simple inherited basis are *albinism* and "dwarfism" of a type known as *achondroplasia.* For some common characteristics such as eye and hair color, minor complications in their inheritance pattern prevent us from using them as examples. For some characteristics such as allergies, there is almost certainly some elementary genetic mechanism involved but it has not yet been precisely defined. Other traits, such as interracial differences in pigmentation, have a definite genetic basis, but are inherited in a more complex fashion. Some characteristics that we believe to have some inherited basis depend to some extent on the person's genetic composition, such as cleft lip, but developmental variables may be more important. Height and weight of an individual also fall into this category. Finally, at the other extreme, there are certain characteristics for which the extent of the genetic component is still under investigation, and in which environmental influences must play a large part, such as intelligence as defined by performance on standard IQ tests. For now, however, we shall consider only a few traits that are particularly useful in illustrating the nature of simple genetic inheritance in humans.

Dominant Inheritance

Let us take as our first example the inheritance of achondroplasia. In this type of dwarfism, the long bones do not develop properly, with the result that the individual is usually quite short, although the trunk and head may approach normal size (Figure 5-1). The intelligence of these persons is unaffected and they mature physically and psychologically; in fact, it has been noted that their personalities tend to be of a very cheerful type.

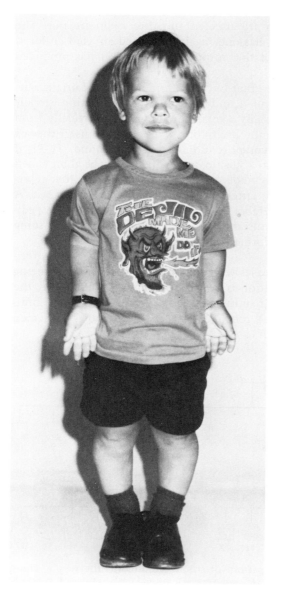

Figure 5-1 An achondroplastic boy whose well-being is clearly evident. (Courtesy of Judith Hall, University of Washington Medical School.)

THE PEDIGREE.   The relationships in a family with this characteristic are shown in Figure 5-2. In a standard *pedigree* chart, the squares refer to males and the circles to females. A horizontal line connecting the two represents a mating, and a vertical line connects them to their progeny, who, as a group, are referred to as *sibship.* The first child is given on the left with successive progeny to the right. Individuals who are affected by the given characteristic are designated by a solid symbol. In order to make it possible to specify certain individuals in the pedigree, the rows of individuals (generations) are numbered with Roman numerals *I, II, III,* and so on, and the individuals in a given row are numbered consecutively with Arabic numerals.

In this pedigree, an affected male (*I-1*) has married a normal female (*I-2*) and they have seven progeny, four who are affected and three who are not. The marriage of one affected child (*II-1*) has once again produced affected children (*III-1* and *III-3*), but the marriage of the normal progeny (*II-7*) has produced only normal grandchildren.

Twins are represented by two lines that have a common point of origin, suggesting their shared birth (as in *II-3* and *-4,* and *II-6* and *-7*). A distinction is made between identical and fraternal twins by a small line projecting vertically from the sibship line to indicate the common origin of the identical twins. If the sex of an individual is unknown, the symbol is indicated as a diamond (*IV-9*); if there are several similar individuals in a sibship, the pedigree may be abbreviated by putting the number inside the symbol, as in *IV-7* and *IV-9,* the first representing two female progeny and the latter two children of unspecified sex. A stillbirth is indicated by a smaller symbol than usual. At any point in the pedigree, a small question mark can be inserted within a symbol to indicate some doubt as to the reliability of the information indicated by that symbol.

THE PROPOSITUS.   Except in those cases where complete populations are being examined for specific diseases, attention is ordinarily drawn to a family by the existence of a particular genetic trait in one individual. Such an affected individual is called either the *propositus, proband,* or *index case* and is indicated on a pedigree by an arrow (as in *III-3*).

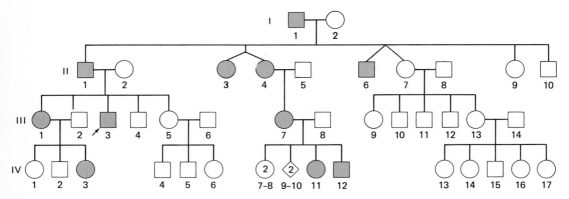

Figure 5-2 A pedigree showing the pattern of simple dominant inheritance, along with the more common symbols used in pedigrees. Detailed explanation of symbols in text.

This pedigree of a family affected with achondroplasia illustrates three special properties of the inheritance of certain defects: (1) all affected individuals have at least one parent who is also affected; (2) about half of the progeny of an affected individual are also affected; and (3) both sexes appear to be affected about equally frequently. These rules characterize *simple dominant inheritance.*

THE RELATION TO MEIOSIS.   It is reasonable to ask why the progeny show the characteristic to the same extent as the parents—would it not make more sense if each child were a blend, intermediate between the two parents? Furthermore, why are only half the children of an affected parent also affected, and not all of them?

The answers to these questions can easily be found if we consider the consequences of the meiotic divisions. The end result of meiosis is the production of gametes, half of which carry one of the two chromosomes of each homologous pair, the other half the other chromosome. Let us suppose that the achondroplastic dwarf parent has a genetic factor which we will call *D* that is responsible for his condition. This factor is present on one of the two homologous chromosomes present in the cells. Which chromosome carries this particular trait, whether in the A or the G group, whether number 3 or number 17, is unknown at present, but this is unimportant to us now. On the other homologue of the chromosome carrying *D* is the normal factor, *d*. This person's genetic composition with respect to the achondroplastic factor is therefore *Dd*; we refer to this as his *genotype.* After meiosis, half of his gametes will carry *D* and half *d*.

Since it is the presence of *D* on one of the chromosomes that makes an individual achondroplastic, all normal persons have a genotype of *dd* and all of their gametes will carry *d*. Thus it is easy to see that when an achondroplastic individual mates with a normal person the chance combination of their gametes will result in half of the progeny being achondroplastic and the other half normal (Figure 5-3). Because the allele *D* manifests its characteristic *phenotype,* achondroplasia, in the *Dd* combination, and *d* does not, *D* is said to be *dominant* over *d*, and *d* is *recessive* to *D*.

Linear Order of the Genes

In certain tissues of the *fruitfly, Drosophila melanogaster,* we find a unique situation in which the chromosomes enlarge by successive replication without cell division, so that they attain a size a thousand times greater than normal, with resulting visible detail that is far greater than can be obtained from human chromosomes at the present time. This makes them ideal for genetic research

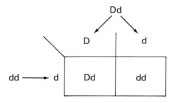

Figure 5-3   A simple checkerboard, or Punnett square, illustrating the standard method for predicting the kinds of progeny from a mating. In this case, one parent carries two *dd* factors so that all of the gametes are *d*. In the other case, one chromosome carries *D* and the other *d* so that half of the gametes carry one and half the other. These symbols on the margins are then combined in the squares, to give the expected proportions of progeny, ½ *Dd* and ½ *dd*.

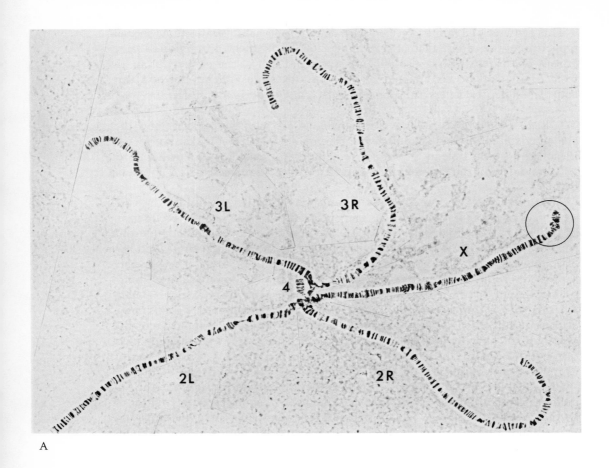

A

Figure 5-4 (*A*) The polytene chromosomes found in the salivary glands of the
fruitfly *Drosophila*, much larger than usual
chromosomes because they consist of thousands of chromosome strands side by side.
The individual bands may be considered to
be the location of specific genetic factors.
(*B*) A line drawing showing how the *Drosophila* chromosomes form this typical polytene configuration. All the chromosomes
come together at the locations of their centromeres, and the two homologues of each
chromosome of the diploid are so tightly
synapsed that the two together appear to be
a single strand. (Courtesy of G. Lefevre,
California State University at Northridge.)

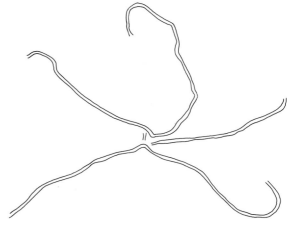

B

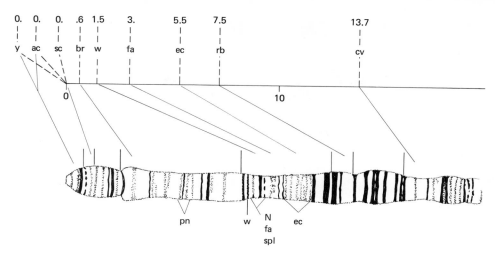

Figure 5-5 A magnification of the tip of the X-chromosome of *Drosophila*, the section encircled in Figure 5-4*A*. The letters in the top row represent symbols for genetic factors (*y* = yellow body color, *w* = white eyes, *rb* = ruby-colored eyes, etc.), and the lines to the section of the polytene X-chromosome show the location of that particular gene.

and experimentation. Because each chromosome is composed of so many strands joined together, they are referred to as *polytene chromosomes* (*poly* = many, *tene* = thread). Figure 5-4*A* shows a set of polytene chromosomes from a cell of the salivary glands of *Drosophila*. They form an unusual configuration because (1) the two homologues are completely fused together so that each arm is actually two homologues and (2) all of the chromosomes are joined together at their centromere regions because of an apparent "affinity" for the chromosome material, *heterochromatin*, that surrounds the centromere.

Figure 5-4*B* is a line drawing showing how the four polytene chromosomes of *Drosophila*, including an X-chromosome and a small dot chromosome, make up this configuration. At higher magnification it can be seen that each chromosome arm is crossed by bands, approximately a thousand on each arm. These bands may be considered to be the genes, those factors located on the chromosomes that are responsible for the development of specific tracts in the individuals carrying them. Figure 5-5 shows the end of the X-chromosome of *Drosophila*, along with a detailed picture of the locations, or *loci* (plural of *locus*), of some genes. In *Drosophila* it is possible to find the exact locus on each chromosome arm of a large number of genes whose functions are known. The limitations of the chromosome detail in humans have prevented any comparably exact assignment of genes within a chromosome, although great strides are currently being made using the newer staining techniques in assigning various types of genes to specific chromosome regions.

Allelic Relationships

In the example of achondroplasia, the gene *D* is responsible for dwarfism and *d* is responsible for normal development in the absence of *D*. These two different forms of a gene are called *alleles*. Thus, *D* is an allele of *d* and vice versa; but

Figure 5-6 A severely affected child, the progeny of two parents with achondroplasia. She probably received the *D* allele from each parent and represents the very rare case of the homozygote for this gene. (Courtesy J. G. Hall, J. R. Dorst, H. Taybi, et al.: Two Probable Cases of Homozygosity for the Achondroplasia Gene. In *Birth Defects: Orig. Art. Ser.*, ed. D. Bergsma. Part IV. Skeletal Dysplasias. The National Foundation–March of Dimes, White Plains, N.Y. Vol. IV [4] pp. 25, 26, and 28, 1969.)

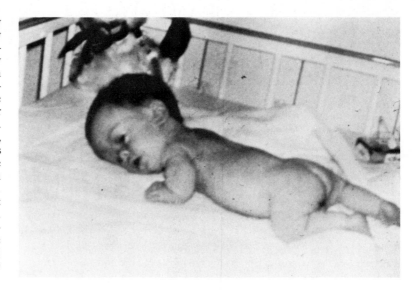

D, responsible for achondroplasia, would not be an allele of the gene for, say, eye color, since that gene would be found at a completely different locus, probably even on a different chromosome. If the two alleles on the two homologous chromosomes carried by an individual are the same (*DD* or *dd*), that person is said to be *homozygous* for those alleles; if they are different (*Dd* or *dD*), he or she is *heterozygous*. It is natural to wonder what the appearance, or phenotype, of the homozygote for the gene *D* for achondroplasia would be. Such homozygotes are rare, because the *D* allele would have to be contributed by each parent —in other words, each parent would have to be affected. In the case of achondroplasia, such children are probably very severely affected (Figure 5-6) and die young, if they survive to birth. For many relatively rare dominant genes, the phenotype of the homozygote is unknown.

In adopting symbols for the designation of alleles, the student has wide freedom. Following the practice initiated by Gregor Mendel in his pioneering work with the garden pea, however, it is customary to use a letter of the alphabet to designate a locus, and to use the capital to denote the dominant allele and the lower case for the recessive. Some confusion can be avoided if different loci are given different letters, and if the letter chosen is the initial letter of a word describing the phenotype, such as *D* for dwarfism or *a* for albinism.

Recessive Inheritance

Dominance requires the presence of only one different allele, of the two present in the cell, to produce a distinctive appearance, or phenotype. The alternate, recessiveness, requires the simultaneous presence of two like alleles for the new phenotype to appear.

In the common forms of albinism, the biochemical reactions that lead to the formation of *melanin*, a dark skin pigment, are blocked so that such individuals

Figure 5-7 A group of Shimopaui Indians including three albinos. (Courtesy of the Smithsonian Institution.)

have extremely light skin and hair, with little or no coloration to the iris of the eye, giving their eyes a pale blue or pink appearance (Figure 5-7). Although this is a relatively rare defect, with only about one in 20,000 people showing it, it is so striking that most adults have at one time or another noticed an albino human. In any case, it is a defect well known to children from popular white-colored varieties of pets, such as mice, rats, guinea pigs, and rabbits.

In order to manifest albinism, an individual must have two *a* alleles, i.e., must be homozygous *aa*. Those who have one *a* allele and one *A* allele, i.e., are heterozygous *Aa*, are normal in appearance, as of course are those who are homozygous *AA*. In many cases, the allele defined as the recessive one may show some slight effect in the heterozygote, but the relationship is conveniently referred to as a recessive one nevertheless.

A typical pedigree showing the inheritance pattern of albinism is given in Figure 5-8. Here we can see that an affected individual appears in a family without either of the parents being affected. When one child in a sibship is affected, there may be several others, depending on the family size, but the overall expectation is not one half, as it is in dominant inheritance, but rather one fourth. Figure 5-9 shows why this is so. The homozygous recessive type appears because both parents are heterozygous, and with two heterozygous (*Aa*) parents we can see that if half the gametes from each of the parents carry *a* then the probability that any one zygote produced will be homozygous *aa* is one half of one half, or one quarter. This checkerboard diagram, introduced in the early days of genetics

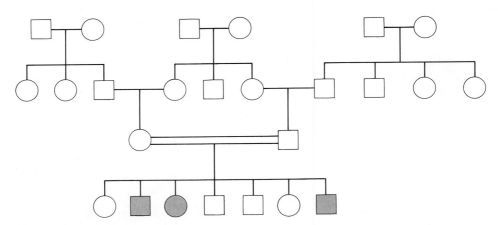

Figure 5-8 A pedigree showing the inheritance of albinism. In this particular case, the two parents of the albinos are first cousins.

Figure 5-9 A checkerboard showing how two heterozygotes, *Aa*, will produce, on the average, one quarter homozygous *aa* children.

	Eggs from Aa	
	A	a
A	AA	Aa
a	Aa	aa

Sperm from Aa

by the British geneticist Punnett and therefore referred to as a "Punnett square," is a simple and accurate way of determining types and frequencies of genotypes expected from parents of given genetic constitutions. Even the most professional geneticist will make use of this system to explain or predict the results of genetic crosses.

With respect to the locus for albinism under consideration here, most people will be homozygous for the dominant allele (*AA*) and it is therefore possible for us to designate the genetic constitutions or genotypes of the individuals in a pedigree with some degree of assurance as in Figure 5-10. Where there is a good chance that a phenotypically normal person may be heterozygous Aa (as would be the case, for instance, if one of his parents were known to be Aa), but it cannot be stated for certain whether he is AA or Aa, then that person's genotype may be given as A–.

CONSANGUINITY. Under ordinary circumstances, the occasional rare individual homozygous for a deleterious recessive allele is produced because each of his parents happened by chance to be heterozygous for the same recessive. However, in those cases where the parents are related, i.e., have a recent ancestor in common, that ancestor might have carried a recessive which, by chance, was transmitted to the two parents. In this case, the probability of a homozygote appearing is very greatly increased.

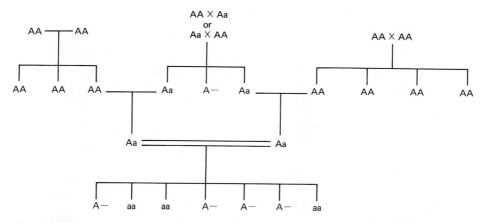

Figure 5-10 The pedigree shown in Figure 5-8 with the genotypes of the members replacing the usual pedigree symbols. Because the recessive allele for albinism is relatively rare, we are justified in assuming that any normal person is homozygous *AA*, unless he is either a parent of a homozygous child or one of a sibship in which either a homozygote or a heterozygote has appeared.

The increased frequency of homozygotes from marriages of related persons, or *consanguineous* matings, is of great importance in human genetics studies, and such unions, where known, are represented in a pedigree by a double line connecting the partners. The appearance of *consanguinity* in a pedigree is itself a strong reason for suspecting recessive inheritance and, equally, pedigrees that involve recessive traits will often prove to have consanguinity in the ancestry of the affected members.

A quick way of recognizing a pedigree involving consanguinity is to note that it is characterized by an area that is completely enclosed by pedigree lines. Figure 5-8 is that of a consanguineous marriage: the two parents giving rise to the sibship with two albino children are first cousins, since they have a pair of grandparents in common, and it was from one of these grandparents that both received the recessive allele.

The Work of Gregor Mendel

What we have described earlier in this chapter is an illustration of the rules of inheritance that were first derived in the middle of the nineteenth century by Gregor Mendel, an Austrian monk (Figure 5-11*A* and *B*). His experiments with the garden pea, *Pisum sativum*, can truly be considered one of the great landmarks of modern biology, and if not the first careful and thoughtful attempt to analyze breeding experiments then certainly the first to achieve success.

Gregor Mendel was born in 1822 in northern Moravia, which was then a part of Austria, now in Czechoslovakia. At the age of 21 he was admitted to the Augustinian monastery at Brünn (now Brno, Czechoslovakia), where he spent the rest of his productive life. After self-training in the natural sciences, he took an examination for a teaching certificate in 1850 and failed. His poor perfor-

A   B

Figure 5-11   Portraits of Gregor Mendel, as a young man (*A*) and late in life (*B*). The props on the left in *B* appear to be symbols of a bishop, a rank Mendel did not attain. (Courtesy of V. Orel, Moravian Museum, Brno, Czechoslovakia.)

mance resulted more from his lack of a formal university education than from poor aptitude; his natural intelligence was well appreciated by those around him.

He then spent several years at the University of Vienna, where he studied all the basic sciences including paleontology. His emphasis was on physics, and his physics course was taught by Professor Doppler (well known for the Doppler effect), and his courses in mathematics, chemistry, and botany were also taught by well-known scientists of the time. A second attempt to pass the examination for a teaching certificate was thwarted by an indisposition at examination time. (This phenomenon has been observed as well among present-day university students.) Although classroom teaching occupied a good part of his time at the monastery, he nevertheless managed to devote considerable time to his research interests. We may surmise that if his talents had been directed toward full-time classroom teaching, as they are with many teachers today, his scientific experimentation would have been greatly curtailed and he might never have started his experiments in plant breeding.

In any case, instead of a simple-minded monk puttering about his small garden and happening upon some basic scientific principles by accident in the best Hollywood tradition, we have the picture of a trained scientist setting up his experiments with knowledge and purpose. In fact, in addition to his work with the garden pea, Mendel carried out breeding experiments as well with the pea weevil, with honeybees, and with mice. He was, in addition, an amateur meteorologist and made daily records of rainfall, temperature, humidity, and barometric pressure until his death. He kept records of sunspots and published a detailed description of a tornado which passed over the monastery. He requested, and obtained, permission to be present at an autopsy performed by local authorities. In 1868, after his experimental work with the garden pea, Mendel became abbot of his monastery. This office led him to excessive involvement in administrative duties, including a bitter controversy with the state on problems of taxation. The harassments of his administrative chores contributed to the phenomenal consumption of more than 20 cigars a day. Despite this, he continued his scientific experimentation to the end of his life. As he grew older, he suffered from a long progressive disease of the kidney with dropsy, uremia, and hypertrophy of the heart. He died in 1884 at the age of 61.

MENDEL'S RESEARCH. The best account of Mendel's work comes from his own paper published in 1866. Although Charles Darwin's epoch-making work on *Natural Selection and the Origin of Species by Means of Natural Selection* did not appear in print until 1859, Mendel states as one of the main reasons for starting the breeding experiments with the garden pea in 1854 that "It requires indeed some courage to undertake a labor of such far-reaching extent; this appears, however, to be the only right way by which we can finally reach the solution to a question the importance of which cannot be overestimated in connection with the history of the evolution of organic forms." If Darwin had been aware of Mendel's work and had appreciated its significance, he would have surmounted one of the major obstacles to his theory of natural selection: how it might be possible for desirable characteristics to be transmitted in full form from one generation to the next without being diluted out in successive generations. In fact, Darwin himself had observed segregation of simple characteristics in the second generation of crosses of the snapdragon, and had made counts that approximated a 3:1 ratio. But, not being sensitive to the significance of this fragmentary observation, Darwin proposed a theory of heredity he called "pangenesis," according to which the germ cells somehow received and incorporated information that came to them from the extremes of the body and then passed this information on to the next generation.

Darwin's theory of pangenesis was never really accepted by the scientific world, which would have to wait until 1900 for the rediscovery of Mendel's paper and the elucidation of the simplest rules of inheritance. In this paper, Mendel had pointed out the necessary criteria for setting up a proper genetic analysis: (1) making specific crosses with well-defined characteristics; (2) keeping track of the results for each generation; and (3) making exact counts, to, as he put it, "ascertain their statistical relations." The organism to be chosen for the experiment, according to Mendel, must possess a variety of constant characteristics from one strain to the next, must be easily crossed experimentally and readily protected from foreign pollen, and must not suffer from problems of infertility of the hybrids. The garden pea had additional advantages: it could be grown easily and had a relatively short period of growth.

Mendel obtained a large number of different strains of pea from nurseries throughout Europe and grew as many as 34 strains in the summer of 1854; the following year he checked to see which of the strains had constant characters, and the following year he selected a total of seven different characters for his final experiments.

Two additional advantages to the garden pea merit mention. When a plant produces seeds, certain characters are found on those seeds (round or wrinkled form; yellow or green color), which already represent the next generation. When Mendel made a cross of members of a line characterized by round seeds with those characterized by wrinkled seeds, the first generation would appear within a few months—in this case, as round seeds on the female parent plant. Furthermore, there would appear not just one progeny per mating but as many as there were seeds on the plant—probably from 30 to 40. Thus, although we generally imagine a series of plant crosses as involving parental plants giving rise to first-generation (F_1) plants and those in turn to second-generation (F_2) plants, each generation taking a year and each plant requiring a sizable amount of garden space, Mendel eliminated the first year's lapse by getting the first generation of a seed character on the parental plant, and was able to count many individuals (seeds) per plant.

This explains how he was able to collect such very large numbers for his ratios, on such a small plot of land and in such a short period of time (Figure 5-12). These advantages become apparent if his original figures are examined to compare the number of experimental individuals involving the two seed characters with the number involving the five plant characters he worked with. The two seed characters were the basis for progeny counts more than three quarters of the time; the five plant characters only one quarter.

MENDEL'S RESULTS. From the crosses of lines with two different contrasting characteristics, such as a tall plant crossed with a short plant, Mendel observed that one of the two characters manifested itself in the first generation. He introduced the terms "dominant" to describe the character that appeared in the F_1 heterozygote and "recessive" to describe the one that did not, and suggested the convention of using a capital letter for the dominant allele and the corresponding lowercase letter for the recessive. From matings of the two first-generation hybrids (F_1), from which he got the 3:1 ratio of phenotypes, he concluded that the factors responsible for the parental characteristics were not lost or changed in the hybrid, but were segregated as individual factors in equal proportions in the germ cells of the hybrid. This is the *principle of segregation,* or Mendel's first law. We have already applied that principle to explain the proportions of affected persons in a sibship for simple allelic differences such as achondroplasia.

In experiments involving not just one pair of characters but two, Mendel showed that the characters behaved independently of each other. For instance, consider that pair of characters affecting the seed color—either yellow or green, with yellow being dominant to green. The dominant allele yellow is symbolized by Y and the recessive green by y. If a strain homozygous for yellow (YY) is mated to a strain homozygous for green (yy), the F_1 seeds will be genotypically Yy, and phenotypically yellow. If two plants from these F_1 seeds are crossed, the F_2 seeds on those F_1 plants will be approximately three quarters yellow and one quarter green.

Round and wrinkled are also seed characters, with the second generation yielding roughly three quarters round and one quarter wrinkled. If a plant taken from a strain characterized by round green seeds is mated to one with wrinkled yellow seeds, all the F_1 seeds will be round and yellow. However, if these F_1 seeds are germinated and a cross is made between two similar plants (or if one such

Figure 5-12 A view of the area in which Mendel performed his experiments. It was his inspired use of seed characters that enabled him to accumulate large quantities of data in such a small space over a short period of time. (Courtesy of V. Orel, Moravian Museum, Brno, Czechoslovakia.)

plant is allowed to self-fertilize), then the seeds produced in the F_2 generation will combine the two $\frac{3}{4}$ to $\frac{1}{4}$ expectations into a $\frac{9}{16}{:}\frac{3}{16}{:}\frac{3}{16}{:}\frac{1}{16}$ distribution. Figure 5-13 shows a simple system making this combination of two independent crosses. The resulting ratio is generally expressed as a 9:3:3:1 ratio. The independence of different pairs of contrasting characters is embodied in Mendel's second law: the *principle of independent assortment.*

In the century since Mendel did his research we have come to understand that his law of independent assortment is based on the independent segregation of different chromosomes at meiosis. The locus for the *Y-y* alleles is on one chromosome, and the locus for the *R-r* alleles is on another. In the F_1 double heterozygote *YyRr*, the alleles separate from each other, following the behavior of the homologues carrying them, so that four different types of gametes are produced: *YR, Yr, yR,* and *yr,* with equal frequencies (Figure 5-14).

In retrospect, it appears that Mendel's numerical results were much closer to theoretical expectations than one might have anticipated on a chance basis. In fact, it has been calculated that results as close to the exact ratios as he got should occur by chance only one time in about 40,000. These too-good results have worried a number of investigators, some suggesting that perhaps Mendel had an overzealous laboratory assistant who miscounted the various phenotypes to make them conform more closely with his superior's expectations. Still another explanation is that, considering Mendel's calling, he may have had some divine assistance not ordinarily available to the average scientist! Since Mendel could

Figure 5-13 One method for deriving the 9:3:3:1 ratio found when two independent pairs of characters segregate independently. One set of the contrasting pairs. Round and Wrinkled, are entered on the top along with their expected proportions in the F_2 $\frac{3}{4}$ and $\frac{1}{4}$. The other contrasting pair, Yellow and Green, are entered at the side margin, along with their expectations. Each class is then combined with each other, to give four phenotypic classes, and the respective marginal frequencies are multiplied to give the proportions expected for each class. From this it can be seen that in an F_2 segregating for both the Round-Wrinkled and Yellow-Green pairs of characters the expected proportions are $\frac{9}{16}$ Round-Yellow, $\frac{3}{16}$ Round-Green, $\frac{3}{16}$ Wrinkled-Yellow, and $\frac{1}{16}$ Wrinkled-Green.

Proportions of Round vs Wrinkled		
	3/4 Round	1/4 Wrinkled
Proportions of Yellow vs Green — 3/4 Yellow	9/16 Round Yellow	3/16 Wrinkled Yellow
1/4 Green	3/16 Round Green	1/16 Wrinkled Green

not have profited from any manipulation of his data and, in any case, such a performance on his part would have been completely out of character judging from his well-substantiated willingness to accept contrary data, biologists have made a number of suggestions as to how such excellent results might have been obtained. Possibly he extended his experiments when the ratios appeared to depart from the expected to make sure that an unusual result was not a bona fide departure from his simple rules. Such an innocent procedure—perfectly rational and legitimate—would tend to eliminate the exceptional cases and make the rest of the data "look good."

So it was that the foundations of genetics were laid. Mendel was able to repeat his results with the garden pea using several other plant species (such as corn and four o'clocks). However, when he attempted to obtain hybrids in the hawkweed (*Hieracium*), he failed completely. It would not be known for more than 30 years that the reason for this failure was that seeds of that species can be

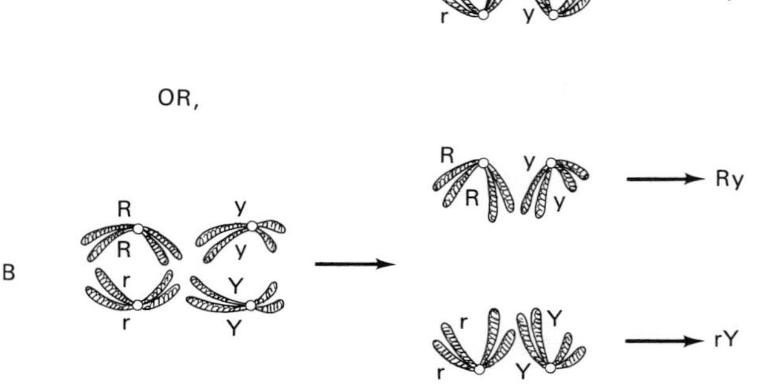

Figure 5-14 The arrangement of chromosomes in a plant heterozygous for two pairs of contrasting characters. Only two pairs of chromosomes are shown—those that are hypothesized to carry the loci of the independent factors. The orientation of the chromosomes at metaphase of the first meiotic division can be such that (*A*) the chromosomes with the dominant alleles of each heterozygous pair (*R* and *Y*) go to the same pole at anaphase, with the recessive alleles (*r* and *y*) going to the other. Equally frequently, however, (*B*) the chance orientation of these independent chromosome pairs will be such that a dominant allele of one pair will go along with the recessive of the other, to give the combinations *Ry* and *rY*. The final result, at the end of the second meiotic division, will consist of four possibilities, *RY, Ry, rY,* and *ry*, each expected equally frequently.

purely maternal in origin and may be produced without either a normal meiosis or fertilization. We remain fascinated with Mendel's short paper, not just because he explained the simplest rules of inheritance, but also because he implied the basic nature of the genetic mechanism in higher plants and animals. Without any knowledge of the role that chromosomes play in heredity—their significance would not be appreciated until almost a half century later—Mendel implied constancy of genes, diploidy of the mature organism, the reduction division and resultant haploidy of the gametes, and the segregation of independent units (the chromosomes) at meiosis. He explained mathematically the consequences of large numbers of generations of matings of closely related individuals, or inbreeding, and how genetic factors might interact to produce unusual ratios and unusual phenotypes. Possibly it was just this elegance and abstractness in his approach and presentation, stemming from his training as a physicist, that confused the botanists of his time and led to their failure to recognize the value of his work until it was "rediscovered" in 1900. If any one scientific work were to be described as being ahead of its time, surely Mendel's *Experiments in Plant-Hybridization* stands as first choice.

Other Examples of Simple Inheritance

It is sometimes felt by the beginning student that the discipline of human genetics must consist primarily of a listing of all known inherited characteristics of man, along with their mode of inheritance. A recent catalogue of inherited human traits* lists 583 defects as "good" autosomal dominants, another 635 as suspected or probable dominants, 466 as "good" autosomal recessives, and another 481 as probable recessives. Clearly even a partial listing in an elementary text would be impossible. However, there are so many important principles to be discussed that, in fact, the discussion of specific inherited traits is largely limited to their usefulness in exemplifying basic principles. Furthermore, the vast majority of cases of simple inheritance known are of interest only to specialists in medicine, biochemistry, or population studies. It is also true that for many physical characters of general interest—handedness, pigmentation variants such as freckles and moles, height, weight, and so on—the mode of inheritance is more complicated, probably involving numbers of interacting loci. Even for eye and hair color, although it is true that the darker pigmentation types are usually dominant to the lighter—brown eyes dominant to blue, brunette hair dominant to blonde—there exist not only well-established contradictions to these rules in some pedigrees but also an array of intermediate types (red hair, green and hazel eyes) for which no simple explanation suffices.

MALFORMATIONS OF THE LIMBS.   Anomalies of the hands and feet are not uncommon, and when they appear they usually show dominant inheritance. In Figure 5-15*A* is a clear example of *polydactyly*, in this case six digits instead of five, along with an x-ray photograph (*B*) of the feet revealing the bone structure. A shortening of the fingers (Figure 5-15*C*) known as *brachydactyly* is caused by

*V. A. McKusick. 1975. *Mendelian Inheritance in Man*, 4th ed. Baltimore: Johns Hopkins University Press.

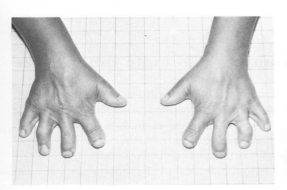

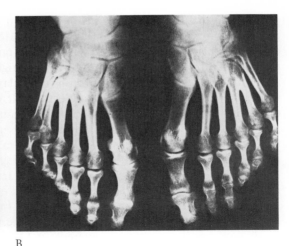

A

B

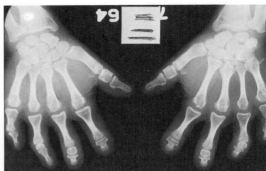

C   D

Figure 5-15 Two common dominant abnormalities of the hands and feet. (*A*) A photograph of a polydactylous pair of feet, along with an x-ray photograph of the same (*B*). (S. B. Pipkin and A. C. Pipkin, *J. Hered.*, **37**:93–96, 1946.) (*C*) A pair of brachydactylous hands, with the x-ray photograph (*D*) showing the shortening of the terminal joints. (D. Hofnagel and P. S. Gerald, Hereditary Brachydactyly, *Ann. Hum. Genet.*, **29**:377–82, 1966. Used by permission of Cambridge University Press.)

the shortening and fusing of the terminal bones (Figure 5-15*D*). Brachydactyly is of historical interest, because it was the first authenticated case of simple dominant inheritance in man, the pedigree having been published in 1905 (Figure 5-16). In another common defect, syndactyly, two or more of the digits of the hands and/or feet are joined together.

The preceding abnormalities are dominant in the sense that they produce a characteristic phenotype in the heterozygote. With the exception of the case of the achondroplastic homozygote described earlier, homozygotes for these alleles are unknown, because the defects are sufficiently rare that heterozygotes have not married to produce homozygous offspring. Another exceptional case is shown in Figure 5-17. In this instance, two first cousins with minor hand and feet abnormalities produced an offspring with extreme defects, undoubtedly caused by homozygosity for the allele carried by both parents.

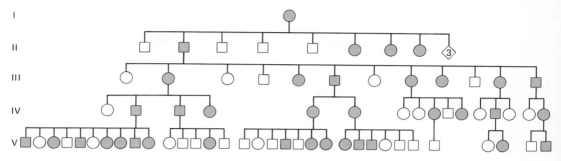

Figure 5-16 The first human pedigree of a simple mendelian dominant, showing the inheritance of brachydactyly. (Farabee Papers, Peabody Museum, Harvard University, **3,** 1905.)

CYSTIC FIBROSIS. The most common serious genetic disease in the United States is *cystic fibrosis*, affecting about 5 out of every 10,000 newborns in the White population. Its incidence in nonwhites (Orientals and Blacks) is much less, amounting to about 4 per million. It is typified by a defect in protein metabolism that leads to degeneration of certain organs such as the pancreas, chronic respiratory infection, and lung destruction. Affected persons usually die before the age of 18 years, but one affected person out of ten reaches adulthood. When they survive to reproductive age, affected females may produce normal offspring but the males are usually sterile.

This disease appears to be the result of homozygosity for a single recessive; both parents of an affected child must therefore be heterozygous. Biochemical tests

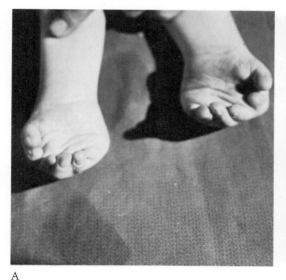

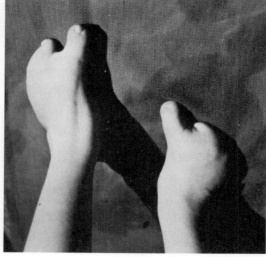

A   B

Figure 5-17 The extremely malformed hands (*A*) and feet (*B*) of an offspring of first cousins, both of whom showed minor foot defects. (J. A. Edwards and R. P. Gale, *Am. J. Hum. Genet.*, **24:**464–74, 1972.)

make it possible to detect otherwise normal heterozygotes, although the tests are not yet sufficiently simple to allow for extensive screening of individuals taken at random from the population. One of these tests involves the inhibition of cilia, cells with fine threadlike projections which beat to produce movements of adjoining fluid, in oysters and other shellfish.

TAY-SACHS DISEASE. *Tay-Sachs disease*, a recessive condition involving degeneration of the nervous system, appears at about the sixth month of life. The degeneration of the optic nerve leads to blindness, followed by loss of intellectual capabilities, progressive muscular weakness, and paralysis. It is usually lethal by the age of 2 or 3. Although the biochemistry of this condition is now fairly well understood, there is no known cure. Heterozygotes are quite normal, but may be detected by a reduced concentration of a specific enzyme.

The disease occurs with a high frequency in the Ashkenazy Jews (those from Central Europe), where it is present in about one out of every 6,000 births. Among non-Jewish births, it has a much lower frequency, about one per 500,000.

SICKLE-CELL ANEMIA. In 1910, a young West Indian Black complained of fever and muscular pain. Upon medical examination he proved to have a previously unrecognized form of anemia, one in which his red blood cells took on an unusual elongated (or sickle) shape (Figure 5-18). In 1928, this was recognized as an inherited condition and in 1949 it was shown that the red cells sickled because they contain an abnormal *hemoglobin*, the protein present in red blood cells and responsible for the transport of oxygen from the lungs to the cells of the body.

More than 2 million people in the world are homozygous for an allele, S, that causes sickle-cell anemia, including 50,000 to 100,000 Blacks in the United States. About one in ten of all U.S. Blacks is a heterozygote (usually with no ill effects whatsoever), but about one in 500 is a homozygote, with sickle-cell anemia. This disease affects the homozygote by producing sickling crises, during

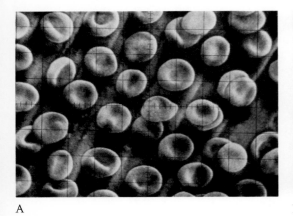
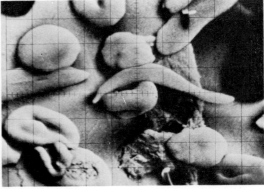

A   B

Figure 5-18 (*A*) Normal red blood cells, magnified $2000\times$. (*B*) Sickled red blood cells, magnified $5000\times$. Both photographs were taken by a scanning electron microscope. (Photographs by Irene Piscopo, Philips Electronic Instruments, Inc., Mount Vernon, N.Y.)

which some cells, with their irregular spindle shape, clog small blood vessels and cut off the oxygen supply to certain tissues. Under these conditions of low oxygen pressure, the cells sickle even more, aggravating the situation and causing extreme pain and tissue damage. Many affected persons die during childhood, but some live to old age.

Sickling occurs because of an abnormality in the hemoglobin molecules which attach to each other end to end to form long fibers that distort the cell shape. Some chemical treatments (urea, sodium cyanate, aspirin, and others) seem to offer some hope for relief by diminishing the capacity of the abnormal hemoglobin to form long fibers. It has been suggested that there may be a lesser incidence of severe sickling in homozygotes in African groups because of the increased (fortyfold) cyanate content of the common foods in their diets—yams, sorghum, millet, and cassava.

The reason for this unusually high frequency in certain populations seems to be that the heterozygote, who is ordinarily unaffected, has an increased resistance to malaria. Certainly there is a high correlation in Africa between regions with a high incidence of the *S* gene for the abnormal hemoglobin and those with high incidence of malaria (Figure 5-19). In addition, it has been shown that sicklers (i.e., homozygotes) have a lower frequency of infection by the malarian parasite than nonsicklers, and that the heterozygotes, when infected, have lighter infections than do those without the *S* gene.

Thus it appears that this allele performs a useful function in malaria-ridden areas by providing the heterozygotes with some protection against the disease. The price paid by the population for this protection is the general debilitation and high mortality of the homozygotes produced when two heterozygotes marry. At the present time, in nonmalarial regions such as in the United States and around the Mediterranean Sea, where some Italian and Greek populations carry the allele, the presence of the *S* allele serves no useful purpose; and, except for some cataclysmic event like the breakdown of civilized society, it is not likely to in the future.

THALASSEMIA.　In addition to carrying some *S* alleles for the sickling hemoglobin, populations around the Mediterranean Sea, and some Italian and Greek populations in the United States derived from them, show a different variant hemoglobin, responsible in the homozygous state for an anemic condition called *thalassemia*. It is believed that the widespread distribution and high frequency of this gene for abnormal hemoglobin are related to the distribution of malaria because of the protection against that disease conferred by the heterozygous state, just as with the gene for sickle-cell anemia.

PHENYLKETONURIA.　With a frequency of about one in 25,000 in Northern European populations, there appear homozygotes for a simple recessive who lack an enzyme responsible for converting the amino acid phenylalanine into the amino acid tyrosine. As a consequence, phenylalanine and its by-products accumulate in the body and retard normal mental development. The retardation becomes evident 6 months after birth and the majority of patients who reach adulthood are clearly subnormal in intelligence, with an occasional rare case reaching borderline intellectual development. In addition, since tyrosine is necessary for the production of the pigment melanin and is not produced by

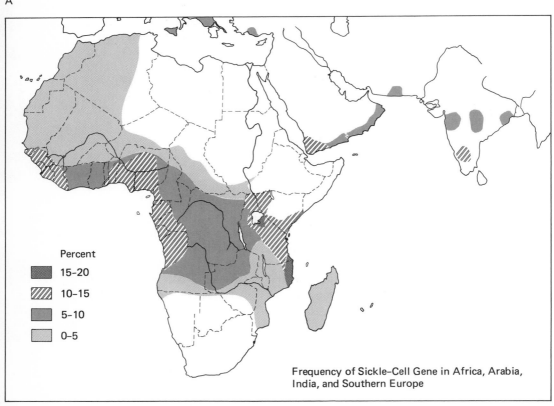

A

Distribution of Falciparum Malaria

Percent
15–20
10–15
5–10
0–5

Frequency of Sickle-Cell Gene in Africa, Arabia,
India, and Southern Europe

B

Figure 5-19 (*A*) The distribution of malaria in Africa and southern Asia. (*B*) The frequency of the sickle-cell allele in the same area. (A. C. Allison, in *Genetical Variation in Human Populations*, Pergamon Press, Inc., out of print.)

the normal intracellular synthesis, melanin is synthesized from the reduced amount made available in the diet. For this reason, affected individuals tend to have light hair and blue eyes.

In this instance, therapy is possible. If a newborn homozygote is placed on a diet with a low phenylalanine content, sufficient to provide the amino acid in the quantity necessary for normal development but not too great that it and its by-products reach high levels in the blood, along with additional tyrosine to make up for the amount missing because of the metabolic block, development may be more nearly normal. Such treated individuals may have normal intelligence, the degree of improvement depending on the speed with which the diet, and monitoring of the blood phenylalanine levels, is started after birth. Because the disease can be readily diagnosed shortly after birth by relatively inexpensive urine or blood tests, with only a small fraction of misdiagnoses, many states have mandatory test programs to determine the possible presence of *phenylketonuria* (*PKU*) in each newborn.

SIMPLE GENETIC CONDITIONS OF ANECDOTAL INTEREST. From time to time, attention is called to the possibility that people of historical interest have suffered from some common genetic disease. The appearance of hemophilia, the disease responsible for excessive bleeding, in the royal families of Europe is well known and is discussed in detail in Chapter 7. Abraham Lincoln, with his tall gaunt bearing, is suspected of having had Marfan's syndrome, which is responsible for longer than normal bones, particularly of the digits; the Greek Ypsilanti family, for whom a town in Michigan is named, was known to have the debilitating muscular disease myotonic dystrophy, causing premature death; Toulouse-Lautrec, the French artist, may have inherited from his first-cousin parents a recessive gene for pyknodystosis, responsible for his short stature and weak bones; King George III may very likely have had porphyria, a metabolic disease that gives rise to great pain and temporary insanity, which would explain his behavior at the time of the American Revolution. Finally, it has been suggested that Christ may have suffered from camptodactyly, a permanent crook in the fingers, based on the appearance of his fingers in a well-known painting of him. Possibly it was the artist's model who had the characteristic!

References

BROCK, D. J. H., and O. MAYO. 1972. *The Biochemical Genetics of Man*. New York: Academic Press.

CERAMI, A., and E. WASHINGTON. 1971. *Sickle Cell Anemia*. New York: The Third Press, Joseph Okpaku Publishing Co.

FREDRICKSON, D. S., and E. G. TRAMS. 1966. Ganglioside lipidosis: Tay-Sachs disease. In J. B. Stanbury, J. B. Wyngaarden, and D. S. Fredrickson, eds., *The Metabolic Basis of Inherited Disease*. New York: McGraw-Hill.

GOODMAN, R. M., ed. 1970. *Genetic Disorders of Man*. Boston: Little, Brown.

GUSTAFSSON, Å. 1969. The life of Gregor Johann Mendel—tragic or not? *Hereditas*, **62**:239–58.

HALL, J. G., et al. 1969. Two probable cases of homozygosity for the achondroplasia gene. In D. Bergsma, ed., *Birth Defects: Orig. Art. Ser.*, Vol. V, pp. 24–34. Baltimore: Williams & Wilkins.

HARRIS, H. 1970. *The Principles of Human Biochemical Genetics*. Amsterdam: North-Holland.

HOEFNAGEL, D., and P. S. GERALD. 1966. Hereditary brachydactyly. *Ann Hum. Genet.*, **29**:377–82.

KNOX, W. E. 1966. Phenylketonuria. In J. B. Stanbury, J. B. Wyngaarden, and D. S. Fredrickson, eds., *The Metabolic Basis of Inherited Disease*. New York: McGraw-Hill.

McKUSICK, V. A. 1975. *Mendelian Inheritance in Man: Catalogs of Autosomal Dominant, Autosomal Recessive, and X-Linked Phenotypes*, 4th ed. Baltimore: Johns Hopkins University Press.

STANBURY, J. B., J. B. WYNGAARDEN, and D. S. FREDRICKSON, eds. 1973. *The Metabolic Basis of Inherited Disease*. New York: McGraw-Hill.

WITKOP, C. J. 1971. Albinism. In H. Harris and K. Hirschhorn, eds., *Advances in Human Genetics*. New York: Plenum.

WOOLF, C. M., and F. C. DUKEPOO. 1969. Hopi Indians, inbreeding, and albinism. *Science*, **164**:30–37.

Questions

Useful terms: pedigree, sibship, propositus, genotype, phenotype, dominant, recessive, polytene, gene, locus, allele, homozygous, heterozygous, consanguinity, segregation, independent assortment.

1. Draw a pedigree, as extensive as you can recall, for your own family. Add a few complications to make it more interesting. Fill in some of the symbols so that the pattern will be that of typical dominant inheritance. Repeat the same exercise, assuming segregation for some simple recessive characteristic.

2. Why was Gregor Mendel's experimental work with the garden pea so successful? How was Mendel able to make so many observations in such a short period of time on such a small plot of land?

3. Describe Mendel's two laws, using actual characteristics in a specific cross.

4. Describe the manifestation and manner of inheritance of the following diseases: (a) cystic fibrosis, (b) Tay-Sachs disease, (c) sickle-cell anemia, (d) thalassemia, (e) phenlyketonuria.

5. Is a dominant allele defined as one that has the same phenotypic effect in both the homozygote (AA) and the heterozygote (Aa)?

6. As Mendel showed, the 3:1 ratio is very important in theoretical genetics. Do we often find a 3:1 ratio in human families where both parents are heterozygous for the same recessive allele? Would this depend on family size?

7. If two independent pairs of alleles are segregating for a set of recessive traits, we would ordinarily expect a 9:3:3:1 ratio. Has this ratio ever been observed in a human family?

8. Discuss the relevance of the following conditions to the concepts discussed in this chapter: albinism, achondroplasia, eye and hair color, polydactyly, syndactyly, brachydactyly, cystic fibrosis, Tay-Sachs disease, sickle-cell anemia, thalassemia, phenylketonuria, Marfan's syndrome, myasthenia gravis, pyknodystosis, porphyria, camptodactyly.

6

Sex Determination

Definition of Sex

Sex may broadly be defined as the combination of the genetic material of two different individuals to form a new individual. From this scientific definition it is clear that there is no restriction on the physical form of the sexes—they may be identical to each other (and often are in primitive plants and animals), and there is no limitation on the number of sexes that may occur in one species, although our preoccupation with the condition in humans and other higher animals leads us to think of two morphologically different sexes, male and female. In fact, in some species of lower animals and higher plants, there may exist a large number of different "sexes." If diploidy is an almost invariable feature of life, sex is even more so, being found in all plants and animals, with the possible exception of a group of primitive plants, the blue-green algae.

The Evolutionary Advantages of Sex

Except for the rare cases of identical twins, it is likely that no two humans who have ever existed on the face of this planet since the dawn of time have been genetically identical. The allelic variation in a large fraction of the tens of thousands of loci that human chromosomes carry, along with the process of recombination that occurs during meiosis in both sexes and the chance union of two dissimilar gametes, gives every zygote its own genetic constitution. From this point of view, every individual represents a natural and unique experiment to test the potentialities of the virtually unlimited number of different gene combinations possible in the species. There is an opportunity for beneficial gene or chromosomal changes that occur independently in the germ line of different individuals to meet in an even more advantageous combination in their progeny. In addition, allelic changes with apparently neutral effects may combine with similar alleles in other lines to produce some new, unusual, and possibly advantageous phenotypes.

On the other hand, consider the consequences if all human females reproduced without the benefit of fertilization of the eggs by sperm, that is, by *parthenogenesis*. Each progeny could contain nothing but the genetic material of the mother, and would in most cases be a carbon copy of her. This would happen for generation after generation, with an occasional spontaneous genetic change altering each line. Such a change, once established, would become a permanent feature of the line. Other maternal lines would likewise slowly diverge in different ways, and after a period of time each maternal line would have its own special characteristics, gradually deviating from what we now consider to be humanness.

The phenomenon of sexuality, then, not only has the great advantage of reshuffling genes to test out new allelic combinations in every individual produced, but, perhaps more subtly, has the effect as well of reaching into the common "gene pool" that characterizes mankind and of constantly reestablishing the common basis for the unique properties of the human species. This maintenance of a common gene pool will of course be a consequence of sexuality not just for humans but for all living species.

The Forms of Sexuality

PROTOZOANS.   In one-celled organisms it is common for the two sexes to look alike. Indeed, the knowledge that sexes exist may be obtained only by experimental testing. If two different populations of one-celled animals, protozoans, are each derived from a single ancestor by repeated mitotic divisions, an individual from culture A and an individual from culture B may fuse in a process of *conjugation* and exchange genetic material. This event will not usually occur between two individuals of the same culture. Such a mating may not produce a new individual immediately, but those progeny subsequently produced by mitosis or *fission* of the mated pairs now have a new genetic constitution. By testing large numbers of independently derived strains, it is possible to find out how many independent, self-sterile groups or mating types (i.e., sexes) exist in that species; it has been found that for one species of paramecium, for instance, there are at least four different mating types, and for another, six.

PLANTS.   In molds, the sexual process takes place between two groups that are morphologically indistinguishable. The only difference between them appears to reside at one locus with two different alleles; these are referred to simply as + and −. Many plants have male and female gametes produced on different individuals—even the ancient Babylonians recognized the necessity for having a male date palm growing near a female date palm to produce dates. In fact, they practiced artificial pollination to insure a good crop. Other plants, for instance tobacco, have both male and female sex organs on the same plant, but are self-sterile; that is, the pollen will not function on a plant of the same constitution or on one of identical genetic constitution with respect to a particular locus. This is a fact of obvious importance to agriculturalists. An orchard of cherry trees consisting of all the same variety of tree would produce no fruit because cherries are self-sterile, and all trees of a specific commercial variety are sterile with each other. The knowledgeable horticulturist will plant several varieties of tree within the same area so that, in the spring, the bees will readily transport pollen back and forth from one strain to another. Since the desirable characteristics of the cherry (flavor, size, and appearance) are purely maternal in origin, the exact variety of cherry contributing the pollen is of no importance. However, to perpetuate the good qualities of a strain, planting seeds would be of no use since the seeds will always be hybrids—instead, cuttings from the desired variety are induced to form roots and develop into young trees with precisely the same genetic constitution as the original variety.

ANIMALS.   A few lower animals are *hermaphroditic*, with the sex organs of both male and female in each individual, as in the earthworm and the snail. Clearly such individuals do not need any specific genetic mechanism of sex determination, since the male and female cells have the same genetic composition. A basically hermaphroditic animal may preferentially differentiate into one sex or the other depending on external conditions, the circumstance which makes it appear as though the sex is in fact being environmentally determined. A classic example of this sort of determination is found in the marine worm *Bonellia*, in which fertilized eggs develop into females when there are no other females in the vicinity, but, if there are females present, into males which live parasitically within the bodies of the females.

In the hymenopterans, bees and wasps, sex is generally determined by the number of chromosome sets. Queens (fertile females) and workers (sterile females) come from fertilized eggs that have the diploid number of chromosomes, 32. The difference between the queens and the workers is not genetic but developmental, depending on the kind of food the immature individuals are given during early development. On the other hand, drones (males) are produced parthenogenically, i.e., without benefit of fertilization of the egg by a sperm, and have the haploid number of 16. The meiotic divisions of the drones are modified so that the sperm have the adult number, 16. Further work has shown that the distinction between female and male has a more subtle basis, that the female has two different alleles, i.e., is *heterozygous*, at a certain locus which makes her female. The male, being haploid, has only one allele (i.e., is *hemizygous*), which causes the developing egg to become a male. If bees are closely inbred experimentally so that a fertilized egg gets two identical alleles, that individual, although diploid, will develop into a male because of the homozygosity for the sex alleles.

Sex Determination in Drosophila

By far the most detailed work on sex determination has been carried out with the fruitfly, *Drosophila*. It was established very early that females have two X-chromosomes and males have an X and a Y, just as humans do, and also that the Y-chromosome is necessary for fertility in the male but has no other effect. When nondisjunction occurs and an XXY egg is produced, it develops into a female indistinguishable from normal, and a single X egg similarly produced, without a Y, develops into an infertile male. The first rule for sex determination in *Drosophila* seemed simple: a zygote with two X-chromosomes developed into a female, one with only one X into a male. Thus it was clear that the X-chromosome was sex determining and, since the properties of a chromosome must come from the genes carried on it, there must exist sex-determining loci somewhere on the X. However, a search for specific sex-determining loci on the X-chromosome, by fractioning it into bits and pieces with x-rays, proved fruitless, although it occupied the attention of a number of investigators for several years. Instead, it appeared that the action of the additional X in producing femaleness was not related to a few specific sex-determining loci but was rather a generalized, nonspecific effect of sections of chromosomes.

TRIPLOIDY IN DROSOPHILA. It could be shown, however, that sex determination does depend on a balance between the autosomal complement and the sex chromosomes. When the number of autosomal sets equals the number of sex chromosomes, the resulting individual is a typical fertile female. Thus a *triploid Drosophila* female with three sets of autosomes and three sets of X's is virtually indistinguishable from a normal diploid female with two sets of autosomes and two X's. A much rarer class, consisting of *tetraploid* females, with four sets of autosomes and four X-chromosomes, also proves to be relatively normal and fertile.

Triploid females of *Drosophila* produce gametes with a wide variety of different chromosome constitutions, since there are three chromosomes of each type instead of the normal two. As a rule, each set of three solves the problem of segregation by sending two of the three to one of the two poles at meiosis and the other to the second pole. Since this occurs independently for all groups of three homologues, a wide variety of chromosome numbers is found in the gametes. All zygotes which have unbalanced sets of autosomes die early in development. For a zygote to survive and develop it must have the same number of each of the large autosomes; the X's, however, can vary in number. In this way, progeny can be produced with different ratios of X-chromosomes to autosomal sets (Figure 6-1). When the ratio is one X to two autosomal sets, maleness results; a ratio of two X's to two autosomal sets results in femaleness; and any intermediate ratio, such as two X's to three autosomal sets, gives an intermediate phenotype called an *intersex*. This has male and female characteristics together in variable degrees. Individuals with three X-chromosomes but only two sets of autosomes are highly abnormal females of very low viability, virtually completely infertile and with gross abnormalities. Such females were originally named *superfemales* because of the high rate of sex chromosomes to autosomes; how-

Figure 6-1 The different kinds of progeny produced when a triploid *Drosophila* female is mated to a normal male. The arrows indicate the type of gametes that will produce viable offspring; these carry either one complete set (1A) or two complete sets (2A) of autosomes, each with either one or two X-chromosomes.

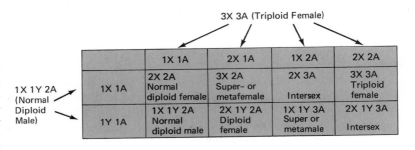

1X 1Y 2A (Normal Diploid Male)	3X 3A (Triploid Female)			
	1X 1A	2X 1A	1X 2A	2X 2A
1X 1A	2X 2A Normal diploid female	3X 2A Super- or metafemale	2X 3A Intersex	3X 3A Triploid female
1Y 1A	1X 1Y 2A Normal diploid male	2X 1Y 2A Diploid female	1X 1Y 3A Super or metamale	2X 1Y 3A Intersex

ever, they are not the kind of individual such a term would imply. Another name, *metafemale*, has been suggested as an alternative. At the other end of the scale are those *Drosophila* with only one X-chromosome but three sets of autosomes. This ratio is less than that of ordinary maleness; such individuals are malelike, but morphologically abnormal and completely sterile. Because of the low ratio they are called *supermales* (or *metamales*). These ratios are summarized in Table 6-1.

The concept that the determination of sex is primarily a matter of balance between the autosomes and either two or one X-chromosome, with the Y imparting fertility to the male but with no other essential function, was thought for a long time to represent the model for all animals with an X-Y sex-determining mechanism. In 1959 it was shown that most human beings characterized by Klinefelter's syndrome were in fact XXY; as we have already seen, these persons are essentially male. The contrary class, the X0 condition, that of the female with Turner's syndrome, was also described cytologically in 1959. These results are quite contrary to the situation in *Drosophila*: in humans the Y-chromosome and not the number of X-chromosomes is the primary sex determinant. This point is so important that it merits reemphasis: it is almost invariably the case that an individual without a Y-chromosome is a female and one with a Y-chromosome, regardless of how many X-chromosomes may be present, is a male.

Table 6-1 The relationship of the sex chromosome/autosome ratio to sex in *Drosophila melanogaster*. Note that as the number of sex chromosomes relative to the sets of autosomes decreases the sexuality shifts from femaleness toward maleness.

	Constitution	
	Sex chromosome/autosome	Ratio
Superfemale	3X:2A	1.5:1
Tetraploid female	4X:4A	1:1
Triploid female	3X:3A	1:1
Diploid female	2X:2A	1:1
Intersex	2X:3A	0.66:1
Diploid male	1X:2A	0.5:1
Supermale	1X:3A	0.33:1

The Puzzle of the Missing X

The X-chromosome in humans is unusual in still another way. From the previous discussions of chromosome imbalance involving the autosomes, either trisomies or monosomies, it is obvious that the addition or deletion of even a small chromosome from the normal diploid complement has severe consequences to the development of the embryo and fetus. On the other hand, this rule does not seem to apply to the sex chromosomes, for the subtraction of an X-chromosome from a female (as in the case of Turner's syndrome) has effects on the phenotype that are perceptible only to one experienced in such diagnoses. It seems reasonable to conclude that the X-chromosome does not follow the same rules as the autosomes with respect to the effects of aneuploidy, particularly since it is a large chromosome, distinguishable from other members of the C group only by special staining techniques. There must be some special explanation for this unusual developmental behavior of the human X-chromosome.

DOSAGE COMPENSATION.   The bulk of the synthetic machinery of the cell is derived from genes carried on the autosomes, which, always occurring in pairs in normal cells, operate normally in the diploid state. The genes on the X-chromosome, however, are faced with a problem, because the male cell with only one X-chromosome must perform at essentially the same level as the female cell with two.

Clearly, the logic of this problem can be approached from either of two directions to make male and female cells synthetically equivalent. Either the single X-chromosome of the male can work twice as hard so that it performs at the same level as the two of the female, or the two X's of the female can work half as hard to perform, jointly, at the same level as the single X of the male. This adjustment of the cell to take into account the inequalities that result from having two X-chromosomes present in one sex and only one in the other is referred to as *dosage compensation*. Of the two solutions suggested, *Drosophila* follows the first course: a single X in the male works much harder and approaches the two-X condition of the female, so that the autosomes in both sexes sense the equivalent of two X-chromosomes. In humans, one of the two X-chromosomes in the female "lies down on the job" and performs no substantial function. Male and female human cells are similar in having only one functional X-chromosome (with minor qualifications to be considered later).

The Significance of the Sex Chromatin

As early as 1891 it was observed that nuclei of cells in the resting stages have an undifferentiated mass of *chromatin*, which was referred to as X. Over the next 20 years it became apparent that this body was a chromosome that played a role in sex determination. The interest in the X-chromosome, as it was henceforth called, then focused on its meiotic behavior, particularly in the male where the homologue was not another X, but a Y-chromosome, and on its developmental effect in sex determination.

X-CHROMOSOME INACTIVATION.   The sex chromatin body (as distinct from the sex chromosome) was described and illustrated by Cajal, a Spanish nerve anatomist in 1909, in his descriptions of nerve cells of dogs, cats, and humans. However, he was not aware of the relationship of the presence or absence of this body to the sex of the individual from which the cells were taken. In 1937 a German cytologist, Geitler, pointed out that in some insects the two sexes differ in their manifestation of the sex chromatin bodies. These observations, however, did not appear meaningful until 1949, when two Canadian workers, Barr and Bertram, reported that in the nuclei of nerve cells of cats the two sexes are characteristically different with respect to the sex chromatin bodies. When a sex chromatin body appears, the cell is from a female; the male never shows any (Figure 6-2). In 1959 Russell and Ohno independently proposed that the sex chromatin resulted from the inactivation of a single X-chromosome, and in 1961 Lyon suggested further that the X that is inactivated may be either maternal or paternal in origin, and that the inactivation occurs during early embryonic development, resulting in groups of cells with the same X-chromosome active. Such genetically similar groups descended from a single cell are known as *clones.*

This idea of X-inactivation is sometimes referred to as the Lyon hypothesis, and the inactivation process as lyonization. However, since substantial contributions to this theory have come from a number of other workers, we shall refer to this phenomenon simply as *X-chromosome inactivation* and to the inactive X visible at interphase, sometimes called the Barr body, as the *X-chromatin.*

How sex chromosome inactivation occurs is not at all clear. After fertilization of the egg, the cells proceed through the cleavage divisions, and at about the eighth or tenth mitotic division, about 12 to 16 days after fertilization, a "decision" is made by each cell that if there is more than one X-chromosome present only one will be functional. Thus, if there are two X's within a cell, one of them, selected in a lottery not yet understood, will become the functional X and the other nonfunctional. This information (which X has been chosen for function-

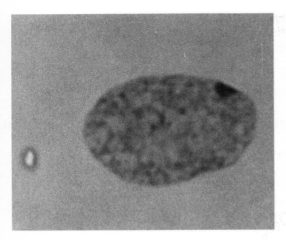
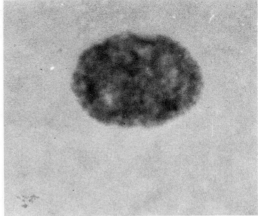

Figure 6-2 The X-chromatin body in a human female (left) and the absence of such a body in the male (right). (Courtesy of J. Geraedts, University of Leiden.)

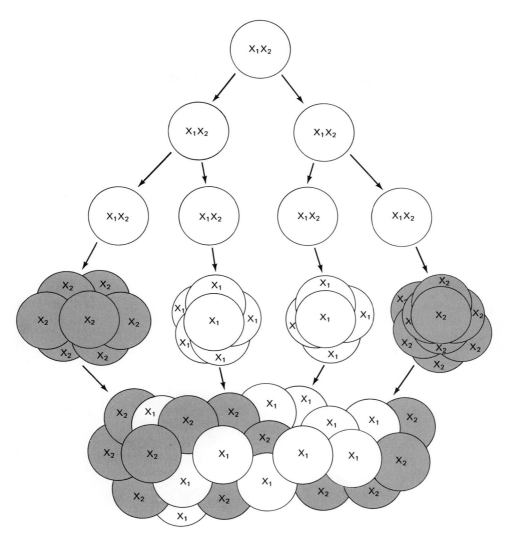

Figure 6-3 Schematic showing how a zygote with two X-chromosomes gives rise to a mosaic individual, with approximately half of its cells expressing one of the two X's and half the other. After several cleavage divisions (only two are shown here for simplicity), one of the two X's, randomly selected, becomes inactive. Further mitoses produce clones of cells with the same X inactive; because these divisions occur in a mass of cells with some mixing of individual cells, a mosaic is produced with grouping of similar cell types, but also with some intermixing.

ality) is transmitted to the daughter cells, and, as the cells go through subsequent mitoses, each cell present at that early stage gives rise to a clone of descendant cells, all of which have the same functional X (Figure 6-3). A female is thus a mosaic, part of her cells having one X functional and part having the other homologue functional.

The Mosaic Female

If there existed allelic differences for skin pigmentation or some similar obvious phenotype on the X-chromosome, X-chromosome inactivation would have been as obvious as any other physical difference because the heterozygous female would have a blotchy appearance, resulting from the natural grouping of the cells from each clone derived from single cells with a specific X inactivated. In this case, appreciation of the unusual developmental role of the X chromatin would not have been delayed until the 1950's. Unfortunately, such clear morphological characters do not exist in humans. We can, however, see sex chromosome inactivation at work in the cat, where such factors do exist. The type of cat that exhibits this characteristic is the well-known *calico* or *tortoiseshell*.

At one locus on the cat's X-chromosome there exist two alleles for coat color, *C*. One of these alleles produces black pigment (C^B) and the other yellow (C^Y). A calico cat is heterozygous for these two alleles with the genotype $C^B C^Y$ (Figure 6-4). The sequence of events that gives rise to a calico cat is shown in Figure 6-3. In different patches of tissue, one or the other X-chromosome is inactivated.

Figure 6-4 The classic visual example of X-chromosome inactivation: the *tortoiseshell cat*. In addition to the black and orange (reproduced here as gray), these two cats also have an independent gene which removes the pigment entirely from large sections, making them partially white; such three-colored cats are *calico*. The cat in *B* has, in addition, a gene for tabby which superimposes a banded pattern. (Courtesy of P. G. N. Kramers, University of Leiden.)

A

B

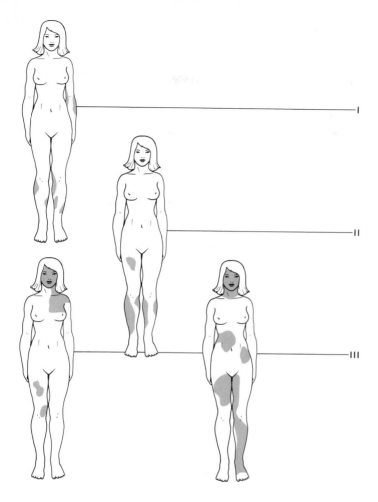

Figure 6-5 A mother, daughter, and two identical twin grand-daughters all heterozygous for an allele reducing the number of sweat glands in the skin. The stippling indicates areas where one X-chromosome (that with the allele for anhidrotic dysplasia) is active and the clear areas represent skin with the other X active. (From A. H. Kline, J. B. Sudbury, Jr., and C. P. Richter; The Occurrence of Ectodermal Dysplasia and Corneal Dysplasia in One Family; *J. Pediatr.*, **55**: 355–66, 1959.)

Those patches of cells in which the chromosome carrying the black allele is inactivated give rise to yellow fur, and those in which the chromosome carrying the yellow allele is inactivated produce black fur. (The white fur, in which color pigment is inhibited, is caused by an independent autosomal gene and is irrelevant to our present discussion.)

Thus, if one were to see a calico cat from a distance, it could be reasonably predicted that the cat is a female. Male calicos occur with a very low frequency and are sterile, since the calico characteristic requires two X-chromosomes and maleness requires a Y, making the male calico the equivalent of a human with Klinefelter's syndrome.

It is possible to obtain evidence for the mosaicism of the human female heterozygous for a few sex-linked characters. A defective condition known as *anhidrotic dysplasia* is responsible for an absence or low frequency of sweat glands, and this condition can be detected in areas of the skin by an unusually low resistance to electricity in affected areas. In Figure 6-5 three generations of heterozygous women have been "mapped," showing the areas where the chromosome

Figure 6-6 Cell cultures from a woman heterozygous for two sex-linked alleles that produce detectably different enzymes. When a mixture of the two is placed in an electrical field, one (*A*) migrates at a slightly faster rate than the other (*B*). The original heterogeneous cell culture from the woman indicated that both enzymes were present, but when nine single cells were used to start nine clones, six of the clones produced only variant *B* and the other three variant *A*, showing that any one cell, and its progeny, will carry only one active X-chromosome. (R. G. Davidson, H. M. Nitowsky, and B. Childs, *Proc. Natl. Acad. Sci. U.S.A.*, **50**:481–85, 1963.)

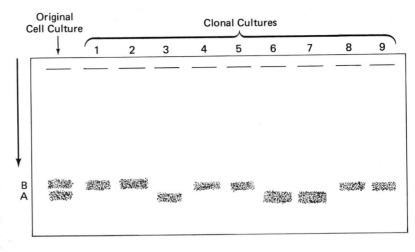

carrying the normal allele is active and where the homologue carrying the allele for the absence of sweat gland is active.

In a second demonstration of this mosaicism, a woman is heterozygous for two detectably different enzymes that modify a sugar molecule, glucose, and appears to carry both enzymes in her cells (Figure 6-6). However, when individual cells are isolated, and their descendants tested, all the clones carry either one of the enzyme types or the other, but not both, indicating that in some cells one X is active and in the rest the other X functions.

The Presence of Multiple X-Chromosomes

In cells with more than two X-chromosomes, there may be more than one X-chromatin body. The usual rule is that only one X remains active and all others become inactivated; a cell with three X-chromosomes would have two X-chromatin bodies.

Because of X-inactivation, the human can tolerate more variation in X-chromosome number than in autosomal number. A 3X (triplo-X) female is not ordinarily distinguishable from a 2X female; such persons are fertile and otherwise normal, except that their psyche may be disturbed as indicated by an IQ slightly lower than normal and a predisposition to mental illness.

In fact, the degree of deviation from the normal sex chromosome complement can be rather extreme, as is seen in Table 6-2. In this table is a list of some of the

Sex Chromosomes	Number of Cases	X-Chromatin Masses (percent)				
		0	1	2	3	4
XX	12	55	45			
XXX	12	24	48	28		
XXXX	3	7	21	49	23	
XXXXX	1	10	20	32	30	8
XY	12	100				
XXY	12	45	55			
XXXY	5	24	50	26		
XXXXY	8	13	35	31	21	
X0	12	97	3			
XYY	1	100				
XXYY	1	55	45			
XYYY	1	100				

Table 6-2 The number of X-chromatin bodies in cells of persons with various numbers of X-chromosomes. (Adapted from W. M. Davidson, Sexual Dimorphism in Nuclei of Polymorphonuclear Leukocytes in Various Animals. In K. L. Moore, *The Sex Chromatin*, W. B. Saunders Company, Philadelphia, 1966.)

chromosome numbers that have been observed (sometimes as mosaics) in individuals with more than one X-chromosome (in at least some cells), both with and without a Y-chromosome. It can be seen that in some individuals the number of X-chromosomes can reach a total of four and in mosaics even as many as five.

On the other hand, it is obvious that the additional X-chromosomes cannot be completely inactivated under all circumstances; otherwise, we would expect an X0 individual to be exactly like an XX. Also the XXY Klinefelter's should be a male normal in all respects if all X's beyond one were completely inactivated. The hypothesis of X-chromosome inactivation must therefore be qualified to account for the observation that there is, in fact, some slight influence of the additional X-chromosomes beyond the first, particularly with regard to intellectual and sexual development. The basis for this influence is not clear. It is possible that inactivation may be the rule for most somatic cells but that there are some special cell lines, perhaps in certain organs, which do not undergo inactivation. Because single cells in the germ cells of the female can be shown to possess both of the enzymes produced by different alleles on the two X-chromosomes, we can argue that X-chromosomes in such cells do not undergo inactivation. A second possibility is that inactivation occurs late enough in embryonic development that the early action of the two X-chromosomes together influences subsequent development. Still a third possibility is that the X-chromosome is not entirely inactivated but remains active in a small segment, and that it is the genetic imbalance of this small segment which, when either absent (in the X0 female) or present in excess (in the XXX female and other multiple X types), is responsible for the differences of these females from the normal XX.

From the preceding discussion, it may appear that the X-chromatin bodies are invariable features of cells with more than one X. Unfortunately, this is far from the case. In one investigation, counts of cells of triple-X females gave 24 percent with no sex chromosome bodies, 48 percent with one, and 28 percent with two. Similar counts of tetra-X (4X) cells gave 7 percent with none, 21 percent with one, 49 percent with two, and 23 percent with three. Other cell types with percentages are given in Table 6-2. Clearly, the presence or absence of X-chromatin in a single cell or in a few cells would not be sufficient to make a definite determination of the X-chromosome constitution of an individual.

The Y-Fluorescing Bodies

Human chromosomes stained with the special dye *quinacrine* generally show a very prominent Y-chromosome at metaphase (Figure 6-7). The brightness is detectable not only in ordinary metaphase chromosomes but also in resting cells and even in mature sperm cells; in the latter two types of cells the fluorescent spot is referred to as the *Y-chromatin body.* This makes it possible to make a specific analysis for sex in addition to the X-chromatin tests. The X-chromatin tests can distinguish among X, XX, and XXX individuals, but not between XY and X0 or between XX and XXY. The detection of one or more Y-chromosomes using fluorescent dyes makes it possible to describe the sex chromosome complement of a cell completely. These techniques provide a powerful tool for determining the sex of an individual without actual karyotyping in those cases where there is some ambiguity, as sometimes happens at the time of birth. Cells scraped from the lining of the mouth are quite useful; the determination is fast and simple, requiring only an immediate staining procedure and microscopic examination. It should be pointed out, however, that there do exist some Y-chromosomes which characteristically do not fluoresce. That this is an innate property of the particular Y-chromosome is shown by its transmission from father to son in family studies.

A   B

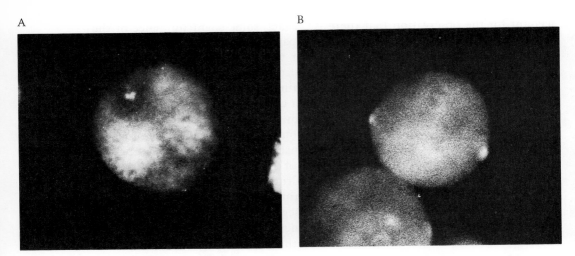

Figure 6-7 The brightly fluorescing Y-chromosome. (*A*) The single spot found in the cells of a normal male. (*B*) The two Y-chromosomes of an XYY male. (Courtesy of B. K. McCaw, University of Oregon Medical School.)

Human Sexuality

When the word "sex" is used with reference to humans, it has implications beyond those covered by our original biological definition. These can be categorized as cellular, organic, or behavioral.

CELLULAR SEX. At the cellular level, there are two aspects of sexuality, the chromosome and the genetic constitution of the individual. Basically it is the XX-XY mechanism that is responsible for the initiation of the two sexes, and under ordinary circumstances any individual with two X-chromosomes can be assumed to be female and the XY can be assumed to be male. However, in addition to the chromosomal basis, the genetic constitution of the individual may play a role. There are individuals who clearly have two X-chromosomes, without a Y, who appear to be males from external phenotype, and there are individuals who are XY chromosomally who appear, phenotypically, to be females. Furthermore, such aberrant individuals may be found concentrated in single pedigrees, indicating that a genetic basis is responsible for their unusual differentiation, rather than a developmental accident.

ORGANIC SEX. The organic manifestation of sex has three aspects: internal sex, external sex, and hormonal sex. During the course of early prenatal development, the presence of the Y-chromosome in the male will ordinarily suppress the development of the ovaries and female ducts and promote the development of the testes and male ducts. In the absence of the Y-chromosome, the reverse happens.

Perhaps the most commonly used criterion of sex is the structure of the external sex organs (Figure 6-8). Not only is this criterion simple, direct, and usually determinable at birth, but also the behavioral patterns the individual is expected to show (instinctively, or by training) will depend on this sex index. In most cases this characteristic is obvious at birth, but in a few it is not and the newborn may be of ambiguous sexuality. In such cases it would be desirable to go to easily discernible cellular aspects in order to make a determination of the individual's basic sexuality.

The third criterion at the organic level is that of the hormonal sex. The changes induced by the sex hormones become most obvious at puberty, when both male and female sexes develop the specific characteristics for which they are most distinctive. In those cases where the development of the secondary sexual characteristics is impeded (as in Turner's syndrome), the application of additional female hormones will promote their development, although the problem of sterility cannot usually be helped. On the other hand, abnormal hormonal effects may change sexual development so that the individual takes on some of the characteristics of the opposite sex. Thus we have the development of breasts in individuals who would otherwise be classified as males. Contrarily, the "masculinization" of women at older ages—the deepening of the voice and increased facial hair—is a matter of common observation.

BEHAVIORAL SEX. There are behavioral traits that are commonly considered to be either masculine or feminine; these can be categorized as psychological

Figure 6-8 An illustration of the universal recognition of organic sex as the most important criterion for differentiating the sexes. (Courtesy of Leonie Pera and G. H. Valentine, *The Chromosome Disorders,* Heinemann Medical Books, London, 1966.)

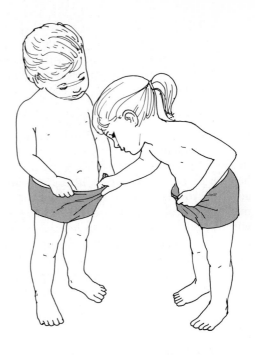

THERE IS A DIFFERENCE!

sex (Figure 6-9). Every individual has characteristics of both sexes to varying degrees, and these may be expressed by actions or attitudes that may not be consistent with the other criteria for sex labeling. In extreme cases, an individual may take on, in toto, the behavior of the opposite sex. Another aspect of behavioral sex is social sex, referring to the influences that individuals may be exposed to early in life that cause them to adopt the attitudes and behavioral patterns expected of them.

OVERALL SEX. In the vast majority of cases, the various indicators of sex are in substantial agreement in one direction or the other. On the other hand, at each step in the development from the zygote to the mature individual there is the possibility of a deflection from the typical pattern; it is not surprising that there should exist such a wide array of differences in sexuality. Our concern is with the biological basis for these differences, when such a basis exists. With respect to behavior, for instance, probably the most interesting and important aspect of sex, little can be said here. This is an area of vital interest to psychologists and anthropologists; the points of view and their interpretations of sexual behavior seem to be as numerous as the investigators themselves. At this point we shall simply note that there exists no evidence for a cellular or a somatic basis for such phenomena as homosexuality and transvestism.

INTERSEXUALITY. In some cases, a human of one sex embodies some of the physical characteristics of the other sex and is, therefore, referred to as an *intersex.*

Figure 6-9 The influence of the behavioral sex. When all other differentiating characteristics have been eliminated, the two sexes may identify themselves in somewhat subtle ways. (Courtesy of Leonie Pera and G. H. Valentine, *The Chromosome Disorders*, Heinemann Medical Books, London, 1966.)

THERE IS A DIFFERENCE!

The hormonal influence may occur prior to birth. The classic example is the so-called freemartin in cattle. When twins of opposite sexes develop in the cow, the hormones of the male (and in many cases some cells as well) reach the developing female embryo and induce that female to become malelike. When newborn, the female may resemble a bull calf, but the external genitalia are those of a female and the internal sex organs are usually poorly developed. This does not occur in humans—a twin of one sex is unaffected by the presence of one of the other sex in the womb during development. However, male hormones may sometimes influence an otherwise normal female fetus to take on male characteristics, as has been observed repeatedly after women have been given such injections to inhibit chronic miscarriage. Such a child may be identified as a male at birth and be raised as a male without question until puberty, when female secondary sexual characteristics start to develop.

HERMAPHRODITISM. Hermaphroditism is part of the larger category of intersexuality. Strictly speaking, a *hermaphrodite* is an individual with both male and female sex organs and, as has been noted earlier, both of the sexes in one individual may be functional. In humans, hermaphrodites usually have ambiguous genitalia, or, if they have genitalia of one sex, have a mixture of the two types of internal sex organs. A contribution to this class might come from mosaics in which the somatic cells are of two different compositions. One of the common types of mosaics, for instance, comes about when a Y-chromosome is lost from a male cell during early cleavage, giving rise to two cell lines, one XY and the other X0. This combination can give rise to an individual who appears to be part male and part female. However, sharp differences are not usually found in such cases, because the direction of differentiation may not depend entirely on the chromosome constitution of the individual cells, but rather may be determined by the levels of hormone in the circulation.

References

Barr, M. L., and E. G. Bertram. 1949. A morphological distinction between neurones of the male and female, and the behavior of the nucleolar satellite during accelerated nucleoprotein synthesis. *Nature,* **163:**676–77.

Davidson, R. G., H. M. Nitowsky, and B. Childs. 1963. Demonstration of two populations of cells in the human female heterozygous for glucose-6-phosphate dehydrogenase variants. *Proc. Natl. Acad. Sci. USA,* **50:**481–85.

Davidson, W. M. 1966. Sexual dimorphism in nuclei of polymorphonuclear leukocytes in various animals. In K. L. Moore, ed., *The Sex Chromatin.* Philadelphia: Saunders.

Jones, H. W., and W. W. Scott. 1958. *Hermaphroditism, Genital Anomalies and Related Endocrine Disorders.* Baltimore: Williams & Wilkins.

Lyon, M. F. 1962. Sex chromatin and gene action in the mammalian X-chromosome. *Hum. Genet.,* **14:**135–48.

Mittwoch, U. 1973. *Genetics of Sex Differentiation.* New York: Academic Press.

Moore, K. L., ed. 1966. *The Sex Chromatin.* Philadelphia: Saunders.

Russell, L. B. 1961. Genetics of mammalian sex chromosomes. *Science,* **133:** 1795–1803.

Welshons, W. J. 1963. Cytological contributions to mammalian genetics. *Am. Zool.,* **3:**15–22.

Questions

Useful terms: parthenogenesis, gene pool, fission, intersex, metafemale, metamale, dosage compensation, sex chromatin.

1. How does the scientific definition of sex differ from the popular view?
2. About a dozen years ago in England, a woman claimed that her newborn son had been produced parthenogenetically after she had gone swimming in the ocean and presumably exposed herself to salt water. Can you find anything wrong with her argument?
3. In what way does the manifestation of sex differ in the lower forms of the plant and animal kingdoms compared to the higher forms?
4. How is it known in *Drosophila* that the autosomes play a part in sex determination, in addition to the influence of the sex chromosomes?

5. What does Turner's syndrome tell us about the mechanism of sex determination in man?
6. What is meant by dosage compensation; why is it necessary?
7. How is the sex chromatin related to dosage compensation?
8. Why is sex chromosome inactivation not obvious in the appearance of humans? Is there any animal in which it is quite clear?
9. Have any human adults been seen with as many as three or four sex chromosomes? Can you explain the difference between these two?
10. Occasionally newborn children are found with ambiguous sex organs, so that it is difficult to determine the sex at birth. Explain how this could be accomplished taking advantage of the sex chromatin bodies and tell why it might be desirable to supplement these tests with a test for Y-chromosome fluorescence.
11. What are the different criteria by which a person may be classified into one or the other of the two sexes? Which do you consider to be more important than the others?
12. Discuss the relevance of the following conditions to the concepts discussed in this chapter: calico or tortoiseshell cat, intersex, hermaphrodite, mosaic, anhidrotic dysplasia.

7

Sex-Linked Inheritance

The sex chromosomes have an unusual pattern of inheritance from one generation to the next. The X-chromosome of the father goes only to his daughters, his Y-chromosome only to his sons. With this in mind, it can be expected that genetic traits determined by genes carried on those chromosomes will also exhibit an unusual pattern of transmission. This pattern is commonly referred to as *sex-linked inheritance*, although the phrase is misleading since the transmitted traits are not strictly linked to the sex of the individual. Rather, they follow the inheritance of the sex chromosomes and therefore affect both male and female progeny, but unequally. Historically, however, the term has been used since the early days of genetic research and we shall continue the usage here.

Y-Linkage

The Y-chromosome has a very important function in determining the differentiation of the embryo into maleness. How many genes on the Y-chromosome are responsible for this task cannot be said at the present time, although in theory it would take only one to trigger the series of biochemical and developmental reactions leading to maleness during the late embryonic and early fetal period. After a little reflection, we might not expect to find very many loci for other characteristics on the Y-chromosome, for a number of reasons. First, it is one of the smallest chromosomes of the complement. Second, we have seen that

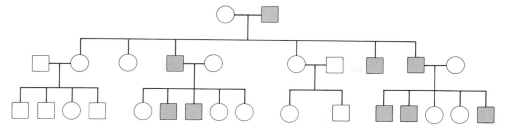

Figure 7-1 Pedigree showing the manner of transmission of a Y-linked, or *holandric*, character. The sons of all affected males are similarly affected; daughters are never affected. Note that this type of transmission parallels precisely the transmission of the male sex.

individuals with two Y-chromosomes do not suffer from very severe abnormalities, and in many cases are indistinguishable from normal males, suggesting that the number of genes duplicated in XYY individuals is relatively small. Finally, the female is clearly a normal individual in all respects, yet lacks a Y-chromosome completely.

It is not surprising, then, except for maleness itself, and one other exception to be discussed in Chapter 18, we are hard-pressed to find any genetic characteristics transmitted by the Y-chromosome. If they occurred, there should be no difficulty in identifying them because they would appear in a pedigree as in Figure 7-1, where every male with that trait would have similarly affected father and brothers, but unaffected sisters. There is one possible candidate for this pattern: in India several pedigrees have been found in which the ear rims of all the males are excessively hairy. This may be a true case of Y-linked inheritance, but it has also been suggested that there may be an autosomal dominant allele occurring with a very high frequency (100 percent in both males and females) in these particular pedigrees. If this allele expressed itself only in males, just as hairiness of other parts of the body—the chest, for instance—is characteristic of males, then we would expect a similar pedigree. Since earlier cases which had been accepted as Y-linked or *holandric* inheritance, such as scaly skin (*ichthyosis*) and webbed toes, are now regarded as spurious, verification of the hairy ear phenotype as evidence for Y-linkage would depend on obtaining pedigree data involving outcrosses of affected individuals to persons from distantly related groups.

X-Linkage

Characteristics determined by alleles known to be located on the X-chromosome, on the other hand, not only are frequent but also represent some of the most interesting and dramatic of those known in the human population. Partial color-blindness involving red/green vision is undoubtedly the one most common genetic anomaly present in the human race, affecting more than 5 million males in the United States. Hemophilia, another X-linked characteristic, although not nearly so common, has a fascinating history in the royal families of Europe.

Some of the simple rules of X-linked inheritance have been known for centuries. Even during the period before the birth of Christ it was vaguely understood that if a male affected with certain defects married a normal woman their children would usually all be normal, but the daughters would then produce affected grandsons. These empirical rules were not really understood until 1910. At that time, T. H. Morgan, an embryologist at Columbia University who had been an eloquent anti-mendelian, harshly critical of the growing number of biologists who were interpreting all inherited traits as simple single-gene effects, observed an unusual event in some cultures of fruitflies he had been growing in his laboratory on rotting bananas just for fun. Some of the flies had eyes that were completely white instead of the normal red (Figure 7-2). In working out the details of the inheritance of this characteristic, he showed that it corresponded exactly to that of such human characteristics as red/green color-blindness and hemophilia. More importantly, however, he triggered an explosion of research effort in genetics, involving many colleagues and students who laid the foundations for our present highly sophisticated knowledge of genetics.

SPECIAL PROPERTIES OF X-LINKED LOCI. It is clear that, because of the gross difference in size of the two chromosomes, most of the loci on the X cannot be

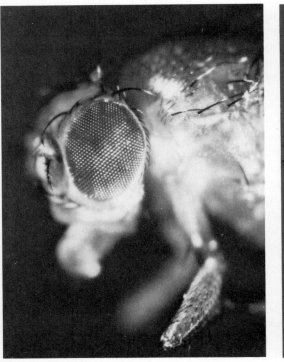 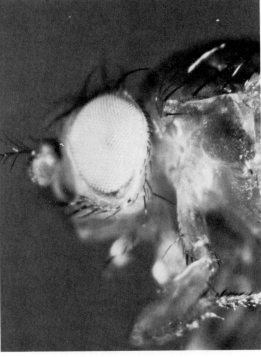

A   B

Figure 7-2 The eye of *Drosophila melanogaster*, at high magnification so that the individual elements of this compound eye are visible. (*A*) The normal or *wild-type* red color; (*B*) the mutant *white*, the first striking mutant to be discovered in this species.

present on the Y, and therefore the male must be effectively haploid with respect to most (if not all) X-chromosome loci. For this reason, the terms "dominant," "recessive," "homozygous," and "heterozygous" do not apply in the usual way to these loci in the male, because those terms are meaningful only in describing the relationship of two alleles to each other. Since the male has only one allele of each X-chromosome locus, he is said to be hemizygous for those alleles. The female, with two X-chromosomes and thus two alleles for each X locus, might be described as homozygous or heterozygous; however, in most of the cells of her body one or the other of her X-chromosomes is inactivated, making her a mosaic for those characteristics for which her two X's have different alleles. If, at a particular locus, a woman has different alleles on her two X-chromosomes, the allele responsible for the phenotype of the mosaic female is referred to as the dominant allele.

The consequences of mosaicism can be illustrated by the example of *hemophilia*, a blood disorder which will be considered in more detail later in this chapter. A male carrying the allele for hemophilia on his X-chromosome produces very little protein of a type called antihemophilic factor (AHF) necessary for normal clotting of the blood, with serious, sometimes fatal, consequences. A homozygous female with two such alleles, one on each of her X-chromosomes, would have similar problems. However, a heterozygous female will have, on the average, half of her cells with the chromosome carrying the normal allele active, while the other half have the other chromosome with the allele for hemophilia active. That half with the active normal allele will produce enough AHF so that her clotting time will be normal, or nearly so. Since she is a heterozygote with one allele normal and the other the allele for hemophilia, but has a normal phenotype, the normal allele is considered to be dominant and the hemophilic allele recessive. For convenience, the terms "homozygous," heterozygous," "dominant," and "recessive" are used with reference to sex-linked characteristics, but it should be kept in mind that this is not necessarily the same cellular relationship that is implied when the terms are used in connection with autosomal alleles.

TRANSMISSION OF X-LINKED LOCI. Figure 7-3 shows what to expect from simple matings of males and females carrying different sex-linked alleles. If a male carrying one allele mates with a female homozygous for a different allele for that locus (Figure 7-3A), all of their daughters will get his allele (and all of his sons his Y-chromosome). If his allele is dominant, then all of his daughters will be affected and none of his sons. If the allele is recessive (i.e., did not show up in the mosaic for the two different alleles), then his daughters will appear normal but will be heterozygous—capable of passing their father's defective allele on to their children—and all of his sons and their children will be normal since they do not receive the X-chromosome from their father.

If a woman homozygous for the defective allele (Figure 7-3B) mates with a normal man, then all of her children will carry the allele since they all receive an X-chromosome from her. If this allele is dominant, all of her progeny will show the defect. If it is recessive, however, the females will not show the characteristic since they are heterozygous and possess the dominant normal allele; the males, on the other hand, being hemizygous (without a homologue to the single X), will express the gene.

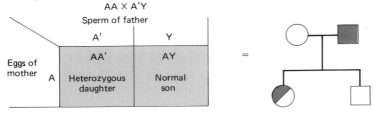

A. Homozygous normal mother X hemizygous affected father

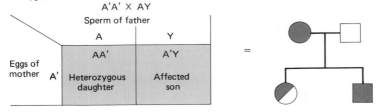

B. Homozygous affected mother X normal father

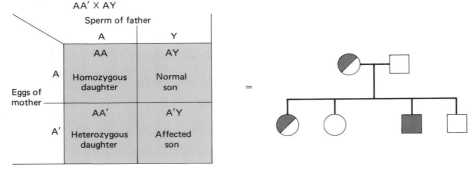

C. Heterozygous mother X normal father

Figure 7-3 The transmission of characters determined by alleles on the X-chromosome when (*A*) the father is hemizygous, (*B*) the mother is homozygous, and (*C*) the mother is heterozygous. The Y-chromosome of the male is indicated by a Y. The "sibships" are not meant to indicate actual families, but instead show the theoretically expected types of progeny in the proportions ideally expected: thus, there are two different types of progeny expected equally frequently in (*A*) and (*B*) but four in (*C*). Heterozygous females are indicated by circles only half filled in: such females would be affected if the character in question were dominant but not if the character were recessive.

When the female is heterozygous (Figure 7-3*C*), four types of progeny may be produced: normal and affected males, and heterozygous and homozygous females. A litter of cats showing the four types from such a mating is shown in Figure 7-4.

Human pedigrees of sex-linked recessive traits commonly consist of combinations of the situations described in Figure 7-3*A* and *C*, in which affected males

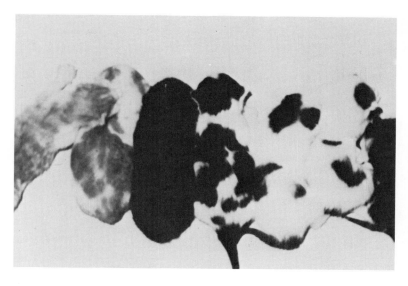

Figure 7-4 Segregation of the sex-linked characters for black and orange in the litter of a heterozygous mother. (The white coloration is caused by extraneous autosomal genes and should be disregarded.) The two orange kittens on the left are males. The kitten on the extreme right is both black and orange (i.e., is calico) and is female. The four black kittens in the center are three males and one female. These four classes correspond to those shown diagrammatically in Figure 7-3C. (Courtesy of H. C. Thuline, Rainier School, Buckley, Washington.)

produce no affected offspring, but only heterozygous daughters, and in which those heterozygous daughters then produce sons half of whom are affected and daughters half of whom are heterozygous.

This pattern of transmission is illustrated by a sex-linked condition causing excessive dryness and scaliness of the skin, *ichthyosis* (Figure 7-5A). In such pedigrees the affected individuals are invariably males (Figure 7-5B), who transmit the recessive to their daughters, who then transmit it in turn to half of their offspring.

Red/Green Color Blindness

By the use of special tests for color perception, it can be readily shown that one male in ten or twelve sees colors in a slightly different way from all other males (and virtually all females). Such males fall into the broad category of the *red/green "color-blind."* When they are questioned closely, several interesting facts emerge. First, they do not regard their "defect" as important. Such affected persons probably went through the first dozen or more years of life without being aware of any great color problems, except for an occasional insignificant error of confusing one minor hue for another. In fact, hundreds of thousands, perhaps millions, of males in the United States are in the red/green color-blind class without being aware of their condition. Second, the "color-blind" person may drive a car; he has no difficulty in differentiating red and green traffic lights.

What, then, is the basis for classifying him as red/green color-blind? Does he actually see red and green, or some other color, or nothing, or what? Exactly what is the genetic basis for his failure to pass the standard tests for color vision? Because this is such a very common condition, we will consider the phenomenon of color vision in some detail.

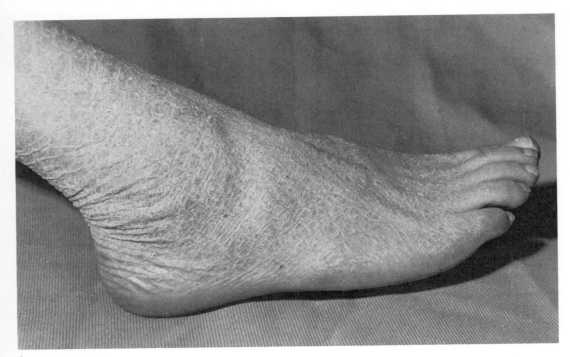

A

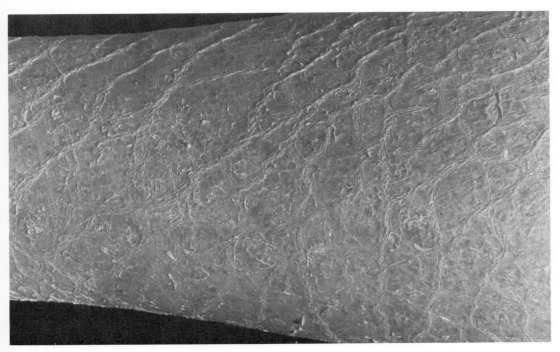

B

Figure 7-5 (*A* and *B*) A sex-linked recessive condition, ichthyosis, causing excessive dryness and scaliness of the skin. (*C*) A pedigree of this condition, taken from a much more extensive Dutch pedigree, showing the simple rules of sex-linked inheritance. (L. N. Went et al., *Ann. Hum. Genet.*, **32:**333–45, 1969.)

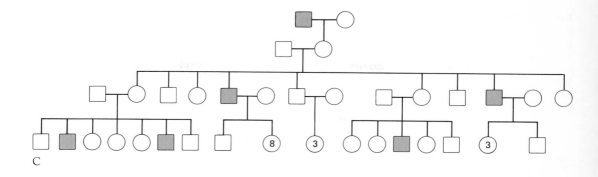

C

THE PHYSICAL BASIS FOR COLOR VISION. Of all the human senses, vision is the most complex and the least understood. As light hits the retina of the eye, an image is formed somewhere in the cerebrum that bears a close relationship to what might be described as reality. This image has many properties that are not at all well understood, including that of color.

Quanta of light, distinguished from each other only by small differences in wavelength, strike the retina of the eye and there are transformed into what we subjectively perceive as color. However, because our nervous system has the remarkable capacity to interpret these wavelengths as colors, we ordinarily think of color as an objective property of the object perceived and not merely a subjective impression occurring within our heads. This point is not self-evident—as eminent an observer as Isaac Newton, for instance, considered color to be an intrinsic property of the light wave.

Just as no one has succeeded in describing the biological basis for human perception of a visual image, so has no one yet formulated a completely satisfactory explanation of color interpretation of light wavelengths. This is not because workers in the field have not given it considerable attention: in fact, there have been more than two dozen theories of color vision since the initial work of Young in 1802 and of Helmholz some 50 years later. We shall adopt the Young-Helmholz theory of color vision with minor alterations here to explain some of the common genetic defects.

TOTAL AND BLUE COLOR-BLINDNESS. At the outset it should be pointed out that total color-blindness, in which a person can perceive only shades of gray, is so rare a phenomenon (with a frequency of less than one per 10 million) as to be of no great interest to us here. Color-blindness affecting the perception of blue is also an extremely uncommon deficiency, often brought about by injury or disease rather than genetic factors. Instead, what is commonly referred to as "color-blindness" is a difficulty in distinguishing certain colors in the range from red to green, in the same way that the rest of the population does. This is not actually a blindness, despite its name, and might be better thought of as color-confusion.

RESPONSE OF COLOR-SENSITIVE CONES. There are commonly assumed to be three sets of color-sensitive eye cells, or *cones*, sensitive to red, green, and blue (or violet). The blue sensitivity is very rarely affected for genetic reasons; defects

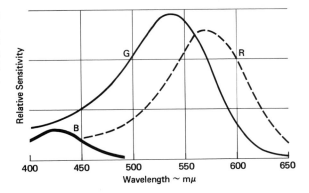

Figure 7-6 Graph showing the relative sensitivity of the three cone types, blue (*B*), green (*G*), and red (*R*), as a function of the wavelength of light, from blue wavelengths on the left through green (500 to 550) and yellow (575 to 585) to red on the right. (G. Wald, *Science*, **145**:1007–16, 1964.)

of color vision that are inherited are almost without exception defects of the red/green system.

The response of the cones to different wavelengths of light is not sharply defined. There is considerable overlapping and the impression of color that we perceive comes from the relative ratio of stimulation of the three kinds of cones by particular wavelengths (Figure 7-6). The discrimination of the eye is so exacting, however, that it can ordinarily differentiate more than 150 colors on the basis of relative stimulation of the three different cone types. Notice that according to this theory there is no cone that responds to yellow. Yellow is perceived when the red and green cones are stimulated equally. This can happen in two ways. If the particular wavelength perceived is at the point where both the red and green cones are stimulated about equally (Figure 7-7*A*), the net effect is yellow—this is why we see yellow between red and green in a rainbow. Also, if red and green lights are superimposed so that both kinds of cones are stimulated fully, the net effect will again be that of yellow (Figure 7-7*B*). This latter principle is used in color television sets, in which the simultaneous glowing of red and green dots adjacent to each other on the screen fuses in our perception as a single color—yellow. It is easy to understand why yellow colors produced artificially

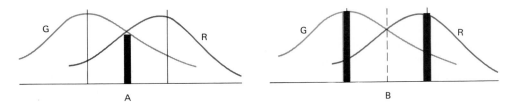

Figure 7-7 Two different ways of producing the yellow sensation by the simultaneous stimulation of green and red cones equally. In (*A*), a single wavelength between red and green is so positioned on the spectral sensitivity curve that both red and green are stimulated equally. This is how a specific wavelength between red and green in a rainbow produces a yellow sensation. In (*B*), two different wavelengths stimulate the red and green cones, respectively, to give a strong but equal response from each. It is in this way that a television set produces a yellow color, a yellow that can appear more gaudy than natural yellow does.

by adding red and green light together and stimulating both cone types maximally may be more vivid than those found naturally, as in rainbows where a single wavelength stimulates the cones only partly.

The red-sensitive cones contain a pigment which absorbs light primarily in the lower wavelengths according to the curve in Figure 7-7, and the green cones have a different pigment which mainly absorbs light in the "green" or middle wavelengths. These two kinds of cones give distinctly different kinds of messages to the brain, which are then interpreted as "red" and "green." Whether a cone will contain predominantly red-sensitive or green-sensitive pigment depends on which of two loci on the X-chromosome is functional in that particular retinal cell. We know that these two loci are located physically quite close to each other on the X-chromosome, because naturally occurring breaks in that chromosome strand very rarely separate them. Why the red-pigment-producing locus of some eye cells is the active one and the green suppressed, and the green locus active in others and the red suppressed, is not known. A good guess is that the initial signal is given by the nerve endings as they develop in the retina of the early embryo: those nerves that will be responsible for red cerebral impressions after birth incite the production of red-sensitive pigment, and those responsible for green perception incite green pigment production.

What happens when one of the two loci is nonfunctional? John Dalton, the British physicist and chemist, better known for the gas law, was severely red/green color-blind, and made the following comment: "My yellow comprehends [= includes] the red, orange, yellow, and green of others; my blue and purple coincide with theirs." Dalton apparently had a nonfunctioning (or poorly functioning) locus normally responsible for red pigment; this reduction in the red sensitivity of his retina completely blurred the distinction between red and green that normal people possess. Similarly, when the locus responsible for the green pigment does not function normally, the reduction in green sensitivity overlaps the red, leading to a parallel confusion.

From this simple description, it is clear that both types of color-blindness are similar to some extent; they can readily be distinguished, however, because persons with the red defect (like Dalton) will show a greatly reduced sensitivity to pure red, but not to pure green, whereas those with the green defect have the reversed characteristics.

MILD COLOR-BLINDNESS. In about sixty percent of the cases, however, the reduction in pigment is much less extreme so that discrimination between red and green is more easily made, but sometimes with difficulty. A common question is: how do the moderately red/green color-deficient see red and green? They will, in fact, see red and green at the proper points in the spectrum, but the limited response of one of the two types of cones tends to obscure the sharp differences between the two colors that normal people are quite aware of. This does not, however, impair their ability to distinguish red from green, as in traffic lights, provided that the lights are bright and not too distant.

TESTS FOR COLOR-BLINDNESS. This confusion in the red/green color-blind is taken advantage of in designing the tests for color-blindness seen earlier in this chapter. The best known of these are those of Professor Shinobu Ishihara, who constructed a series of plates with numbers made up of colored dots. The normal person associates all reds together and all greens together, regardless of shade or mixture, and sees one number. The color-blind, being less sensitive to the distinctions between red and green, will associate the lighter shades with each other, regardless of whether they are yellow, pink, or light green, and so see an entirely different number. Different plates can be designed to determine whether it is the red or green cone that is deficient, and whether the red/green color-blindness is partial or complete.

Such tests are now routinely included in the series of eye tests given to driver's license applicants and it is not unusual for a male in his late teens or early twenties to discover for the first time that he has a color vision deficiency, but has quite successfully and unconsciously compensated for it in his early years. Generally speaking, such color-deficient persons will be able to discriminate among red, green, and amber sufficiently to qualify for a driver's license. In fact, studies of 160 color-blind drivers show that they have accidents no more often than other drivers and that this is true for both mildly and severely affected red/green color-blind drivers.

DISTRIBUTION OF COLOR VISION DEFECTS. The frequency of defects of color vision differs from one ethnic group to the next. In general, the technologically "primitive" peoples (hunters and food-gatherers) show a very low incidence of color-blindness, whereas those from more technologically advanced groups tend to show a higher frequency. It has been suggested that this is related to the greater survival value of individuals for whom color vision is important for existence, such necessity being relaxed as people move from primitive survival methods to more advanced societies. One difficulty with this interpretation, however, is that the lower frequency of color defects in more primitive populations results from a reduction in the milder forms of color deficiency, the more severe types remaining relatively constant from one population to the next. A list of the frequency of common color-blindness in various ethnic groups is given in Table 7-1.

In smaller numbers males have been tested to determine whether the color deficiency involves the red or the green cone, and whether the extent of the deficiency is mild or severe. Table 7-2 gives the frequencies of the various types of red/green color deficiencies in European populations.

Hemophilia

Perhaps the most infamous inherited congenital disease is the "bleeding disease," *hemophilia*, caused by a sex-linked recessive allele which results in the failure of the blood to clot properly. It appeared in the royal families of Europe via mutation occurring either in Queen Victoria (1837–1901) or in one of her parents. We can be fairly certain that it did not occur in that family prior to that time, because of the detailed records that have been kept on the royal family and the great likelihood of detecting a disease like hemophilia, considering the roughness of the times and the lack of medical care in the centuries prior to Queen Victoria's reign.

Of her children, three daughters (Victoria, Alice, and Beatrice; see Figure 7-8) were heterozygous for the gene—hence unaffected carriers—and only one son,

	Number	Color-Blind (percent)
By Country		
Belgium	9,540	7.5
Britain (naval recruits)	123,414	7.5
France	6,635	9.0
Germany	6,863	7.7
Netherlands	3,168	8.0
Norway	9,047	8.0
Russia	1,343	9.2
Switzerland	4,036	8.3
By Locality		
Hausa, Nigeria	380	2.1
Miao tribesmen, Thailand	312	2.2
Eskimo "full-bloods"	297	2.5
Eskimo "half-bloods"	132	6.6
Maoris, New Zealand	571	2.6
Chomorros, Mariana Islands	246	3.3
Tibetans	241	5.0
Thailand, northern	2,128	5.2
Bantus, Uganda	1,294	4.7
Papuans, New Guinea	4,820	3.9

Table 7-1 Rates of red/green color vision deficiency in selected samples. (Adapted from R. H. Post, *Humangenetik* **13**:253–84, 1971.)

Leopold, was hemophilic. All of her other children were free of the defective allele. Because the present royal family of England is descended from one of the normal sons, Edward VII, it can be safely assumed to be free of the disease. However, Alice and Beatrice married into other royal families of Europe and transmitted the gene to their progeny, so that the next several generations of European royalty had an undue proportion of hemophiliacs (Figure 7-9).

Perhaps the best-known case involved the wife of Czar Nicholas II of Russia, Czarine Alix, who was the granddaughter of Victoria. She had a son, Czarevitch Alexis, who was hemophilic. It is of historical interest to note that the monk Rasputin claimed to have supernatural powers that could cure Alexis, and succeeded in becoming a powerful figure in the Russian royal court. It was partly a result of Rasputin's influence that the Russian monarchy became disorganized

		Type of Cone Affected	
		Green (percent)	Red (percent)
Severity of Defect	*Mild*	53	8
	Severe	19	20

Table 7-2 Frequencies of the four common types of red/green color-blindness in the European population. (Adapted from R. Cruz-Coke, *Colorblindness, An Evolutionary Approach*, 1970. Courtesy of Charles C Thomas, Publisher, Springfield, Ill.)

Figure 7-8 Some of the royalty of Europe surrounding Queen Victoria, as they posed for a group photograph during a family wedding in 1894. Queen Victoria (*A*) was the initial heterozygote; her youngest daughter, Princess Beatrice (*B*), and two grand-daughters, Princesses Alix (*C*) and Irene (*D*), all received the defective allele from her. Beatrice transmitted the allele to the Spanish royal family and Alix to the Russian. (Courtesy of the Gernsheim Collection, Humanities Research Center, University of Texas.)

and eventually suffered a complete collapse. Although there were certainly additional sociological factors of greater weight, Rasputin and thus hemophilia can be said to have been in part responsible for the downfall of the Russian czardom and the institution of the present soviet government.

The term *hemophilia* is, strictly speaking, a misnomer, since its literal transla-tion is "liking for blood," a disposition which might better describe a vampire than the victim of a well-known sex-linked defect. It has been recorded as far back as the Hebrew law book, the Talmud, in which it was noted that when excessive bleeding occurred during ritual circumcision the other male members of the affected family were to be exempt from the ritual, implying an awareness of the pattern of inheritance. The disease was first definitively described by a U.S. physician, Dr. John Otto of Philadelphia, who in 1803 outlined its clinical manifestations as well as noting that only males were subject to the disease, and the females of the same family were unaffected but could still transmit it to their male children.

MEDICAL ASPECTS. The primary characteristic of hemophilia is the greatly increased clotting time of the blood. Whereas normal blood may clot within 5 to

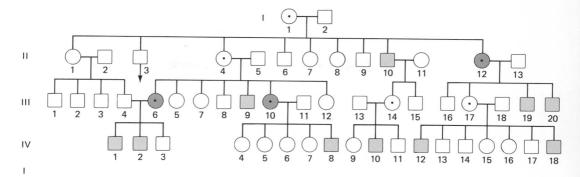

Figure 7-9 An abbreviated pedigree of Queen Victoria and her descendants, showing the distribution of hemophilia in royal families of Europe. Some of the more important (from our point of view!) persons in this pedigree are

- I1. Queen Victoria.
- I2. Prince Albert.
- II3. King Edward VII of England.
- II4. Princess Alice of Hesse Darmstadt.
- II10. Prince Leopold, Duke of Albany.
- II12. Princess Beatrice of Battenberg.
- III6. Princess Irene of Hessen.
- III10. Princess Alix, wife of Czar Nicholas of Russia.
- III14. Princess Alice, wife of Alexander, Prince of Teck.
- III17. Princess Victoria, wife of King Alfonso of Spain.
- IV8. Czarevitch Alexis of Russia.
- IV10, 12, 18. Lord Trematon, Prince Alfonso, and Prince Gonzalo, respectively; all three died after automobile accidents.

10 min when placed in a test tube, hemophilic blood may take from 30 min to several hours to clot. The hazards of this condition can be easily appreciated. A cut that might be completely innocuous to a normal person could bleed for days in a hemophiliac. Minor injuries that are not usually even noticed in unaffected people, such as bruises, can produce serious hemorrhaging in affected persons. Simple operations such as dental extractions can be fatal. Internal bleeding, especially in such vulnerable areas as the soft palate of the mouth, represents a special hazard because of the difficulty of stopping the flow of blood inside the body. Strains and sprains of the arm and leg joints can result in accumulations of blood inside these relatively delicate areas, inhibiting movement.

Twenty-five years ago, only one quarter of all hemophiliacs reached the age of 16 years. Today, more than 95 percent survive beyond this age. Treatment for hemophilia consists of reducing the coagulation time, which can be done by transfusion of fresh, normally clotting blood, or by injection of antihemophilic globulin derived from normal blood. The techniques for this latter treatment have become sufficiently standardized that affected individuals can maintain their own supply of globulin in refrigerators and inject themselves at regular intervals. It is, nevertheless, a serious disease with many potential dangers, especially during youth when injuries are more frequent, and these can accumulate with additional side-effects to decrease the affected person's general well-being.

CHRISTMAS DISEASE.   During a World War II bombing raid in England a small boy was injured; he was subsequently diagnosed as having hemophilia, but of a different sort than the commonly recognized one. His surname was Christmas: the hemophilia was therefore called Christmas disease. The blood component lacking in affected persons is not the same as that absent in ordinary hemophiliacs; although the two types of hemophilia are sex-linked, they are definitely alleles of different loci.

There exist, in fact, eleven different inherited diseases known to affect clotting of blood; of these, only classical hemophilia and Christmas disease are sex-linked, the others being autosomal, and for the most part recessive. The clotting factor lacking in classical hemophilia (or hemophilia A), in addition to being referred to as AHF, is also called factor VIII. The factor lacking in Christmas disease, or hemophilia B, is called PTC for the platelet and thrombocyte blood cells of which it is a component, or, in international symbolism, factor IX. Hemophilia A is six times more common than B.

Muscular Dystrophy

There are three major types of muscular dystrophy: one of these is inherited as an autosomal recessive, another as an autosomal dominant, but the type described by Duchenne affects juveniles and is inherited as a sex-linked recessive. This is by far the most frequent type, occurring at the rate of about three males per 10,000 births. The abnormal development of the muscles shows up very early, sometimes at birth, and progresses rapidly, generally leading to death before the age of 20 years.

Frequencies of Alleles in Populations

It seems fair to assume that the chromosomes that males carry are a random selection of all the X-chromosomes present in the human population. In fact, the totality of all these chromosomes must be one third of all X-chromosomes in existence, with females possessing the other two thirds of them. If they are a random sample, then we can come to some simple conclusions about the relative frequencies of various X-linked alleles by examining the relative frequencies with which they occur in the male sex. Thus, if in the population of the United States, 6 percent of all males are color-blind because of a defect at the "green" locus, then it is reasonable to conclude that 6 percent of all X-chromosomes, irrespective of whether they occur in males or females, carry the allele for green color-blindness.

Now, we can further make the simple statement that the chance that any particular chromosome, whether it be found in the male or the female, will carry the gene for color-blindness involving the green locus is 6 percent or .06. We can equally simply state that the chance that an X-chromosome will carry the normal allele is 1 − .06 or .94.

Why are so many more males color-blind than females? The female would have to have both X-chromosomes with the recessive allele to show the phenotype, and this is less likely than a male having only one.

If we know that the frequency of the *c* allele is .06 and the frequency of its alternative *C* is .94, then we can ask what the relative frequencies of the different genotypes might be on the basis of simple algebra. This approach suggests that we can consider each person as being a random selection of X-chromosomes (two for the female, one for the male) drawn from a much larger pool of chromosomes, in which the chance that any chromosome of a given type will be drawn is given by its overall frequency in that pool. Let us consider the probabilities of the different combinations when we take two chromosomes from this large supply of X-chromosomes in order to constitute a female. The chance that any one chromosome will be normal, with the *C* allele, is .94. The chance that two in succession will be normal is $.94 \times .94 = .8836$. What is the chance that one of the X-chromosomes will be normal and the other abnormal? This can happen either of two ways: either the first will be normal and the second abnormal ($.94 \times .06 = .0564$) or the first will be abnormal and the second normal ($.06 \times .94 = .0564$). The total probability, then, of one normal and one abnormal chromosome (that is, a heterozygote) will be $2 \times .0564$ or .1128. Finally, the chance that both chromosomes will have the recessive *c* is $.06 \times .06 = .0036$, which is less than 1 percent. Since this condition can be described as recessive (two alleles of type *c* being necessary in the female for color-blindness to express itself), we have a very much reduced frequency of expectation for color-blind women than for color-blind men. Population surveys, in fact, confirm the level predicted by simple algebra.

This simple approach for estimating gene frequencies in populations was first proposed by the British mathematician Hardy and the German physician Weinberg in 1908, and is often referred to as the *Hardy-Weinberg rule*. In its more complex mathematical formulations, it can be difficult to understand, but in the exercise we have carried out here there is no difficulty in comprehending the basic principles. We will have more to say about this later.

Our rules for the expression of genes carried on the X obviously break down for some of the syndromes we have already discussed. Females who have Turner's syndrome, with a single X chromosome, obviously have a probability of carrying X-linked defects equal to that of males. This theoretical expectation is borne out by observation of the frequency of red/green color-blindness in Turner's syndrome females, which is approximately as high as the male frequency. This would not be the case, of course, if the nature of the defect put the female at a severe disadvantage (e.g., hemophilia). If a medical practitioner should come upon the case of a female with a rare sex-linked disorder, he might well immediately look for symptoms of Turner's syndrome. This would be particularly true in the case where the father was known not to have the disorder—for example, for a female affected with muscular dystrophy of the Duchenne type.

References

MCKUSICK, V. A. 1962. On the X-chromosome of man. *Quart. Rev. Biol.,* **37:** 69–175.

MCKUSICK, V. A. 1965. The royal hemophilia. *Sci. Am.,* **213:**88–95

MORTON, N. E., C. S. CHUNG, and H. A. PETERS. 1963. Genetics of muscular dystrophy. In G. H. Bourne and Ma. N. Golarz, eds., *Muscular Dystrophy in Man and Animals.* New York: Karger.

OHNO, S. 1967. *Sex Chromosomes and Sex-Linked Genes.* Berlin: Springer.

POST, R. H. 1971. Possible cases of relaxed selection in civilized populations. *Humangenetik*, **13**:253–84.

RUCH, T. C., and H. D. PATTON. 1965. *Physiology and Biophysics*, 19th ed. Philadelphia: Saunders.

STERN, C., W. R. CENTERWALL, and S. S. SARKAR. 1964. New data on the problem of Y-linkage of hairy pinnae. *Am. J. Hum. Genet.*, **16**:455–71.

STRAATSMA, B. R., M. O. HALL, R. A. ALLEN, and F. CRESCITELLI, eds. 1969. *The Retina: Morphology, Function and Clinical Characteristics.* UCLA Forum in Medical Sciences. Berkeley: University of California Press.

THOMPSON, M. W., B. LUDVIGSEN, and G. MONCKTON. 1962. Some problems in genetics of muscular dystrophy. *Rev. Can. Biol.*, **21**:543–50.

WENT, L. N., et al. 1969. X-linked ichthyosis: Linkage relationship with the Xg blood groups and other studies in a large Dutch kindred. *Ann. Hum. Genet.*, **32**:333–45.

WRIGHT, W. D. 1967. *The Rays are not Coloured. Essays on the Science of Vision and Colour.* London: Adam Hilger Ltd.

Questions

Useful terms: sex-linked inheritance, holandric, hemophilia, hemizygous.

1. Draw a pedigree showing the manner of inheritance of a genetic defect with its locus on the Y-chromosome.
2. Since the Y-chromosome is essential for male fertility, could genes responsible for male infertility be transmitted with the Y-chromosome?
3. Brown enamel of the teeth is a well-known sex-linked dominant trait. Draw a pedigree showing how it will be inherited (a) if the father of a family carries the defect, (b) if the mother is heterozygous for it, and (c) if the mother is homozygous for it.
4. In what sense are red/green color-blind persons actually color-blind? How common a defect is this? What is the genetic basis for such color deficiencies?
5. What is the frequency of color-blindness among females? Explain why it is so much different from that in males.
6. If a female should receive an X-chromosome carrying an allele for red deficiency from her father and one for green deficiency from her mother, will her color vision be affected? Explain your answer.
7. What is hemophilia? Why is it so very well known compared to other sex-linked defects?
8. If a relatively uncommon disease such as muscular dystrophy of the sex-linked type occurs in a young girl, what might you suspect?
9. Discuss the relevance of the following conditions to the concepts discussed in this chapter: ichthyosis, red-green color-blindness, hemophilia, Turner's syndrome, Duchenne's muscular dystrophy.

8

Antigens and Antibodies: The Blood Groups

The periodic bouts that we have with viral and bacterial diseases, illnesses of several days with gradually increasing severity, are usually followed by an apparently miraculous recovery, often unaided by any medication, antibiotics, or medical care. These experiences serve as recurring reminders of our dependence on our own *immune system* to combat external invaders. Any person who has wandered into a patch of poison ivy or poison oak may have become painfully aware of the existence of the body's immune system. Those who suffer from hay fever dramatically exhibit an *immune reaction.* The case of the unfortunate person who has become sensitized to as common an event as a bee or wasp sting, and who subsequently dies after another single sting is not an uncommon news item.

Infectious Disease

Humans have always been at the mercy of infectious disease; throughout time it has been the most common cause of death. During the Middle Ages, the plague decimated entire populations, paralleled by the more recent introduction of diseases by White men into areas occupied by peoples not previously exposed. Just as the Indians were wiped out in Yucatan by the first epidemic of yellow fever introduced by Europeans in about 1650, so were previously unexposed Europeans who later migrated to the West Indies and West Africa similarly stricken with yellow fever. The deaths from these epidemics were not limited to children and the weak; in fact, sometimes these diseases seemed to favor robust, mature, and apparently quite healthy adults. Even in recent times epidemics have occurred: it is estimated that in 1918 and 1919 influenza killed a total of 20 million people worldwide, 12 million in India alone.

As time goes on, more and more of these scourges are being eliminated— poliomyelitis, typhoid fever, cholera, smallpox, diphtheria, and scarlet fever, to name a few. The colored poster nailed to the entrances of homes warning of the presence of a patient with a serious communicable disease, common a few years ago, is now virtually unknown in the Western world. To be sure, knowledge of the biological nature of the causes of disease, the use of vastly improved sanitation, and the development of effective medicines have all been important in the struggle, but there can be no doubt that the application of the most elementary techniques involving the immune reaction has played a leading role.

Some of the characteristics of the immune system are technically complex, but the most important are not. Every person is familiar with the widespread occurrence of childhood diseases such as chickenpox, measles, and mumps, and the fact that once a person has had such an illness it will almost certainly never return. One classic example of the long-range memory of the immune system comes from the measles epidemic that ran through the Faroe Islands in 1781, 1846, and again in 1875. In each instance large numbers of people were affected, but in the later epidemics persons who had been alive during the previous one and had contracted the disease at that time were not again affected, despite the intervening decades.

Manipulation of the human defense mechanism began with the experiments of the English physician Edward Jenner in following up a suggestion made by a woman in charge of a dairy. She pointed out that those of her milkmaids who came down with the rather minor illness cowpox were never stricken with smallpox. Jenner inoculated a small boy with cowpox germs and continued his observations with a set of "experiments" designed to prove the effectiveness of vaccination against smallpox by the use of cowpox vaccine. His scientific paper describing the results of his experiments was rejected by the journal to which he submitted it, and he spent the remainder of his life embroiled in the controversies surrounding his ideas. Since that time, immunization against smallpox and tetanus has become commonplace, and vaccines that protect against other less common diseases (such as cholera and plague) are available when conditions require their use. The most recent development, the antipolio vaccine and the use of weakened polio strains for immunization, has brought the incidence of this dread disease down to a minute percentage of its former frequency.

On a more personal level, we are reminded of our own immunity characteristics when our blood is typed and we are designated as O-positive or A-negative, for instance, and are vaguely aware that this information is of great importance if our blood is to be transfused into another person (or, less fortunately, if we ourselves require a transfusion). The adverse effect of the *Rh factors* in newborn infants has received considerable attention in the daily press and every woman who has been tested and found to be Rh-negative is forewarned that her progeny could suffer as a result. Finally, the press reports case after case of organ transplantation from one human into another, many of them successful and many not; in the latter cases we know that the failure is often due to *rejection* of the transplanted organ, a property of the immune system.

The ABO Blood Groups

The art of the blood transfusion, which has been instrumental in saving literally millions of lives over the past 60 years (developed just in time to have a sizable impact on the chances of survival of the wounded during World War I), evolved slowly from an historical standpoint. The reasons for its late arrival include a prior lack of information about the circulatory system, and about the necessary sterile techniques, and an absence of the tools (hollow needles, etc.) necessary to perform the operation properly. The major conceptual obstacle, however, was the lack of any understanding of the blood groups.

HISTORICAL DEVELOPMENT.   Possibly the earliest attempt to "transfuse" blood took place in 1492 when Pope Innocent VIII, in a coma, was administered the blood of three young men. He died, along with the three donors. (It seems likely that he was given the blood to drink rather than as a true transfusion.) In the 1660's, the physician to Louis XIV of France performed a successful transfusion of about half a pint of lamb's blood into a boy who was weak and feverish from excessive bloodletting, a common practice then thought to have curative effects. However, subsequent transfusions were unsuccessful and that physician was charged with murder. Blood transfusions, which had such a high fatality rate at that time, were banned in France, England, and Italy.

At the beginning of the nineteenth century there was a resurgence of interest in this technique after it was shown to be effective in some cases of anemia. Almost 500 transfusions were performed during the nineteenth century through-out Europe, the majority with human blood and the rest with blood from ani-mals. Again, however, there were severe unexpected reactions in such a large proportion of the patients that the technique was abandoned once more.

LANDSTEINER'S INITIAL DISCOVERY   The primary cause for this incompatibility of the blood of different persons was discovered by the immunologist Karl Landsteiner, who showed in 1900 that human blood can be classified into sev-eral distinct groups. Subsequent work quickly established the number of such groups to be four.

We can demonstrate the existence of these four groups quite easily. Let us take samples of blood from 100 persons, number the samples carefully, and mix a drop of each sample with a drop from every other sample. After a few minutes we will see that in about a third of the mixtures clumping, or *agglutination,* oc-curs (Figure 8-1), but the rest remain fluid. At first it may appear that agglutina-tion is taking place at random between different samples. However, a closer inspection (and some thought) reveals something more: that the samples fall into groups, and that samples do not agglutinate when mixed with others of the same group, but do when mixed with those of other groups. Furthermore, the number of such "self-compatible" groups is four.

THE BASIS FOR AGGLUTINATION   To understand why the cells in some mixtures agglutinate and others do not, we must consider the way by which the body protects itself against foreign invaders. When certain substances, usually large complex molecules, enter the bloodstream, the body is able to recognize them as alien and reacts by producing highly specific protein molecules, called

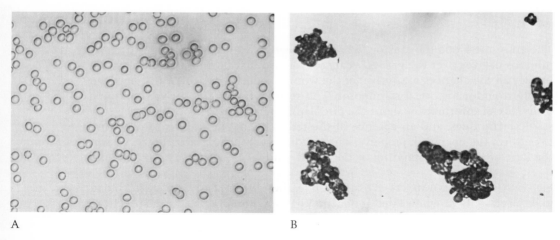

A B

Figure 8-1 A comparison of normal red cells (*A*) of group A with cells that have been agglutinated (*B*) by the addition of the anti-A antibody. Note that the clumps are formed as cells associate with each other. This is not to be confused with clotting, which is caused by the formation of fibrin in the blood fluid. (Courtesy of J. Puro, University of Turku.)

antibodies, against them. Any substance which triggers antibody formation is called an *antigen*. The particular antibody produced by a given antigen is able to form a chemical union with that antigen; when the antigen is located on a red blood cell, the antibody joins two cells together, one at each end of the molecule, causing the cells to agglutinate (Figure 8-2). Each of us possesses a large number—thousands—of compounds with antigenic properties as part of our natural make-up; others may be introduced into our system from outside the body. Similarly, the body normally carries hundreds of different kinds of antibodies, most of them having been previously induced in response to exposures earlier in life to foreign substances, and it is able, on call, to produce antibodies to new antigenic stimuli.

Figure 8-2 Detail showing how a specific kind of antibody molecule, with reactive sites at each end, can link two red cells together. If the antibody molecules attach themselves over the entire surface of the cell, large numbers can be held together in single clumps. (Adapted from F. Haurowitz, *Immunochemistry and the Biosynthesis of Antibodies*, Interscience Publishers, New York, 1968.)

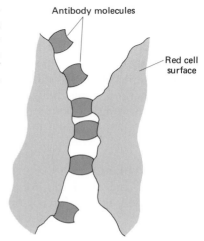

Antibody molecules

Red cell surface

THE SERUM AND CELL COMPONENTS.   Clearly, our first experiment was inadequate to indicate the role of the antigens and antibodies in the agglutination reaction. For this we must refine our technique by separating the red blood cells, or *erythrocytes*, from the fluid, or *serum*, of the blood. This is easily accomplished by putting tubes with the blood samples into a centrifuge and spinning the cells to the bottom. Now we have two components, the cells and the serum, for each of the four blood groups, and we can test all four classes of erythrocytes against all four classes of serum. The results from such combinations are shown in Table 8-1; a positive (+) sign means that an agglutination reaction has occurred, and a negative (−) sign represents no reaction. It appears that this table does not make any sense because the + and − signs appear to occur at random. We shall see, however, that there is a simple explanation for the pattern of + and − signs.

Human erythrocytes may carry either one of the two antigens, A or B, or both together, or neither. The four groups into which we have classified our blood samples are named according to whether their erythrocytes carry neither antigen (group O), antigen A (group A), antigen B (group B), or both antigens (group AB). The serum may contain antibodies anti-A, anti-B, both together, or neither. These antibodies occur naturally and do not have to be induced specifically as antibodies generally do. When the erythrocytes of a person of each of the four groups are added to antibodies of either A or B, the results are seen in Figure 8-3.

Each person possesses, as part of his *ABO* antibody constitution, those antibodies for which he does not have the corresponding antigens. The antigen-antibody compositions of persons of each of the four groups are listed in Table 8-2 and schematically represented in Figure 8-4. We can now reinterpret the results shown in Table 8-1 on the basis of the antigen-antibody components found in each of the four blood groups based on Table 8-2 and Figure 8-3. These are given in Table 8-3. A positive agglutination reaction occurs only when the serum carries the antibody corresponding to the antigen on the red cell.

TRANSFUSION PROBLEMS.   When the erythrocytes of a person carrying either of the two antigens are transfused into a person with the corresponding antibody, the transfused cells are subject to agglutination. The agglutinated red blood cells no longer flow freely to carry oxygen to the other cells of the body, and agglutinated blood masses lodge in the smaller blood vessels, cutting off the flow of what nonagglutinated blood might be left, often with fatal consequences.

		Erythrocytes			
		1	*2*	*3*	*4*
Serum	1	−	+	+	+
	2	−	−	+	+
	3	−	+	−	+
	4	−	−	−	−

Table 8-1   The occurrence of agglutination (+) or not (−) when serum of persons from the four different blood groups is added to the four types of erythrocytes.

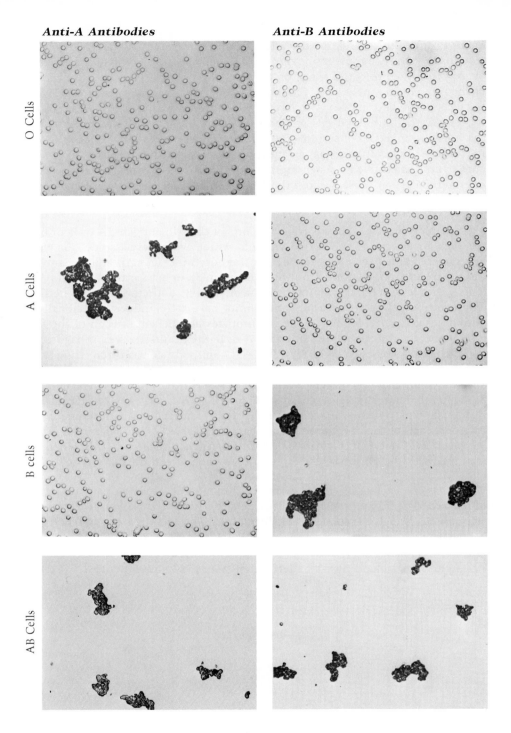

Figure 8-3 Agglutination response when erythrocytes of all four groups are mixed with antiserum containing either anti-A or anti-B antibodies. (Courtesy of J. Puro, University of Turku.)

Blood Group	Antigen on the Red Cells	Antibody in the Serum
O	Neither	Both A and B
A	A	Anti-B
B	B	Anti-A
AB	Both A and B	Neither A nor B

Table 8-2 The presence (or absence) of the A and B antigens on the erythrocytes of individuals of the four blood groups, and the presence (or absence) of the antibodies to those antigens in the serum. Note that no person will simultaneously carry both a given antigen and the antibody against it, but will carry the antibodies for any lacking antigen.

1. Antigenic sites on erythrocyte

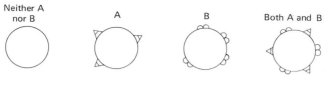

2. Antibody specificities:

3. Constitutions of the four different groups

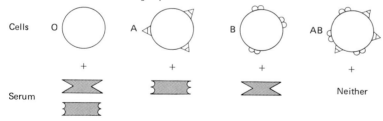

4. Consequences of mixture of AB cells and O serum

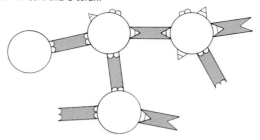

Figure 8-4 The complementary relationship between the specific antigenic binding sites on the cell and the corresponding antibody sites. The cells from the four groups have different surface configurations (*1*) corresponding to the antigenic structure (*2*). Each of the four groups has the antibodies for which cells have no reactive sites (*3*). When cells of one group are mixed with the serum of another, agglutination will occur if the antibodies of the mixture find the complementary site on the cells (*4*); in the case illustrated, the cells have both A and B sites and both antibodies are present.

Table 8-3 Reinterpretation of Table 8-1 in view of the actual antigen-antibody compositions of each blood group. When a given antigen and its corresponding antibody are found in the same square, a positive agglutination reaction results.

Serum	Erythrocytes			
	Group O (neither antigen)	Group A (antigen A)	Group B (antigen B)	Group AB (both A and B antigens)
Group O (both anti-A and anti-B antibodies)	−	+	+	+
Group A (anti-B antibody)	−	+	−	+
Group B (anti-A antibody)	−	−	+	+
Group AB (neither)	−	−	−	−

To illustrate the cases where difficulties may be anticipated, consider the results of transfusing blood from a person of group A into four recipients, each of a different blood group. When the recipient is of group A, his erythrocytes will have the same antigens as the donor's, and his blood serum the same antibodies, so that no complications will result. The group AB recipient will possess no antibodies of either group in his bloodstream, so again no reaction will occur. On the other hand, both group B and group O blood sera contain anti-A antibodies, so the incoming group A erythrocytes will be agglutinated.

One might think that the erythrocytes with a given antigen (A, for instance) in the recipient would be agglutinated by the corresponding antibodies (as anti-A) from the donor's blood serum, but in practice this reaction is found to be relatively unimportant because the transfused antibodies are quickly diluted out and absorbed by other tissues in the recipient. For this reason, compatible transfusions may not be reciprocal: O blood transfused into an A, B, or AB recipient will ordinarily present no problem, whereas A, B, or AB blood transfused into an O recipient will. Since it is the erythrocyte composition of the donor and the serum composition of the recipient that will determine the compatibility of different combinations in transfusion, Table 8-3 can be used to indicate likely transfusion reactions if the word "erythrocyte" is replaced by "donor," and the word "serum" by "recipient." In medical practice, however, it would be unwise to rely on presumed incompatibility based on such a chart. The customary procedure is to make a laboratory test of the donor's and recipient's blood to make sure that no agglutination will occur, either because of ABO blood group incompatibility or for any other reason, such as an unsuspected antigen-antibody incompatibility involving other factors.

INHERITANCE OF THE ABO BLOOD GROUPS. Up to this point in this book we have been concerned with those loci at which is located one "normal" gene,

Phenotype	Genotype
O	OO
A	AA
	AO
B	BB
	BO
AB	AB

Table 8-4 The phenotypes and corresponding genotypes of individuals of the four different ABO blood groups.

vaguely defined as a gene commonly found in the average person, which has an allele, either dominant or recessive, considered abnormal because it is relatively rare. For the ABO blood groups there are three major alleles, any one of which may be found on the particular chromosome that carries this locus. We shall designate the three alleles by the italicized letters, *A*, *B* and *O*.

When a person has the *A* allele he will produce antigen A, and when he has the *B* allele he produces antigen B. Thus the homozygotes *AA* and *BB* are in groups A and B, and, because *O* can be considered inactive, *AO*, *BO*, and *OO* are in groups A, B, and O, respectively. The heterozygote *AB* is in group AB. Since the *A* and *B* alleles express themselves completely in heterozygotes, they are referred to as *codominant* alleles, and such sets of more than two allelic possibilities are referred to as *multiple alleles*.

The list of possible genotypes in the normal diploid individual, along with their corresponding phenotypes, is given in Table 8-4.

From this listing we can immediately see what progeny might result from the mating of any two parents. If they are both group O (*OO*), for example, all of their children must also be O, since there is no possibility of their acquiring any other allele. On the other hand, if one parent is A (of the heterozygous genotype (*AO*) and the other similarly B (*BO*), then all four blood groups are possible among the progeny: *AB, AO, BO, OO*. A complete listing is found in Table 8-5.

Table 8-5 Progeny expected from matings of each of the four groups in all combinations. For clarity the possible genotypes are included.

Father \ Mother		O — OO	A — AA, AO	B — BB, BO	AB — AB	Mother not specified
O	OO	O	O, A	O, B	A, B	O, A, B
A	AA AO	O, A	O, A	A, B, AB, O	A, B, AB	O, A, B, AB
B	BB BO	O, B	O, A, B, AB	O, B	A, B, AB	O, A, B, AB
AB	AB	A, B	A, B, AB	A, B, AB	A, B, AB	A, B, AB
Father not specified		O, A, B	O, A, B, AB	O, A, B, AB	A, B, AB	

MEDICOLEGAL APPLICATIONS. The discovery of these genetic factors occurring with high frequency in the population made it possible to apply genetic principles to legal problems. In the most common application, blood grouping may determine the question of accused paternity. If an accused male has a genetic consitution inconsistent with those of the child and the mother, then he is excluded from possible paternity. Thus, in one case, a woman of group O accused a man of fathering her child. The child proved to be of group A and the man of group B. Clearly he should be exonerated in this instance, since an A child of an O mother must receive the *A* allele from the father, who must then be either A or AB. On the other hand, if the man had actually been A or AB, this evidence would hardly serve to convict him since a sizable fraction of the population of males are either A or AB. The legal use of blood-group evidence varies from one state to another in the United States, and from one country to another, but, generally speaking, evidence excluding the male from culpability is considered compelling, whereas evidence consistent with the possibility of guilt is not.

In other applications, blood grouping has been used to determine maternity, when an infant is wrongfully claimed by a woman to be her own, and to untangle cases of mixups of infants in maternity hospitals. Children who were thought to be ordinary twins have, in several cases, been shown to have different fathers. Identification by blood grouping has helped to identify imposters in inheritance cases. Since the properties of the antigen do not disappear after drying, grouping can be of service in criminal cases where the perpetrator has left a few drops of blood behind. Furthermore, dried blood has been used by anthropologists interested in the consitution of ancient personages: thus more than 300 Egyptian mummies have been classified according to their ABO groups.

The genetic identification of individuals is not limited to the ABO blood groups. There are more than a dozen groups now known, some with unusual allelic variability. These will be taken up in Chapter 20. Furthermore, the recent discussion of distinctly different staining types of chromosomes offers still another means of identification.

The Rh Factors

Prior to 1940, there were several mysterious diseases of newborns, apparently unrelated, with severe symptoms, sometimes ending in death. In some cases the child, or more often the stillborn infant or aborted fetus, was characterized by an extreme excess of fluid that made it appear swollen. In others, the skin turned yellowish (technically, *jaundiced*) shortly after birth. Another congenital disorder was called *erythroblastosis fetalis*, because of the large number of erythroblasts (immature red blood cells) in the fetus's blood. Such children had an excess of blood-forming tissue which spread over the entire body, causing swelling of the liver, spleen, and other organs. We now know that most of these cases have a similar cause and accordingly are grouped under the common name *hemolytic disease of the newborn* (HDN).

A number of medical workers in the 1930's made notable advances in attributing the causes of these maladies to an antigen-antibody reaction, but the solution to the problem came from two directions. Landsteiner and Wiener demonstrated in 1940 that when the blood of a *Rhesus* monkey was injected into rabbits the rabbit blood developed antibodies that could agglutinate not only *Rhesus* blood but most human blood as well. From a sampling of New York City residents, Landsteiner and Wiener found that 85 percent had an antigen on their erythrocytes that would react to this anti-*Rhesus* antibody. The antigen, found in both humans and monkeys, was referred to as the Rhesus antigen.

A year earlier, Levine and a coworker had pinpointed a specific antigen ("X") as the cause of HDN in the successive fetuses of a patient, and as the cause of a severe agglutination reaction in that patient after a blood transfusion from her husband. The Rhesus antigen was subsequently shown to be identical to X. This blood group is now referred to as the Rh group; individuals carrying the antigen are called Rh-positive.

THE PATTERN OF INHERITANCE.   Extensive research by immunologists, geneticists, and medical practitioners has shown that the genetic basis for Rh inheritance is a very complex allelic system—one of the two most complex known in humans. These will be discussed in Chapter 20. However, it is possible to characterize the system simply, by assuming the existence of two alleles, *R* and *r*. A person who is *Rh-negative* (Rh⁻), i.e., who lacks the Rh antigen, is homozygous for the recessive allele *r*. *Rh-positive* (Rh⁺) persons are either homozygous *RR* or heterozygous *Rr*. The inheritance then follows the simple rules for all simple recessives.

THE CAUSE OF HDN.   Unlike the ABO groups, in which each person will carry the antibodies for those A or B antigens that are absent, Rh antibodies are not found in humans unless they are specifically induced. However, Rh⁻ people who lack the Rh antigen may be good producers of anti-Rh antibodies if Rh⁺ cells are introduced (one drop of blood is sufficient!) into their bloodstreams. In fact, about 70 percent of all Rh⁻ persons injected with Rh⁺ red blood cells will develop a good level (or *titer*) of anti-Rh antibodies.

If a pregnant Rh⁻ woman has anti-Rh antibodies in her bloodstream, they may pass through the placenta into the circulatory system of the fetus. If the fetus is Rh⁺ (having inherited an *R* allele from the father), the antibodies going from the mother into the child's circulatory system can destroy the fetal erythrocytes, causing extensive damage. Figure 8-5 illustrates the course of events leading to HDN caused by the Rh factors. Of all Rh⁺ children born to Rh⁻, but sensitized, mothers, about one third will die if not treated.

The mother may have developed the anti-Rh antibodies in response to an earlier transfusion. The ABO compatibility would have been checked, as well as a crossmatch made to test for the likelihood of an immediate antigen-antibody reaction, but a crossmatch is not sufficient to eliminate the possibility that the donor is Rh⁺ and the recipient Rh⁻ but lacking antibodies at that time. The mother may also have developed anti-Rh antibodies from an earlier pregnancy, if some Rh⁺ cells from that previous child had entered her bloodstream during the birth trauma and had immunized her against Rh⁺ blood. The first child, responsible for the immunization, would not be affected because the mother would not yet have formed Rh antibodies.

DETECTION AND TREATMENT OF HDN.   The realization that HDN resulted from an antigen-antibody reaction made it possible to take measures to circumvent or at least decrease its ill effects. First of all, the pregnant woman is tested to determine whether she is Rh⁻ or Rh⁺. If Rh⁺, there is no problem. If she proves

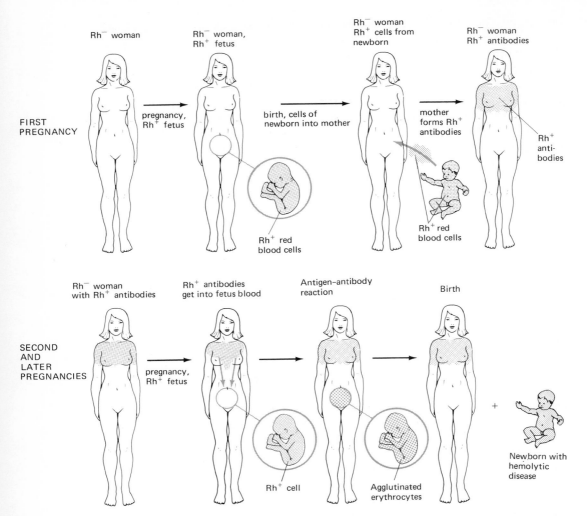

Figure 8-5 The series of events leading to HDN. An Rh⁻ woman is exposed to the Rh⁺ antigen. If she develops Rh⁺ antibodies and produces an Rh⁺ fetus, the antibodies may pass from her circulation into the fetus, destroying erythrocytes in the fetal circulation and causing the premature loss of the fetus or the birth of a child with HDN.

to be Rh⁻, then her husband should be tested. If he is also Rh⁻, again there should be no problem since two Rh⁻ parents cannot produce an Rh⁺ child. The woman's serum may be tested periodically during pregnancy to note any rise in the antibody titer that would cause a miscarriage or a stillbirth, with the possibility of the induction of a premature birth in order to save the child. Nevertheless, in the 1950's about 20 percent of all erythroblastic babies died before birth and another 10 percent shortly afterward. *Amniocentesis*, the process of puncturing one of the membranes surrounding the fetus (the *amnion*) for the removal of fluid from the womb, was later developed to diagnose the status of a potentially erythroblastic child by measuring the amount of the toxic pigment *bilirubin* being released by the destruction of the fetal red blood cells. In severe cases, premature delivery might be the best hope for saving the child.

In a still more heroic technique, the blood of the fetus may be replaced in utero with Rh⁻ blood which will not react to the mother's antibodies. The obvious hazards of such a transfusion may be minimized by an operation involving an abdominal incision through which a limb (say, a foot) of the fetus is drawn and exposed for the transfusion.

Immediately after birth, when the oxygen demands of the infant suddenly increase and any deficiency of red blood cells becomes more acute, a whole-body transfusion (*exchange transfusion*) may be made, in which all of the normal positive blood of the newborn is replaced with nonreacting Rh⁻ blood, so that the antibody is gradually removed as the child slowly manufactures more Rh⁺ erythrocytes. It has been estimated that in one Boston hospital exchange transfusion has reduced the death rate of erythroblastotic children from brain damage from 25 percent to 5 percent. Exposure of the newborn jaundiced infant to bright fluorescent light ("phototherapy") destroys the toxic yellow pigment bilirubin and can be used in addition to exchange transfusion. In one controlled study, 45 infants suffering from HDN required, among them, 69 exchange transfusions when treated with phototherapy for 4 days, whereas another 78 infants not given phototherapy required 224 exchange transfusions.

PREVENTION OF Rh ANTIBODY FORMATION.   An effective method of circumventing the Rh problem was developed both in the United States and in England in the early 1960's. Two teams, working for the most part independently but with some intercommunication, devised a successful procedure for reducing the Rh hazard based on a simple injection that would inhibit the formation of Rh antibodies in the mother in the first place. Because the immunization of the Rh⁻ mother ordinarily occurs during the birth of an Rh⁺ child, when some Rh⁺ cells get into her bloodstream, it is possible to prevent the mother from manufacturing antibodies at that time. That fraction of the blood that carries antibodies, the *gamma globulin*, is extracted from a person who is producing Rh⁺ antibodies, concentrated, and injected into Rh⁻ mothers of Rh⁺ children within 72 hr after birth. Antibody formation is suppressed because the injected antibodies coat the Rh⁺ fetal cells in the mother's circulation and prevent them from inducing more antibodies. Only one half of 1 percent of such treated mothers are immunized (Table 8-6), as opposed to 10 percent of those Rh⁻ mothers of Rh⁺ children who are left untreated.

The time limit of 72 hr for the treatment is based on the period of time arrived at during initial tests of the effectiveness of the procedure. However, the 72 hr period was chosen for a curious nonmedical reason. The original clinical tests of antibody suppression in the United States were made on male inmates of Sing Sing prison in New York, and it was decided by the investigators, who wished to inject the external antibody some period of time after the injection of erythrocytes (to parallel the procedure that might follow birth), that two visits to the prison separated by 72 hr would minimize the possibility that the investigators' regular schedule could be incorporated by inmates into an escape plan, or used as an excuse to riot!

DIFFICULTIES IN THE ERADICATION OF HDN.   This treatment is based on the prevention of antibody formation in the mother in the first place. If an Rh⁻

Table 8-6 Results of clinical tests to prevent Rh immunization by treatment of mothers with Rh+ gamma globulin. (J. C. Woodrow, Some Aspects of Immunogenetics, in *Selected Topics in Medical Genetics*, ed. C. A. Clarke, Oxford University Press, London, 1969.)

Place	Controls		Treated	
	Total	*Number immunized*	*Total*	*Number immunized*
U.S.A. and Canada	814	73	984	1
West German group	756	29	487	2
Liverpool group	320	35	315	1
Edinburgh	101	9	87	0
Sweden	45	3	43	0

mother has been immunized either by a previous birth or by an earlier transfusion of Rh+ blood, it will not be effective. It also has the disadvantage of requiring the injection of Rh+ gamma globulin after the birth of every Rh+ child. The decision to accept this injection is often at the discretion of the mother, who may not be willing or able to accept its cost (in the United States, about 30 or 40 dollars). This is particularly true for therapeutic abortions, in the course of which the Rh− mother may become immunized by an Rh+ fetus, but where neither the psychological nor medical conditions may be such as to arouse proper concern about the consequences of an accidental immunization. In any case, in the United States, less than half of the women who should be receiving this vaccine to prevent the birth of children with HDN are, in fact, receiving it, and, for these various reasons, some 6,000 deaths from this cause still occur yearly in spite of the fact that the disease is theoretically preventable.

OTHER BLOOD GROUPS. Two blood groups, ABO and Rh, have been discussed in detail in this chapter because they are the two most interesting and most important to humans. However, more than two dozen groups are well known and these will be considered in later chapters.

References

BROWNLIE, A. R. 1965. Blood and the blood groups. A developing field for expert evidence. *J. Forensic Sci. Soc.*, **5:**124–74.

CLARKE, C. A. 1968. The prevention of "rhesus" babies. *Sci. Am.*, **219:**46.

EDELMAN, G. M. 1970. The structure and function of antibodies. *Sci. Am.*, **223:**34–42.

HAUROWITZ, F. 1968. *Immunochemistry and the Biosynthesis of Antibodies.* New York: Wiley.

McCONNELL, R. B. 1969. Genetics and diseases of the gastro-intestinal tract. In C. A. Clarke, ed., *Selected Topics in Medical Genetics.* London: Oxford University Press.

QUEENAN, J. T. 1967. *Modern Management of the Rh Problem*. New York: Harper & Row.

RACE, R. R., and R. SANGER. 1968. *Blood Groups in Man*. Oxford: Blackwell.

SNELL, G. D. 1964. The terminology of tissue transplantation. *Transplantation*, **2**:655–57.

WIENER, A. S. 1943. *Blood Groups and Blood Transfusion*. Springfield Ill.: Thomas.

ZIMMERMAN, D. 1973. *Rh—The Intimate History of a Disease and Its Conquest*. New York: Macmillan.

Questions

Useful terms: antibody, antigen, erythrocyte, serum, codominance, multiple alleles, agglutination, erythroblastosis fetalis, HDN (hemolytic disease of the newborn), amniocentesis, amnion, gamma globulin.

1. Can you recall any instance in which your own immune system was clearly incited to action?
2. Do you know of any cases in which the immune system of a person has been responsible for an exaggerated reaction?
3. How would you demonstrate that there are four major blood groups in humans? When the blood of a person of one blood group is transfused into the body of a person belonging to a different group, there may or may not be a severe agglutination reaction. How does the antigen-antibody constitution of each of the two individuals determine the extent of the agglutination?
4. How do we know that the ABO blood groups are inherited?
5. Given all possible kinds of marriages that can occur between persons of the four blood groups, what kinds of progeny might they have?
6. Going back to the question in Chapter 6 about parthenogenesis, if a child were to be produced parthenogenetically, should the blood groups of the mother and child be similar?
7. Suppose that during routine testing of a family the two parents are both found to be AB, and among their children one is of group O. What is the simplest explanation for this?
8. Describe the course of events that leads to erythroblastosis fetalis. Why did the discovery of the Rh blood groups not lead to the elimination of HDN as a cause of infant mortality?
9. What common treatment is now being used as an effective means for decreasing the incidence of erythroblastosis? Under what circumstances is it not effective?
10. Discuss the relevance of the following conditions to the concepts discussed in this chapter: erythroblastosis fetalis, hemolytic disease of the newborn (HDN).

9

The Immune System: Transplantation and Related Problems

It has only been since the mid-1950's that transplantation of organs from one human to another has been shown to be a safe and practicable operation—sometimes. Since the unusual (for that time) success in Boston of transplanting kidneys from one identical twin to another, kidney transplants have exceeded the 10,000 mark. It is now accomplished daily with little fanfare in countries all over the world and is generally considered to be a successful operation. Several years ago, considerable attention centered on the very dramatic and well-publicized heart transplants, which unfortunately have not proved to be as successful. In this chapter we shall look into the problems surrounding these procedures and consider the biological basis for the reaction of one body against an organ from another—the phenomenon known as rejection. Some of the organs commonly involved in transplantation operations, as well as those concerned with the function of the immune system, are shown in Figure 9-1.

Transplantation

The widespread practice of transplantation has already given rise to a host of new problems. Usually the supply of organs of all types is limited and there are more persons near death awaiting transplants than there are transplantable organs available. In many cases some kind of decision must be made as to who will receive the transplant (and live) and who will not (and die). This decision is necessarily influenced by subjective judgments regarding the worth of the person

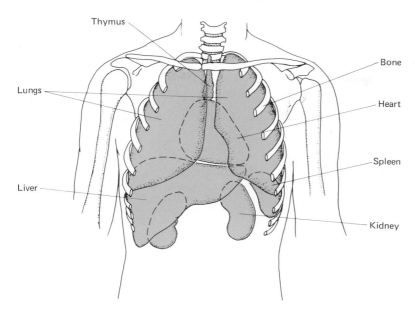

Figure 9-1 The location of a few of the transplantable organs and of those involved in the rejection reaction.

to the particular society in which he happens to be living, his potential contribution to society, the expectations of future behavior, and so on. Every day scores of such life-and-death decisions are being made by doctors and by committees who are only too conscious of the serious consequences of their judgments. In many cases it is not possible to select the "best" recipient but the decision as to which person will be the recipient upon the availability of an organ may depend on compatibility tests. In such cases those with the responsibility for choosing the recipient may feel relieved that other impersonal factors enter into the final decision.

HEART. The first human heart transplant was made in 1967; it was followed the next year by more than a hundred similar operations in the United States (54), Canada (14), France (10), South Africa (2), and 21 other countries (Table 9-1). The limited survival of the average heart recipient, with about half succumbing within a year and only about 10 percent living as long as 2 years after surgery, the average life-span being about 300 days, has led to a decrease in medical and public enthusiasm for this operation. Of course, for terminally ill patients whose survival time could be increased by such a transplant, the procedure may still be a very desirable one.

Since it is necessary to remove the donor's heart at a time when it is still capable of functioning, the donor must be declared dead for reasons other than heart impairment, such as permanent and irreversible brain damage in an accident. This has given rise to new legal and ethical problems with respect to the definition of death, since cessation of heartbeat has historically been used as the primary indicator of death.

Acute rejection of transplanted hearts usually occurs in the first several months after the operation, and is followed by a long-range chronic rejection that may not come until 2 or 3 years later. More of the recipients die of rejection

Table 9-1 The distribution of human heart transplants by year and country. The numbers in parentheses indicate numbers of patients alive on January 1, 1976. (By permission of J. Bergan, Director of the Organ Transplant Registry, and *Journal of the American Medical Association;* unpublished data.)

Year	World Totals	U.S.A.	Canada	France	South Africa	Other Countries*
1967	2	1	0	0	1	0
1968	101 (2)	54 (1)	14	10 (1)	2	21
1969	47 (2)	34 (1)	1	0	4 (1)	8
1970	17 (3)	16 (3)	1	0	0	0
1971	18 (3)	13 (2)	1	0	3 (1)	1
1972	18 (5)	15 (5)	0	0	2	1
1973	33 (6)	21 (5)	1	8 (1)	1	2
1974	29 (12)	17 (9)	0	9 (3)	1	2
1975	31 (19)	23 (16)	0	5 (2)	2 (1)	1
Total	296 (52)	194 (42)	18	32 (7)	16 (3)	36

*Argentina, Australia, Belgium, Brazil, Chile, Czechoslovakia, England, Germany, India, Israel, Japan, Peru, Poland, Spain, Switzerland, Turkey, U.S.S.R., Venezuela.

than from any other cause, with infection ranking second. The third most frequent cause of death is the immediate failure of the transplanted heart itself; 10 of 21 patients in this third category died within the first day and another seven on the second day.

It has been seriously questioned by some medical practitioners whether heart transplantation does, in fact, appreciably extend the life of the average individual with a serious heart problem. The medical teams skilled in such surgery will have at least several and possibly a dozen or more potential recipients from whom one must be selected for the operation. Clearly, those in poor health or of advanced age are poor risks in the medical sense and may be bypassed (consciously or not) in favor of patients with a better chance of surviving the surgical trauma. Since there is necessarily a waiting time between the selection of a suitable recipient and the availability of a heart to transplant, the poor risks may die a natural death first. For these and similar reasons, a patient with better than average survival possibilities may be chosen. Since no controls can be run in this sort of unhappy "experiment," it is not possible to say how long such a person may have survived had he or she never been given the transplant but instead had received the same intensive medical care that a transplant recipient receives. There can be no doubt that in some cases the patient would have lived longer without the the transplant.

LIVER, LUNG, AND BONE MARROW TRANSPLANTS. The first human liver transplant was carried out in Denver in 1963, and by September of 1975 (Table 9-2)

Table 9-2 Totals of five different organ transplants, throughout the world, up to January 1, 1976. (By permission of J. Bergan, Director of the Organ Transplant Registry, and *Journal of the American Medical Association*; unpublished data.)

Year	Heart	Liver	Lung	Pancreas	Kidney
1953–1961					123
1962					67
1963		6	2		157
1964		4	0		359
1965		7	3		453
1966		3	1	2	561
1967	2	8	6	1	832
1968	101	39	6	6	1,245
1969	47	46	7	7	1,538
1970	17	31	2	9	1,990
1971	18	15	4	1	2,904
1972	18	23	3	5	3,486
1973	33	24	2	5	3,828
1974	29	30	0	7	3,620
1975	31	18	1	4	2,756
Total	296	254	37	47	23,919

244 such operations had been performed throughout the world. In one type of operation the patient's liver is removed and replaced, and in another type a second liver is transplanted as an auxiliary to the first. Of 182 transplantations, 145 were of the first type and 37 were of the second. One patient has survived more than 4 years, four patients between 3 and 4 years, two between 2 and 3 years, and 14 between 1 and 2 years after surgery. Of 16 patients at the University of Colorado in Denver, six were still surviving after a period of from 1 to 2½ years after transplantation. This organ continues to be one of the most difficult to transplant successfully for a long-range survival.

Up to 1975, some 37 transplantations of lungs had been performed, but of these only three patients survived as long as a month with the grafted lung still functioning.

Bone marrow is made up of the primitive cells that give rise to blood cells, along with the mature differentiated cells, found in the interior of the bones of the body. When these cells are defective, as when the patient has a disease of the immune system, antibodies may be produced in insufficient quantities, or not at all. In such cases, bone marrow cells of normal persons may be transplanted, usually after an attempt is made to destroy the original defective cells.

It is usually difficult to determine whether the donor cells become established in the recipient, unless the donor and recipient cells differ in their chromosome constitution, antigenic properties, or some other identifiable cell characteristic. In a group of 50 patients in whom the graft took, 55 percent survived longer than 90 days, whereas in a parallel group of 65 in whom it did not (or at least in whom it could not be shown to have taken), only 20 percent survived that long.

The totals of different organ transplants, along with information on survival times, are given in Table 9-3.

KIDNEY.   Of all the organ transplants, those that achieve best success involve the kidney. There is precise information about the details of the operation and its outcome in more than 20,000 cases. Of the recipients, about half are still alive with the graft functioning. Another sixth are still alive with a nonfunctioning graft, and the remaining third are dead. The average survival time of those patients still alive with a functioning graft was 900 days, but this figure includes many cases in which survival will be indefinite, so as time goes on average survival may be expected to increase. Almost two thirds of all grafts are made using cadaver donors; this figure, however, varies from one country to the next. Slightly more than half of the grafts in the United States are from this source, whereas in Australia almost all are from cadavers. Figure 9-2 shows the survival of the graft when the organ comes from each of these sources.

It should be kept in mind that the failure of a kidney graft does not necessarily mean that the patient is doomed, as is usually the case for heart transplant patients. In many cases, the recipient may have one kidney that is still adequately functioning, or may be kept alive by a kidney dialysis machine which removes the blood and purifies it as kidneys do, once every few days. In fact, patients undergoing dialysis at home have survival rates equal to those of the recipients of transplants from living related donors, and have better survival rates than recipients of kidneys from cadavers.

Table 9-3   Total cases reported to the American College of Surgeons/National Institutes of Health (to January 1, 1976), showing the survival of recipients of grafted organs. (By permission of J. Bergan, Director of the Organ Transplant Registry, and *Journal of the American Medical Association;* unpublished data.)

Update on Organ Transplantation in the World	*Heart*	*Liver*	*Lung*	*Pancreas*	*Kidney*
Recipients of transplants	288	242	37	46	21,437
Alive with functioning grafts	52	28	0	1	approx. 10,850
Longest survival with functioning graft	7.1 yr	6 yr	10 mo	3.5 yr	19 yr*

*Identical twin.

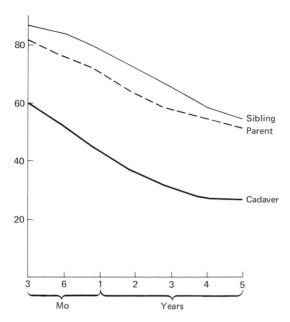

Figure 9-2 The duration of kidney transplant function depending on the source of the kidney. The failure of transplant function does not, of course, doom the patient who has other means—such as dialysis machines—available. (By permission of J. Bergen, Director of the Organ Transplant Registry, and *Journal of American Medical Association*, Second Scientific Report ACS/NIH Organ Transplant Registry, 1972.)

One of the reasons that kidney transplants are more successful than other organ grafts is that, since each individual has two kidneys, it is very often possible to obtain a more compatible one for transplantation from a close relative, who may not be reluctant to part with one of two kidneys. Another important reason is that this particular organ does not elicit the same massive rejection reaction that other organs, such as liver or lung, do. On the other hand, one of the problems in kidney transplantation is that the original disease, which the transplant was intended to alleviate, sometimes appears in the grafted kidney at an average of 2 years after surgery and as late as 6 years. (Similarly, in heart transplants, the primary disease requiring the transplant can affect the grafted heart, although this may not always be obvious, since the transplanted heart has had nerves severed. This makes it impossible for the patient to feel the evidences of heart lesions, although they can be detected by electrocardiograms and other coronary observations.)

OUTCOME OF TRANSPLANTATION OPERATIONS. In any case, it is clear from these statistics that, except for kidney transplants, the outlook for transplantation recipients is not very good in terms of longevity (Table 9-3). It is certainly not as good as one might conclude from the glowing accounts reported in the daily press. Such accounts very often describe the initial case and the operation at length on the front page, but subsequently comment on the fate of the recipient in an inconspicuous item inside, if at all.

The Cellular Basis of the Immune Reaction

Surgical techniques have advanced beyond the point where the success of a transplant would depend primarily on the skill of the surgeon. Instead, the fundamental problem is that of rejection, and, in order to understand it, we must consider in some detail the cellular basis of antibody formation.

ANTIBODY PRODUCTION. As we have discussed earlier in connection with the blood groups, when very large molecules, generally proteins and complex sugars, enter the bloodstream, certain white blood cells are induced to react specifically against these foreign chemicals by forming antibodies; the molecules which incite antibody production are called antigens. Antigenic molecules are found in every living cell from bacteria to human tissue, and there are a large number that are identical in all living cells; against these there will be no antibody production. On the other hand, from one species to the next, there are chemical differences which can be distinguished by the immune system, a good example being the cell surface antigenic differences of the blood groups. It is this system for recognizing foreign substances that lies at the basis of the antigen-antibody reaction in its many different manifestations, such as rejection and allergy.

The way in which a white blood cell manufactures an antibody specific against a foreign antigen is only partly understood. There are two basic problems: (1) how can a cell which has never contacted a particular antigen be induced to manufacture antibodies specific to that antigen, in great quantities? and (2) how is it possible for the immune system to carry the potential for producing many hundreds, perhaps thousands, of different antibodies?

The earliest theory was the "lock and key" hypothesis, according to which the antigen came to rest on the surface of a white cell, which somehow sensed its molecular configuration and then produced many molecules with a complementary configuration. These, then, could react specifically with that antigen because of the complementarity of the surfaces (Figure 9-2). Another idea was that within each individual there exist a very large number of cells, each with a predetermined specificity, covering the range of just about all of the important foreign antigens that might be introduced, so that when a specific antigen enters the bloodstream the corresponding type of white cell is "awakened" and stimulated to undergo rapid mitosis to form large numbers of cells and large quantities of antibody specific against it. A more recent theory suggests that the loci responsible for the production of gamma globulins have the capacity to produce large numbers of variant forms, which they do constantly. When a foreign antigen reacts with such a test antibody, a change occurs on the surface of the cell and causes that cell to undergo rapid proliferation on the one hand and produce much more of that particular antibody on the other. In any case, whatever the precise mechanism is, it is characterized by two outstanding features: the ability to react to an antigen by the production of specifically reactive antibodies, and a sensitization of the immune system to any subsequent exposure to the same antigen.

CELL RESPONSE. The cells which respond to antigenic stimulus are of two types, both derived from a common ancestor. One type of cell is found in the lymphoid tissue, in the thymus gland, the lymph nodes, the tonsils, and spleen and is referred to as the *T-cell* (Figure 9-3); the other type is found primarily in the bone marrow and is referred to as the *B-cell*. When a foreign antigen enters the body, the molecules adhere to the surface of a large white cell, the *macrophage* (Figure 9-3). Both the B- and the T-cells come into contact on the macrophage.

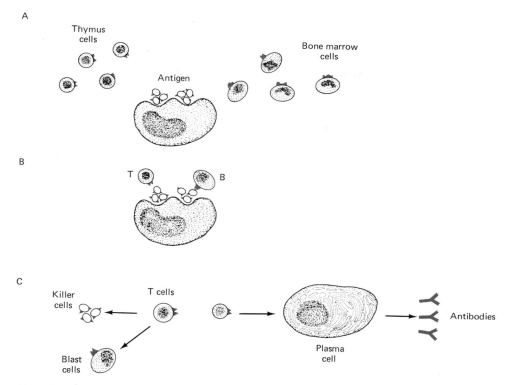

Figure 9-3 The interaction of antigens and the lymphocyte system in producing humoral and cell-mediated immunity. (*A*) The antigen molecules come to rest on the surface of the macrophage. (*B*) The T- and B-cells interact with the antigens (and with each other), resulting (*C*) in the stimulation of the B-cells to differentiate into plasma cells that manufacture antibodies, while the T-cells form killer cells that attack the foreign antigen and blast cells that remain in a primitive state and form the long-term memory of the immune system. (Adapted from Immunological Unresponsiveness, by W. O. Weigle, *Hospital Practice,* **6:**121, New York, 1971, and *Immunobiology,* R. A. Good and D. W. Fisher, eds., pp. 123–34, Sinauer Associates, Inc., Stamford, Conn., 1971.)

The T-cells produce two cell types: killer cells which migrate to the source of the antigenic stimulus, giving the white discoloration characteristic of infections, and an additional type, primitive cells which "remember" the specific antigenic stimulus and are primed to react quickly in the event of a second stimulus much later. It is this latter class of "remembering" cells that goes into action when we are given a booster shot. They can be quiescent for a long period of time, as is seen when we become immune for life after having had a childhood disease. The B-cells differentiate to produce a large white blood cell called the *plasma cell,* and these plasma cells produce the specific antibody against the antigen (Figure 9-3). The antibody is released into the circulatory system and for this reason is referred to as the *humoral* (or fluid) response, as opposed to the whole cell or *cell-mediated* response of the T-cells.

SURVIVAL VALUE OF IMMUNE RESPONSE.   It is obvious that the immune system has a very high survival value; that is, any change in our genetic constitution that makes us better able to ward off a disease would be found with a high frequency in the survivors of an epidemic of that disease, and would eventually become established as a permanent feature of our biological make-up. On the other hand, an infectious organism can "mutate" to produce essentially brand new antigenic types every few years, creating a new epidemic each time, as the influenza virus does. Those new types must still be capable of evoking an immune response, or the human race would have been in danger of extinction throughout its history. A highly virulent strain of a common infectious organism would suffer the disadvantage of jeopardizing its own existence as it kills off its hosts.

These considerations might not apply, however, to organisms to which man has never been exposed previously, and against which he may not have any immune protections. When the astronauts returned from the first visit to the moon, they were kept in isolation for a long period of time in order to eliminate the possibility that they had, by chance, become infected with some unknown organism to which the human race has no immune defense and which could therefore cause an enormous loss of life. When H. G. Wells wrote *War of the Worlds*, he ended his story with the extinction of the alien invaders by Earth-born microorganisms against which they had no resistance. He could equally well, however, have ended it in the reverse fashion with the human race made extinct by microorganisms brought here by the Martians. In any case, if one is ever approached by a creature from outer space it would be well to keep one's distance, if for no other reason than to minimize the possibility of an uncontrolled infection. Furthermore, there may be a strong argument for holding in isolation for a month or two those people who claim to have had any physical contact with such beings. . . .

Overcoming the Immune Reaction

Except for cases of identical twins, every individual is almost certain to have some antigenic differences from any other individual. Not all of these are important in transplantation rejection, as we shall see later, but, briefly, when a transplant is made from one person to another, some of the antigens of the transplanted tissue will evoke an antibody response, with the production of specific killer lymphocytes against that foreign tissue and the gradual destruction of the transplanted tissue. The solution to this problem is to avoid or suppress the immune reaction. This can be approached in a number of ways.

Certain kinds of tissue are relatively immune to rejection. Corneal transplants, for instance, are quite often highly successful operations. This may be because the cornea does not release antigens to the recipient's system, or because the absence of blood vessels in the cornea makes it impossible for those lymphocytes primarily responsible for tissue rejection to reach the cornea. The testes are also immune, protected apparently by the associated circulation which is isolated from the main blood supply. Two other kinds of tissue that are relatively inert are bone (from which all live cells are removed prior to transplant) and blood vessels. In both these cases, it is the ability of the grafts to function without the presence of live cells that makes it acceptable to the recipient.

RADIATION.   One of the first techniques developed for suppressing the immune reaction was to apply a dose of radiation of a fairly high intensity, sufficient to kill or to suppress the mitotic activity of the primitive stem cells which give rise to lymphocytes. Although this is effective up to a point, it can also be very dangerous. In fact, this is a common cause of death in persons accidentally exposed to high doses of radiation, because the immobilization of the immune system leaves the patient vulnerable to simple infectious disease. Of course, in the case of bone marrow transplants, where an attempt is made to kill off the patient's normal lymph stem cells and replace them with those from another person, this may be an effective procedure.

DRUGS.   There are a number of drugs available which suppress the immune reaction: actinomycin, azathioprine, methotrexate, and others. A balance must be struck in the application of these drugs, so that a sufficient concentration is given to suppress the immune reaction, but not enough to completely suppress it and allow the patient to be defenseless against disease. As an overall total, five sixths of all patients who die a short time after a transplant operation die of a subsequent infection, particularly of the lungs. Not only is there danger from infection when the immune system is suppressed, but also the normal control of malignant processes is reduced, so that a transplant recipient is more susceptible to cancers and other malignant growths.

A recent advance in the suppression of the immune system for transplantation involves the use of *antilymphocyte serum* (ALS). This serum carries antibodies (antilymphocyte globulin, or ALG) which react against lymphocytes and suppress the proliferation of those that would attack the transplanted organ.

OTHER TECHNIQUES.   Because of the importance of the problem of the immune reaction to transplantation, a large amount of research effort is going into its solution, and a number of new approaches are currently being tested.

First, let us consider one obvious approach that, unfortunately, will not work. A preliminary test might be made, whereby the recipient of an organ would have a small patch of unessential skin transplanted from the potential donor to his body. Whether that skin transplant is rejected or takes, and, if rejected, how quickly, might then serve as an indication of the probable fate of the organ transplant.

This kind of test, however, would itself sensitize the recipient, almost certainly assuring the subsequent failure of the transplant, unless by some remote chance the two individuals were antigenically identical.

THE MIXED LYMPHOCYTE CULTURE TEST.   Although it is obviously not possible to test for incompatibility in the actual recipient, tests can be made in culture to determine the similarity of the loci of the two individuals. A basic feature of the immune reaction involves the proliferation of lymphocytes in the presence of foreign antigens. In the *mixed lymphocyte culture test*, lymphocytes of the two persons involved are mixed together and cultured. Evidence of growth can be taken as a clear indication that there are dissimilar antigens differentiating the two lymphocyte samples and that a graft between the two individuals would be in jeopardy. An absence of mitosis, on the other hand, suggests that the two sets of lymphocytes do not recognize any foreign antigens from each other, and predicts a more viable transplant. Furthermore, the degree of mitotic activity may serve as an indication of the number of antigenic differences.

LYMPHOCYTOTOXICITY TEST. Another test for compatibility between donor and recipient is the *lymphocytotoxicity test*. If antibodies to a specific antigen are added to a culture of lymphocytes carrying that antigen, the antibody may succeed in killing the lymphocytes. If a large number of sera carrying different antibodies are added to both donor and recipient lymphocyte cultures, the cultures will respond in similar ways to all of the different antibody-containing sera if both sets of cells are antigenically similar. On the other hand, if the two lymphocyte samples are quite different, they will react differently to the various antibody preparations. These reactions are more easily observed if special staining techniques that differentiate between living and dead cells are used. Because this test, unlike the mixed lymphocyte culture test, depends on the specific antigens carried on the lymphocytes reacting with the corresponding antibodies in the sera, it can be used for tissue typing and for determining the specific antigens carried by different individuals.

The Genetics of Compatibility Systems

Compatibility of two sets of tissue is referred to as *histocompatibility* and incompatibility as *histoincompatibility*. That there is a genetic basis for these reactions has already been implied in the actual figures for rejection, which show that success depends on the closeness of relationship between the donor and recipient, being highest between identical twins.

To understand its genetic basis, we may take advantage of the mouse, which has a system not too different from that of humans and with which we can make transplantations at ease, without the same degree of concern for the consequences as we would have for humans. This problem is best approached by using mouse strains that have been highly inbred for many generations, usually by brother-sister matings, until they are homozygous at virtually all loci. Inbred strains of different origins will be homozygous for different alleles; thus, one inbred strain might be of composition $A_1 A_1 B_3 B_3 C_8 C_8 \ldots X_2 X_2$ and a second be of composition $A_2 A_2 B_4 B_4 C_1 C_1 \ldots X_8 X_8$, etc. Then, if we consider two such inbred strains, we first notice that any transplant made within a strain (an *isograft*; see Table 9-4 for terminology) takes readily and permanently without any difficulty but that a transplant between two individuals of different strains (an *allograft*) does not take. We can now breed an F_1 hybrid and make grafts in all possible combinations of parents and hybrid progeny to complete this simple analysis. The results are shown in Table 9-5. From this table we can see that a graft from either parent or from the hybrid will take when the recipient is a hybrid but that a graft from the F_1 hybrid will not take on either of the parents, although it will on sibs.

This experiment tells us that rejection has a genetic basis; furthermore, we can surmise that there are codominant alleles involved because the progeny exhibit antigenic properties of both parents. These interactions are easily understood if we simply imagine that one strain is of composition $A_1 A_1$, the other

Table 9-4 Terminology of graft-host relationship, showing the probability of success depending on the closeness of the relationship. The adjective corresponding to each name is given underneath in parentheses.

Relationship Between Donor and Recipient	Probability of Success	Term	Synonym
Same individual	$+++$	Autograft (Autogenic)	
Genetically similar	$++$	Isograft (Isogenic)	
Unrelated but of same species	$+, -$	Allograft (Allogenic)	Homograft
Different species	$----$	Xenograft (Xenogenic)	Heterograft

$A_2 A_2$, and the hybrid $A_1 A_2$, and that a graft will take only if the recipient has the antigens present in the donor tissue.

How many loci are involved? We can make an estimate of this number by crossing two F_1 individuals and obtaining a large number of F_2. What proportion of the F_2 segregants have tissue that will take when grafted into the original parental strains? If there were only one locus involved, with each inbred strain having its own specific alleles, then in the F_2 we should have one fourth of the segregants like one of the two parents, and one fourth like the other; therefore, grafts from 25 percent of the F_2 onto each original parental strain should take. On the other hand, if there were a very large number of loci, different in the two strains, then the chance that we would get a genotype in the F_2 precisely like one of the parents would be negligibly small, and we would expect virtually zero success. Clearly, we can work out the precise expectations for various numbers of loci, based on an observed frequency of takes between the F_2 segregants and the parental stock. From such analyses involving many inbred strains of the mouse, it can be shown that there is one locus which is particularly potent and there are a dozen or more additional loci of lesser importance, each of these loci having a large number of alleles. There is some evidence in humans for an antigen

Table 9-5 The success of transplants between members of two highly inbred strains and their progeny. The grafts take readily within a given strain (1) but do not between two strains (2). The progeny will accept grafts either from their parents or from their sibs (3), but neither of the parental strains will accept grafts from the progeny (4).

Donor	Recipient		
	Male parent strain	Female parent strain	Progeny
Male parent strain	$+$ (1)	$-$ (2)	$+$ (3)
Female parent strain	$-$ (2)	$+$ (1)	$+$ (3)
Progeny	$-$ (4)	$-$ (4)	$+$ (3)

produced by a locus on the Y-chromosome, but this must be of minor significance in antibody production since male tissue can be grafted to female tissue as readily as female tissue can.

Similar tests obviously cannot be performed on humans, but extensive studies in families, combined with the two-cell culture tests described earlier, show that there are several loci important to transplantation. The first of these is the ABO locus: the potential donor should not carry antigens produced by this locus for which the recipient has the corresponding antibody. It is important to keep this in mind because it limits the availability of certain kinds of potential donors. However, ABO compatibility considerations are so taken for granted in transplantation studies that they are often not explicitly mentioned, but merely implied.

The major histocompatibility "locus" in humans is referred to as the *HL-A* (human leukocyte, A locus). This is not a simple locus but consists of at least four "subloci" located close to each other on the chromosome and generally transmitted together as a unit. Two of these are particularly important in histocompatibility reactions and show great allelic diversity; one of these is called HLA-A, the other HLA-B. The nomenclature of the specific antigenic properties at each of these subloci is sometimes cumbersome; we shall use only a few examples designated simply by numbers: A1, A2, A3, A9, A10, and A11 at the HLA-A locus, and B5, B7, B8, B12, and B13 at the HLA-B locus. Each chromosome will have one representative of each group, so that a specific chromosome might carry A1B5 or A10B13, or any of a very large number of possible combinations; the other homologue might carry A2B6 or A9B8. These designations of the sets of two specific alleles carried on a given chromosome are called *haplotypes*. Each person will have, of course, two haplotypes, and using a slash, the standard symbol to separate the genetic constitution of one chromosome from that of its homologue when several loci or subloci are under consideration, might be of constitution A1B5/A2B6, or A10B13/A9B8 (or any one of a very large number of other combinations). Figure 9-4 shows the relationships of two parents, each with two different haplotypes as they produce several progeny. Note that, since there can only be four different combinations, if there are as many as five children, at least two must be alike with respect to the HL-A locus. However, the probability of compatibility between sibs is actually less because of the ABO groups, as well as other less well-defined antigenic properties.

If we take the number of allelic differences at the first sublocus to be 10 and the number at the second to be 25 (realizing that these numbers are probably underestimates of the actual allelic diversity), there are theoretically 10×25 or 250 haplotypes possible on any one chromosome. In the diploid state, each of these might be found with any other to give 250×249 or 62,250 combinations, but since each combination will have a duplicate ($A_1 A_2 = A_2 A_1$ the number of different heterozygous combinations will be 62,250/2 or 31,125 combinations. Adding to this the 250 possible homozygotes gives us an overall total of 31,375 genotypes possible. A more complicated calculation will show that these genotypes correspond to 17,875 phenotypes. Of course, not all of these will occur with equal frequency, since the alleles found at each of the two subloci themselves do not occur with equal frequencies. Nevertheless, this gives an indication of the magnitude of the task of finding two compatible individuals taken at random from the population—as would be the case, for instance, between a

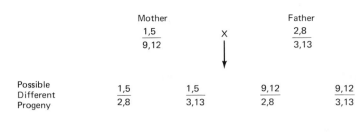

Figure 9-4 Hypothetical scheme showing possible constitutions of two parents, with the four different kinds of progeny. The first number of each pair represents a specific allele at the HLA-A sublocus and the second number, the allele at the HLA-B sublocus. One haplotype found on one chromosome is given above the line and the other haplotype, on the homologue, is placed below the line.

patient awaiting a heart transplant and a potential donor suffering irreversible brain damage in a automobile accident.

When the transplantation antigens were first being investigated, it was hoped that the identification of all of the antigens at this locus would make it possible to approach complete success in matching donors with recipients, just as identification of the erythrocyte cell surface antigens of donor and recipient makes it possible to insure success in virtually all blood transfusions. This has, unfortunately, turned out not to be the case. Although compatibility of the HL-A antigens, along with ABO compatibility, plays a predominant role in the success or failure of transplants, the different antigens appear to have different strengths for reasons that are not yet completely clear, and there appear to be other uncontrollable factors that modify the rejection reaction. In one study in France, for instance, 96 percent of all kidney transplants were successful when the HL-A antigens were identical in both donor and recipient. When they were partially identical only 71 percent were successful, and when they were altogether different less than 50 percent were successful.

In addition, there is evidence for other loci affecting compatibility, and the probability of a take cannot be predicted with certainty on the basis of the similarity of the ABO blood groups and the known HL-A alleles. As a matter of practical procedure it is necessary to make one of the direct tests in cultures such as the mixed lymphocyte test to assess the probability of a take.

Other Antigen-Antibody Reactions

THE GRAFT VERSUS HOST (GVH) REACTION. Ordinarily our concern is with the formation in the recipient of antibodies and killer lymphocytes specific against the antigens present in the grafted tissue, causing its rejection. However, in those cases where the transplant involves lymphocyte-forming tissue, the reverse can occur. If a transplant of such tissue is made after all of the recipient's own has been destroyed, or if the recipient was born lacking the capacity to form them, it would seem a simple matter to transplant comparable tissue into the defective individual in order to restore her or his immune capacity. Rejection should, in principle, not be a problem because the recipient lacks just that tissue responsible for rejection. However, what happens when this is done is that the transplanted tissue itself forms antibodies against the recipient's antigens, with the result that the graft attacks and destroys the host tissue. When a newborn mouse is injected with a sizable number of spleen cells from an incompatible animal, a poorly developed individual, a runt, is the result; it is said to have *graft versus host* or *GVH disease.*

ACQUIRED TOLERANCE.   Under ordinary circumstances, the body will not form antibodies against its own antigens. If a mouse is injected with cells of another strain very early in development, that mouse will not develop antibodies against the foreign antigens, hence the term *acquired tolerance.* This is the case provided that the injected cells are not *immune competent;* that is, provided that they are not capable of producing antibodies or killer lymphocytes themselves, in which case the recipient would develop GVH disease. One suggestion for the extension of transplantation compatibility is to inject newborn infants with incompatible tissues from several different origins, so that they will acquire a tolerance for the antigens on those tissues. In theory they would then be more likely to accept an otherwise incompatible graft later because of their acquired tolerance.

ANAPHYLAXIS AND ALLERGY.   One dramatic manifestation of the responsiveness of the immune system is found in the phenomenon of *anaphylaxis* and its related problem, *allergy.* Anaphylaxis is the sudden and violent, sometimes fatal, reaction of an animal to a second or subsequent exposure to an antigen after the initial sensitization.

A more moderate form of this severe reaction is found in allergy, in such common disorders as hay fever, asthma, and eczema, which can be responsible for fever, hives, rashes, headaches, cramps, and other symptoms. Different *allergens* may produce different effects. Ragweed pollen ordinarily affects the nose and eyes, whereas allergic reactions to foods are more commonly seen as hives or skin rashes.

Allergy may be treated successfully by first identifying the specific allergen by exposing the allergic person to a wide variety of potential offenders, and then by removing the offending factor (sometimes a household pet or an overstuffed article of furniture), or, when this is not possible, by desensitization by a series of injections of small doses of the specific allergen. In addition, since one of the characteristics of the allergic reaction is the liberation of *histamine,* a powerful substance that causes contraction of the smooth muscles, relaxation of the capillary walls, and a fall in blood pressure, *antihistamines* are very effective in relieving the symptoms of allergy.

AUTOIMMUNITY.   It is a truism that under ordinary circumstances a normal individual will not form antibodies against antigens naturally present in his own body. There are, however, some cases in which this rule breaks down; in these instances *autoimmunity* is said to exist. Some cells may be normally isolated from the rest of the body and if an accident, disease, or other misfortune should allow the antigens of such cells to be released and stimulate the immune system, then antibodies may be produced against those antigens (and those cells) leading to an autoimmune reaction.

It has been suggested, as another cause of autoimmunity, that some change may appear in those genes which produce specific antibodies, so that the antibodies are no longer directed against the original antigen but against some antigen found normally in the body. Still another possibility is that an antigen introduced from the outside may induce the production of an antibody which also reacts with antigens normally present in the body.

Some diseases that have been shown to originate in an autoimmune reaction include pernicious anemia, rheumatoid arthritis, disease of the thyroid, and, in some cases, multiple sclerosis, chronic gastritis, and infantile eczema. This list continues to grow as our knowledge of autoimmunity increases.

A striking example is found in injuries to the eye, where damage to one eye may cause the formation of antibodies specific to antigens ordinarily held captive within it, with the result that these antibodies sometime later may attack the other eye and damage it as well. It is a not uncommon medical observation that a person suffering serious injury in one eye may, months later, lose the sight of the second uninjured eye.

Males who have undergone sterilization by vasectomy (which prevents the release of mature sperm, so that they must instead be absorbed by the body) may develop antibodies to those sperm. These antibodies are of two types: those that agglutinate the sperm and those that immobilize them. Whereas only about 1 percent of all males have antibodies in their serum against sperm prior to vasectomization, more than half of all vasectomized males show sperm-agglutinating antibodies after 6 months and more than 61 percent show them after a year. Furthermore, roughly 40 percent of all vasectomized males have in their serum, 1 year after the operation, antibodies that immobilize sperm.

No doubt the most common form of autoimmunity is that which develops in response to malignancies. When malignant tissues produce new antigens, the antibodies and killer lymphocytes formed in response may be effective in destroying the growth. It is likely that every individual who has survived beyond middle age has successfully fought off a cancerous growth without being aware of its presence. Thus, in addition to combatting external infections, this internal struggle may be one of the most advantageous aspects of the immune response.

Immunological Deficiency Diseases

In 1952, an account appeared in a pediatric journal of a boy who was originally diagnosed as having acute rheumatic fever because of a painful left knee joint; after treatment with antibiotics he recovered completely but subsequently relapsed. Over the next 4 years he suffered from severe infections 19 times, each time being only temporarily cured with antibiotics. Because of his high rate of infection, it was thought that perhaps he might have produced an unusually high level of antibodies in his serum.

Antibodies may be detected and identified as proteins of the gamma globulin class by *electrophoresis*, a technique in which a mixture of heterogeneous molecules is placed in an electrical field and the molecules are allowed to migrate to the poles at their own characteristic speeds. (The name "gamma globulin" is derived from the fact that it is the third major group in such a separation, alpha and beta globulins being the first and second, with the word *globulin* referring simply to the large "globule" structure of the molecule.) Electrophoresis of the proteins in the boy's serum showed a blank spot at the level where the gamma globulin fraction was expected. In fact, at the first clinical test it was thought that the electrophoresis apparatus had broken down. When later tests confirmed the original results, this deficiency of gamma globulin was named *agammaglobulinemia.*

In the classic case of agammaglobulinemia, now known to be sex-linked, the lymphocytes are normal and cellular immunity is present, but free antibodies are absent, and, therefore, so is humoral immunity. Although individuals so afflicted are unusually susceptible to many diseases (diphtheria, typhoid), they are easily able to withstand some viral infections such as measles and chickenpox, suggesting that the mechanism of defense may be humoral in the former set and cell-mediated in the latter. Furthermore, it has been found that the gamma globulin fraction containing the antibodies may not be completely absent from the serum, as a crude preparation by electrophoresis would suggest, but may simply be greatly decreased in quantity.

If children with agammaglobulinemia are not diagnosed at an early age, before the time when the protection provided by maternal antibodies that had crossed her placenta into the fetus is disappearing and the child's own immune system should be taking over, the child will very likely die. Unusual susceptibility of a young child, particularly a male, to common infections should give rise to a question about the possibility of this defect being present.

These individuals may be protected by injection of gamma globulin, since such serum fractions obtained from normal people will contain protective antibodies against the common diseases. Although gamma globulin therapy may reduce the frequency of acute infections, it affects chronic infections and progressive pulmonary disease hardly at all. Therefore, it is necessary to continue regular treatment with antibiotics, and the most effective treatment appears to be a combination of antibiotics and gamma globulin. One unfortunate aspect of this treatment is that whatever immune defenses the patient has may produce antibodies against the gamma globulin being introduced, since gamma globulin, being protein, can also act as an antigen.

The patient may also be protected by the transplantation of bone marrow from another individual. When transplantation is attempted, however, it becomes obvious that the immune reaction is not completely absent, because transplantations induce a typical, although somewhat delayed, immune response. In some cases, bone marrow transplants have been successful, more often they have not. Not only is the delayed rejection a hazard, but also the transplanted lymphoid tissue may develop antibodies against the host (graft-versus-host or GVH disease). However, it is worth noting that among patients receiving bone marrow transplants, those with an immune deficiency disease experience a higher frequency of successes than those without such a disease.

Another approach involves identifying the potentially affected child prior to birth (usually after an older sib has proved to be affected). The delivery of the child is then made under strictly germfree conditions. If the child is kept germfree by the most stringent precautions and is isolated in a special chamber, bacteria and viruses, causes of the most common infections, will not reach the child. When the disease is of the type in which the immune system is slow to develop, the child may eventually be released into the outside world. If the disease is permanent, however, release from this germfree isolation may be postponed indefinitely until an effective cure is found.

References

BACH, F. H. 1969. Histocompatibility in man—genetic and practical considerations. *Prog. Med. Genet.*, **6:**201–40.

BACH, F. H. 1970. Transplantation: pairing of donor and recipient. *Science*, **168:**1170–79.

BURNET, M. 1953. *Natural History of Infectious Disease.* Cambridge: Cambridge University Press.

CASTELNUOVO-TEDESCO, P., ed. 1972. *Psychiatric Aspects of Organ Transplantation.* New York: Grune & Stratton.

Death after transplantation. An analysis of sixty cases. 1967. Editorial. *Am. J. Med.*, **42:**327–34.

FUDENBERG, H. H., et al. 1972. *Basic Immunogenetics.* London: Oxford University Press.

GAIL, M. H. 1972. Does cardaic transplantation prolong life? *Ann. Intern. Med.*, **76:**815–17.

MILLER, G. W. 1971. *Moral and Ethical Implications of Human Organ Transplants.* Springfield, Ill.: Thomas.

SIGERIST, H. E. 1943. *Civilization and Disease.* Ithaca, N.Y.: Cornell University Press.

The Third Scientific Report of the ACS/NIH Organ Transplant Registry. 1973. *JAMA*, **226:**1211–16.

The Twelfth Report of the Human Renal Transplant Registry. 1975. *JAMA*, **233:**787–96.

Questions

Useful terms: GVH (graft versus host), acquired tolerance, immune competent, histocompatibility, isograft, allograft, autograft, xenograft, homograft, heterograft, antigenic, haplotype, anaphylaxis, autoimmunity, electrophoresis.

1. In what ways has organ transplantation raised new ethical problems?
2. How has the legal definition of death become an issue since the advent of organ transplantation?
3. In what ways may the procedure of transplantation be said to be perfected and in what ways not?
4. What are the two primary mechanisms by which the immune reaction responds to the presence of a foreign antigen?
5. How may the immune response have any survival value to the human species?
6. Is there any way that the immune reaction may be overcome or bypassed? Describe two different tests that can be applied to increase the likelihood of a "take" of an organ transplant.
7. What is the graft versus host (GVH) reaction?
8. How can it be shown that the compatibility of a graft depends on the genetic relationship of the donor and recipient?

9. Describe what is known about the histocompatibility locus in humans. How much variability is there in the human population at this locus?
10. What is the difference between anaphylaxis and allergy?
11. Rheumatoid arthritis is one of a dozen or so diseases that are believed to be caused by autoimmunity. Explain what is meant by this.
12. If one reads in the newspaper of a child who is protected from the outside world because of an immune deficiency disease, that child is likely to be a male rather than a female. Can you suggest a reason for this?
13. A person with an immune deficiency will have a reduced capability to reject transplants. Why, then, is not a bone marrow transplant a simple solution to repair the innate defect in such a case?
14. Discuss the relevance of the following conditions to the concepts discussed in this chapter: graft versus host disease (GVH), allergy, agammaglobulinemia, autoimmune disease.

10

Mutation and Gene Expression

Mutational changes may be classified as either *somatic* (body) or *germinal* (germ line). Somatic mutations, affecting the body cells at some point after the cleavage divisions have begun, may be of great importance to the individual involved, but are of little importance to the species since only germinal mutations affecting the gametes can be transmitted to the next generation. It is this latter category that is invariably meant when the term "mutation" is used without further qualification.

Somatic Mutation

Cell division is a remarkably exact process, with mistakes such as nondisjunction occurring only rarely. The replication of the genetic material of a cell at interphase is even more precise: errors of replication that change the phenotype of the individual are almost never seen.

   This is not for lack of opportunity for such changes to occur or be observed. For example, consider albinism. More than one person in a hundred is heterozygous for albinism, of genotype *Aa*. During the hundreds of mitoses that occur from fertilization of the egg to the completion of the embryo, one might expect that in a few cases the normal *A* gene would fail to replicate itself properly and instead produce an inactive allele, similar to *a*, which would then give rise to a clone of cells effectively *aa*, or albinotic tissue. Such an event would be obvious because there are millions of cells exposed to the surface (in the skin, hair, and eyes) where such a clone would be easily noticed as a light patch. And this is not the only recessive gene found with a high frequency in heterozygotes in the human population. Nevertheless, such *mosaic spotting,* suggesting the occurrence of body or somatic mutations, is very rare.

Occasionally a person is found with eyes of two different colors, usually one brown and the other blue, green, or hazel. The frequency of such somatic mosaics is very low compared to the known frequency of heterozygotes *Bb* (brown), in which the cell or cells in the early embryo giving rise to one eye might have become genotypically *bb* (blue) by the mutation of the *B* allele to *b*. However, since there are other possible explanations for mosaic tissue in heterozygotes, such as the loss of the dominant allele in a clone of cells giving rise to one eye, thereby unmasking the recessive, the frequency of mutational events must be even lower than the low incidence of somatic mosaics would suggest.

Germinal Mutation

When Gregor Mendel got his garden pea strains from nurseries, he could not have understood why they differed from each other, although judging from his own words in the introduction to his classic paper he was clearly aware of the evolutionary importance of such characteristics.

Their significance was emphasized by Hugo de Vries, a professor of botany and curator of the botanical gardens at the University of Amsterdam, and one of several workers who "rediscovered" Mendel's paper in 1900. As de Vries bicycled from his home to the university, he observed that the evening primrose, *Oenothera lamarckiana,* was characterized by some unusual variations from one individual to the next. This plant was American in origin and had been brought to Europe for cultivation in gardens, but a number of plants escaped and rapidly proliferated in some areas. De Vries brought several of the plants into his garden to study and found that not only did these variant types breed true but also they often produced yet other types that were true-breeding generation after generation. De Vries referred to these changes as *mutations* and in 1901 published a book on mutation theory in which he postulated that these mutational changes were the ones necessary for evolution.

In any case, in its broadest sense we can consider mutation to be any sudden alteration in the genetic composition of a species. Included in this broad spectrum are changes in chromosome sets, or polyploidy, changes in chromosome number, or aneuploidy, and changes that affect the arrangement of chromosome parts either between or within chromosomes. The fourth category of mutation refers to changes in the gene itself, gene mutations. Generally the word "mutation" refers to the category of gene mutation. As the result of the elucidation of the structure of the DNA molecule in the early 1950's by Watson and Crick, and the genetic analysis of fast-growing haploid viruses and bacteria by sophisticated physical and chemical methods by a vast number of others, the chemical basis for gene mutation is now understood. These aspects will be taken up in a later chapter.

Problems in Identifying New Mutations

With experimental plants or animals it is not too difficult to detect new mutational changes as they occur. A laboratory line of animals that has produced no unusual offspring over a large number of generations may be assumed to have suffered a new mutation if a new transmissible type suddenly appears. Indeed, it is common practice to arrange experimental matings so that any new mutations that may occur will be obvious.

DOMINANT VERSUS RECESSIVE ALLELES FROM MUTATION.   When a new genetic type suddenly appears as a single affected individual from unaffected parents, the immediate question is whether the affected person is the result of a new dominant mutation, or of the fortuitous homozygosity for two similar recessive alleles for which the parents happened to be heterozygous. This question can be resolved if the new phenotype can be unambiguously identified as a previously known one with a specific type of inheritance or if sufficient pedigree information becomes available to discriminate among the alternatives. If the phenotype is caused by homozygosity for a recessive, then it is most likely that the two alleles were transmitted from both parent ancestors. It is possible, but less likely, that one of the two alleles is newly mutated and much less likely that both represent new mutational events.

DEVELOPMENTAL MIMICS.   In humans, however, the process of identification becomes much more difficult. If an individual appears with a new characteristic, the first question that arises is whether that change is, in fact, a genetic one or whether it is simply a developmental change, perhaps induced by disease, drugs, or some unknown condition in the mother during pregnancy. Phocomelia, for instance, the condition in which the arms or legs are so short that in extreme cases the hands or feet appear to be attached directly to the trunk, may be induced prenatally by thalidomide but is also well known as a genetic defect. Making a distinction between genetically and environmentally caused changes can be difficult, sometimes impossible. Such changes of a nongenetic nature that mimic known genetic types are known as *phenocopies* (Figure 10-1).

DEVELOPMENTAL VARIABILITY.   It is invariably true that a specific genetic condition may produce some variation in the characteristic phenotype as a result of developmental and genetic variables which must all collaborate in the production of the new individual. In the simple case of polydactyly, for instance, the affected individual has the dominant gene for an excess number of digits; however, the two hands may look quite different from each other. Sometimes one appears quite normal, whereas the other has additional digits. In these cases x-ray photography often reveals extra bones in the hand that appears normal. This type of variation is called *variable expressivity*. The variability may even extend to the extreme where the phenotype does not appear at all in some of the persons with the proper genetic make-up. When a genotype always expresses itself phenotypically, we say it is *completely penetrant*; when it does not, it is said to be *incompletely penetrant*.

Figure 10-1 A pheno-copy in an Ecuadorian Indian. The child, who at first sight appears to be an albino, is in fact the victim of kwashiorkor, a disease of protein deficiency. (Courtesy of C. J. Witkop; original photograph from the International Committee for Nutrition for National Development, National Institutes of Health, Bethesda, Maryland.)

Even such a simple characteristic as eye color may cause problems because of incomplete penetrance. In a small frequency of genotypically brown-eyed individuals, the brown pigment develops poorly or not at all, and the person appears phenotypically to be blue-eyed. However, with a blue-eyed spouse, such a person may produce one or more brown-eyed children. This could lead to some concern on the part of the parents if they have been told categorically that the gene for blue eyes was a simple recessive and that, therefore, two blue-eyed parents can never have brown-eyed progeny (unless, of course, a mutation from blue to brown occurs!).

DIFFERENT LOCI WITH SIMILAR PHENOTYPES.   Even in those cases in which a well-recognized phenotype is identified with a particular type of inheritance, there may be room for error. There may be some difficulty in identifying the phenotype in question; there are a large number of human anomalies that resemble each other quite closely. For instance, about half a dozen different defects can be misclassified as achondroplasia by the casual observer; most of them are caused by recessives for which both parents happen to be heterozygous. Under these circumstances, identification of a new dominant mutation becomes difficult to distinguish from chance homozygosity of a recessive.

Such an apparently straightforward case as albinism has its problems. It is usually considered to be the result of homozygosity for a recessive at a single locus but, in fact, the phenotype can be produced by changes in at least two different loci. We know this to be the case because occasionally two albinos marry and produce normally pigmented children, expected only if the cross is of the composition $aaBB \times AAbb \rightarrow AaBb$.

This should not be too surprising. An extensive series of biochemical steps, each promoted by its individual enzyme, is necessary for the conversion of the simple molecules to the more complex pigment melanin. Each of these steps is controlled by a different genetic locus, and mutations at each of these loci, then, may interfere with the production of the final product (Figure 10-2).

Although the end result is albinism in both cases, the two types can be differentiated biochemically by enzyme tests in the hair bulbs. In these cells, normal people have an enzyme, tyrosinase, which speeds the changes of tyrosine into the pigment precursor, DOPA. The hair bulbs of some albinos have this enzyme; those of others do not (Figure 10-3). This clearly indicates that albinism may result from the interruption of melanin formation at a minimum of two different stages, and this implies that at least two different loci are involved. Albinos may be of either of these two types, which occur with about the same frequency (Figure 10-4).

All of the problems listed above hamper the determination of rates at which new mutations appear in the human population. In addition, of course, there always exists the possibility of mistaken paternity: an affected child of normal parents apparently possesses a newly mutated dominant allele, but in actuality was fathered by a man who was himself affected. This possibility can often be ruled out on the basis of other similarities, especially the blood groups.

Determination of Mutation Frequencies

If we limit ourselves to simple dominant mutations with complete penetrance and invariable expressivity, we find that there are still other problems. In the first place, attention is quite naturally focused on changes that occur with some frequency, since we cannot make a count of nonexistent cases! However, we may rightfully feel uneasy that we will be selecting for loci that are more mutable than average. Clearly, if there are many loci that do not mutate to dominant alleles (and most loci do not), they will be excluded from our investigation. It goes without saying that our estimate of the average rate at which genes mutate to dominant alleles will then be spuriously high.

THE MUTATION RATE. The frequency with which normal alleles mutate in the course of one generation is called *mutation rate*, and is indicated by the Greek Letter μ (*mu*). For dominant alleles, by far the simplest approach is to make an actual count of the new mutations that have appeared in the course of a generation. For instance, we might check all of the children born at maternity hospitals in the United States in the last 10 years for a specific defect, to see whether or not their parents had the same defect. After an effort is made to eliminate the spurious cases, a certain number of individuals can be classified as probably carrying new dominant mutations. A correction must be made at this point to account for the fact that each person has two alleles, and so our population of alleles under observation is twice as large as the number of people.

When these steps are taken, the rates come out in the vicinity of several such new mutations per 100,000 alleles in each generation; that is, the mutation rate is roughly between 10^{-4} and 10^{-5} (between one in 10,000 and one in 100,000) per gene per generation. However, because of the problems mentioned previously, including phenocopies, misclassification, incomplete penetrance, and unconscious selection of the data for those cases with abnormally high mutation

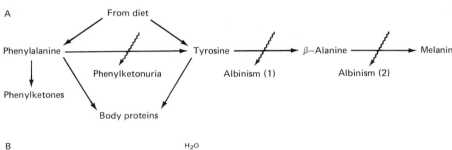

A

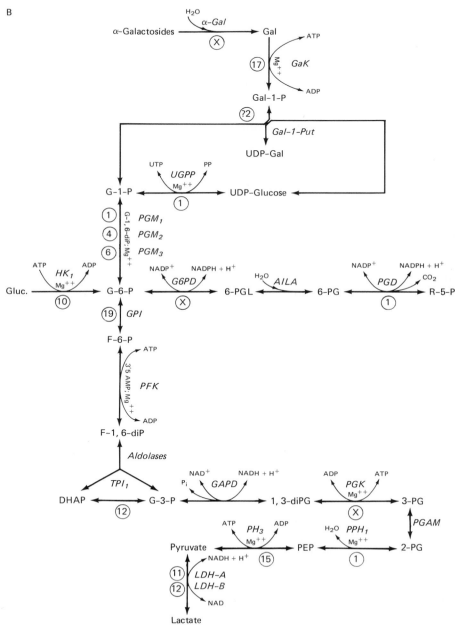

B

rates, it seems safe to say that on the average a human locus will have a probability of spontaneous mutation of about one per million per generation. Inaccurate as this estimate must be, it is still worth making because the numerical value will be used in evaluating mutation risks in populations, particularly where those populations are exposed to agents, such as radiation, that markedly increase the mutation rate.

INDIRECT CALCULATIONS.   For loci that produce recessive mutations, the procedure outlined above cannot be used because it is usually not possible to know whether a homozygote is homozygous because of the occurrence of a new mutation in one of the parents, the other being heterozygous, or because of heterozygosity of both parents. In virtually all cases it will be the latter. A different system for estimating the mutation rate has been devised, based on the following argument: We can assume that the human population is at *equilibrium* with respect to its genetic composition; i.e., that over the last several thousand years, as new alleles have been added to the human gene pool by mutation, other similar alleles have been lost from the pool by inviability, infertility, or other aspects of "unfitness" of the affected individuals. At the present time, a balance exists, more or less, between these two opposing factors. We can reasonably assume that the recent advances in medical science have not yet had sufficient impact to alter the overall balance.

A nice analogy for this state of equilibrium is given in Figure 10-5. This shows water being poured into a container with a hole in the bottom. As the water level in the container gets higher, the water pressure at the opening in the bottom becomes greater and the rate of loss is increased until the amount being lost is equal to the amount coming in. When the level is constant, i.e., is at equilibrium, we need to know neither the amount of fluid in the container nor the size of the hole at the bottom, but simply that the amount leaving the container must be equal to the amount coming in.

Thus, if we assume that the human population is in equilibrium at the present time with respect to allele frequencies, then the number of mutant alleles being lost in each generation is a measure of the number entering the gene pool by

Figure 10-2   *A*. The series of enzymatic steps involved in the production of melanin from phenylalanine and tyrosine. Both of these amino acids are supplied by the diet; in addition, tyrosine is produced within the body from phenylalanine. When the normal gene responsible for the specific enzyme which causes this conversion is not present, phenylalanine is not converted to tyrosine, but accumulates to produce an excess of metabolic products (phenylketones) so that the individual suffers from phenylketonuria. If the normal course of reactions from tyrosine to melanin is interfered with, albinism results. Since the series of reactions may be interrupted at any one of a number of points and each step is controlled by a specific enzyme produced by a specific locus, the genes controlling different steps are nonallelic even if the end result (in this case albinism) appears much the same.
B. Part of the biochemical pathway of carbohydrate metabolism in man. The enzymes are represented by their locus symbols in italics and, when known, the number of the chromosome on which that locus is found is encircled. (Courtesy of P. Meera Khan and D. Bootsma. 1974. Chromosome localization of enzyme information in man. *Proc. XI Int. Cancer Congress*, 56–63.)

Figure 10-3   *A.* The hair of two albinos, both lacking pigment. *B.* After addition of the pigment precursors, one of the hair bulbs manufactures pigment readily (left) and the other does not. (Courtesy of C. J. Witkop, from Albinism, from *Advances in Human Genetics*, Eds. Harris and Hirschorn, Plenum Press, New York, London, 1971.)

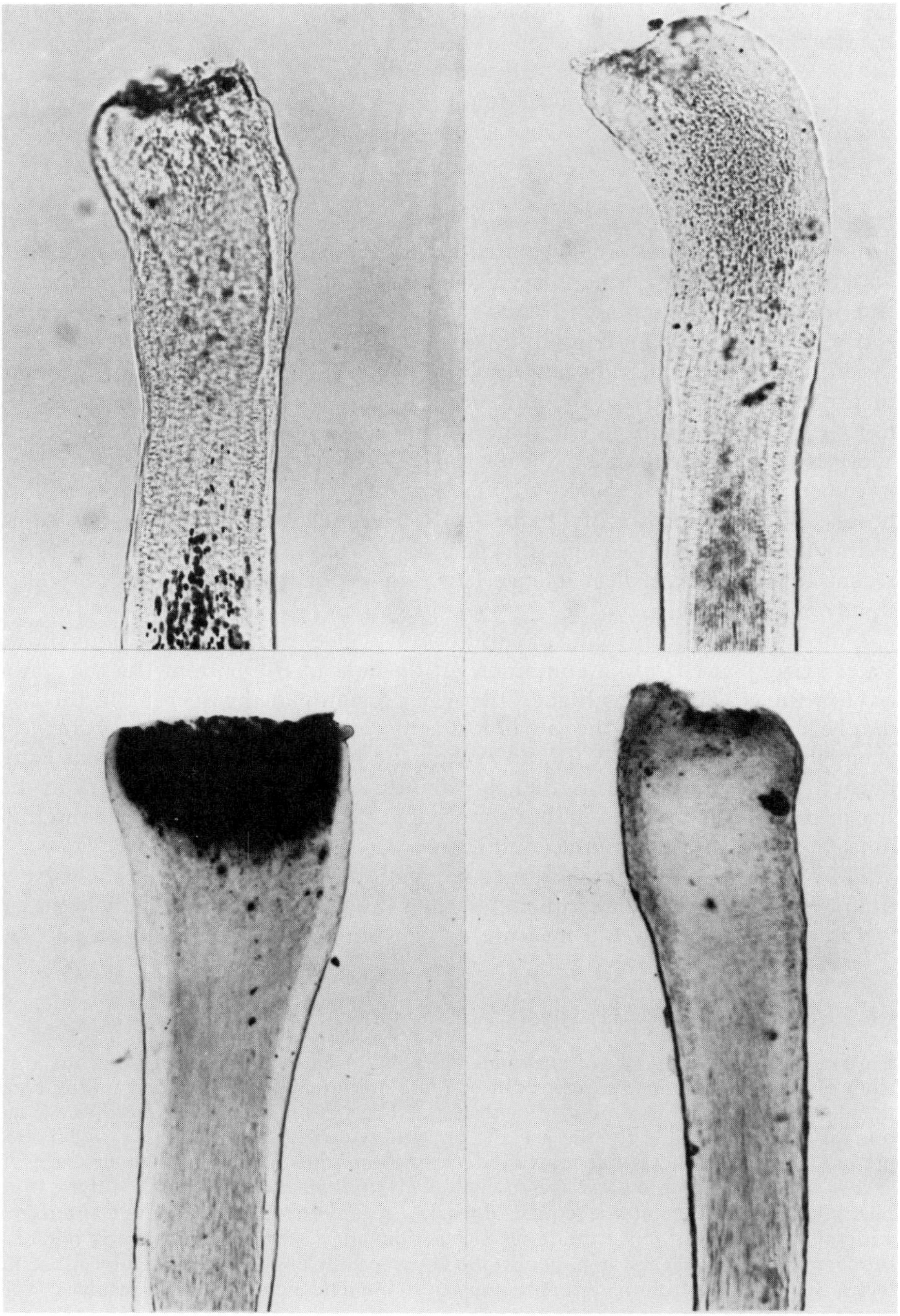

Figure 10-4 The two kinds of albinos (unrelated) produced by recessives at the two different loci differentiated by the test shown in Figure 10-3. (Courtesy of C. J. Witkop.)

mutation. Genes are ordinarily lost by a decreased productivity of the individuals carrying them.

For estimates of mutation rates of recessives, homozygotes are examined to determine what, on the average, their reproductivity has been compared to normal. For the normal comparison, sibs are preferred, since they will have been subjected to roughly the same external influences from family and society with respect to reproduction as the affected homozygotes have been. The difference between the reproductivity of the affected persons and those normal persons chosen for comparison is the loss of reproductivity (L).

The following is a typical illustration. The number of mutant alleles introduced into the population in each generation equals the mutation rate (μ) times the total number of individuals (N_T) times 2 (since each individual is diploid).

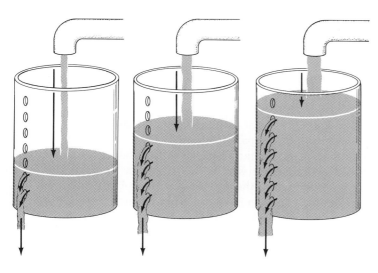

Figure 10-5 Analogy showing the state of equilibrium reached when the outflow equals the input in a simple water system. (From *Principles of Human Genetics*, Third Edition, by Curt Stern. W. H. Freeman and Company. Copyright © 1973.)

The number of alleles lost equals the number of affected individuals in the population (N_{Aff}) times the loss of reproductivity (L) times the number of alleles lost with each reproductive loss. For a simple dominant gene, these relations can be stated algebraically as

$$\mu \times N_T \times 2 = N_{Aff} \times L \times 1$$

or

$$\mu = L \times N_{Aff}/2N_T$$

For recessive genes, in which the reduced fitness of the homozygote causes the loss of two alleles instead of just one, the appropriate relations are

$$\mu = L \times N_{Aff}/N_T$$

Thus, if there are found to be forty individuals with a specific identifiable dominant allele (achondroplasia, for instance) in a population of six hundred thousand persons, and such persons produce only one quarter as many progeny as their unaffected sibs, μ is calculated as $.75 \times 40/6 \times 10^{-5}$, or 5×10^{-5}. For rare recessive sex-linked genes, the relations are like those for a dominant, but only a third of all X-chromosome alleles are found in the male and subject to loss, the other two thirds being present in the female as heterozygotes, and so protected against loss. The loss of reproductivity for all persons, both male and female, carrying such alleles will be only one third as great as is indicated by the loss of reproductivity shown by the homozygous males. For that reason a factor of one third is introduced into the equation to give

$$\mu = L \times N_{Aff}/3N_T$$

All that is needed is some knowledge about the frequency of the defect in the population (N_{Aff}), about the loss of reproductivity caused by the defect (L), and about how the defect is inherited, in order to apply the proper equation. In cases as severe as muscular dystrophy, for instance, the value of L must be very close to 100 percent, or 1. Thus, if we know that the frequency of Duchenne's muscular dystrophy is about 15 per 100,000 in the population, we can immediately estimate the mutation rate of the normal allele to that recessive to be one third as great, or 5×10^{-5}.

Tables 10-1, 10-2, and 10-3 list the mutation rates calculated for a number of dominant, recessive, and sex-linked alleles. For the reasons given previously, these rates are undoubtedly very much higher than the rates for the average, run-of-the-mill gene. One estimate suggests that three quarters of all normal alleles have rates less than one per million and that as many as half may have rates of one per 10 million, or less.

MUTATION RATES PER GAMETE.   What is the chance that, by mutation, a child will have an allele that neither of the parents carried? This is very simply calculated. If we accept the spontaneous mutation rate as 10^{-6} per locus per generation, and the number of loci in the haploid set as 2×10^4, then we have a product of 2×10^{-2}, or roughly 2 percent. For the diploid, we multiply this by 2, coming out with a figure of 4 percent. That is, a newborn child stands a chance of one in twenty-five of carrying a new mutation that was not carried by either of the two parents. Since each of the values entering into such calculations is subject to considerable error, the calculation might be in error by a considerable factor but, hopefully, not as much as a factor of ten too much, or too little.

Table 10-1 Estimates of spontaneous mutation rates at some human autosomal loci: dominant diseases. (From L. S. Penrose, in *Recent Advances in Human Genetics*, ed. L. S. Penrose, Churchill Livingstone, Edinburgh, 1961.)

Trait	Mutation Rate per Gamete per Generation ($\times 10^{-6}$)	Region	Source	Date
Epiloia	8	England	Gunther and Penrose	1935
Chondrodystrophy	70	Sweden	Böök	1952
Aniridia	5*	Denmark	Møllenbach	1947
Microphthalmos without mental defect	5	Sweden	Sjögren and Larsson	1949
Retinoblastoma	4	Germany	Vogel	1954
Partial albinism and deafness	4	Holland	Waardenburg	1951
Multiple polyposis of the colon	13	U.S.A.	Reed and Neel	1955
Neurofibromatosis	100	U.S.A.	Crowe, Schull, and Neel	1956
Arachnodactyly	6	Northern Ireland	Lynas	1958
Huntington's chorea	5	U.S.A.	Reed and Neel	1959

*This estimate differs by a factor of 2 from that given by the author, but it is based on his material.

Characteristics of Mutations

DOMINANCE VERSUS RECESSIVENESS.   Whether a mutant allele is dominant or recessive may be a matter of the gravest importance to the individual who carries it, to say nothing of his parents, sibs, and offspring. This is clearly true for an allele with devastating developmental defects, in which the fortuitous presence of that single allele may mean the difference between life and survival to a ripe old age, if recessive, and death or incapacitation at an early age, if dominant.

Let us consider a few of the simplest ways in which an allele may appear to be either dominant or recessive. One of the commonest kinds of human defects is a biochemical deficiency in which an enzyme is either absent or relatively inefficient so that it does not produce the normal amount of its product. A gene which produces an enzyme or some other protein that is an essential part of the structure of the cell is referred to as a *structural gene*. In the case of albinism and many dozens of well-known human deficiencies, when there is a normal gene present on one of the two homologues and a mutant on the other the normal one will produce sufficient enzyme for the individual to be normal, for all practical purposes. In these cases, the defective gene will appear to be recessive.

Table 10-2   Indirect estimates of spontaneous mutation rates on the assumption of recessive inheritance. (From L. S. Penrose, in *Recent Advances in Human Genetics*, ed. L. S. Penrose, Churchill Livingston, Edinburgh, 1961.)

Trait	Mutation Rate per Gamete per Generation ($\times\ 10^{-6}$)	Region	Source	Date
Juvenile amaurotic idiocy	38	Sweden	Haldane	1939
Albinism	28	Japan	Neel et al.	1949
Ichthyosis	11	Japan	Neel et al.	1949
Total color-blindness	28	Japan	Neel et al.	1949
Infantile amaurotic idiocy	11	Japan	Neel et al.	1949
Microcephaly	49	Japan	Komai et al.	1955
Phenylketonuria	25	England	Penrose	1956

It can be stated as a general rule that mutant genes that are responsible for enzymatic deficiencies will behave as recessives in their inheritance. This does not mean that the heterozygote *Aa* is absolutely identical to the homozygote *AA*: in fact, for many traits classified as simple recessives the heterozygote *Aa* may be distinguished from the homozygote *AA*.

Furthermore, most mutants are characterized by *pleiotropy*, or manifestation of the genetic defect in several morphological ways: thus, albinism is pleiotropic in that such individuals are commonly nearsighted and have *nystagmus*, a rapid involuntary oscillation of the eyeballs, in addition to decreased skin and hair pigmentation. Some aspects of the phenotypes of some genetic conditions may be detectable in the heterozygote, i.e., dominant, and others not, i.e., recessive. Thus the early lethality of achondroplasia is recessive, appearing only in the

Table 10-3   Estimates of spontaneous mutation rates at some human sex-linked loci. (From L. S. Penrose, in *Recent Advances in Human Genetics*, ed. L. S. Penrose, Churchill Livingstone, Edinburgh, 1961.)

Trait	Mutation Rate per Gamete per Generation ($\times\ 10^{-6}$)	Region	Source	Date
Hemophilia	20	England	Haldane	1935
Hemophilia	32	Denmark	Andreassen	1943
Hemophilia	27	Switzerland and Denmark	Vogel	1955
Pseudohypertrophic m.s. (muscular dystrophy)	60	Northern Ireland	Stevenson Blyth and	1958
Pseudohypertrophic m.s. (muscular dystrophy)	47	England	Pugh	1959

homozygote, but the shortening of the limbs is readily apparent in the heterozygote and is therefore dominant.

On the other hand, consider the consequences if a change takes place in a gene that causes it to send an incorrect message to another gene at another locus, so that the latter will function at a time when it should not, or not function when it should. In the lower organisms, genes that control other genes are called *operator* and *regulator* genes; similar gene activities must exist in the higher animals, in which development and differentiation are more complicated. A mutation of such a gene may very well behave as dominant since only one member of the two alleles present in the cell need give the incorrect signal. This may be the explanation of why many of the defects of embryological development (polydactyly, syndactyly, etc.) are dominant in inheritance.

CLASSIFICATION OF ALLELES AS DOMINANT OR RECESSIVE. The dominant or recessive labels we attach to alleles are a matter of convenience, generally dictated by the appearance of some clearly obvious or important aspect of the phenotype in the heterozygote, and not by the detection of any difference between the homozygote normal AA and the heterozygote Aa! Thus, although we may be able to detect heterozygotes for the allele for cystic fibrosis by sensitive biological tests, the allele is properly referred to as recessive because it is the homozygote who is at serious risk of early death. Perhaps the application of these descriptive words to any genetic condition should be made with more consideration to the impact on affected persons and their relatives than on the medical or biochemical details of the action of the allele. Certainly, in one case where this rule was not followed, and medical terminology was applied instead, that of sickle cell anemia, the result has been endless confusion, needless apprehension, and in some cases outright injustice (Chapter 23).

Generally speaking, those anomalies recognized as recessives are more severe than those classified as dominant. A severe recessive may exist hidden in heterozygotes for many generations until a couple who happen to carry the same defective allele become parents and produce one or more affected children. The presence of several affected in a sibship with normal parents immediately gives a clue as to the nature of its inheritance. A dominant allele with similar severity would eliminate itself quickly because of the low viability of the individuals carrying it. If it were not so severe in its effect, it might take more time, but eventually it would disappear from the population for social and other reasons. Therefore, mutant genes transmitted from one generation to the next in the human population are not randomly selected from all possibilities, but are restricted to the recessives and to the more viable dominants.

The Genetic Load

Among the mutations present in the population, a large number will be recessives with lethal effects on the developing embryo and fetus, or, if they do not cause death at such an early age, may be responsible for death or debilitation early in postnatal life. It would be worth knowing how many such deleterious genes the average person carries: Is modern humanity so protected by medical science and the advantages of modern society that we have accumulated large numbers of such deleterious genes? Or are we relatively free of "bad" genes? Further, do different groups of people differ in the number of such genes they carry? Clearly, any answers to these questions, however rough, are of considerable importance.

Obviously we cannot make direct breeding tests for recessive lethal or semi-lethal genes, but we can take advantage of one commonly occurring type of inbreeding, that of first-cousin marriages. First cousins have a pair of grandparents in common. If either of the two grandparents carries a defective recessive, it stands a good chance of becoming homozygous in any child who is a product of the consanguineous marriage (Figure 10-6). To make this statement meaningful, however, it is necessary to state it quantitatively.

Figure 10-6 diagrams a first-cousin marriage with one of the common grandparents (*I-3*) heterozygous for a specific allele. The chance that one of his progeny (*II-2*) will receive it is one-half and the chance that the grandchild (*III-1*) will receive it from his parent is also one-half. The chance that the grandchild will receive the allele designated in the grandparent is therefore $\frac{1}{2} \times \frac{1}{2}$ or $\frac{1}{4}$. A similar argument applies to the other grandchild, *III-2*. Therefore, the chance that the two grandchildren, who are first cousins, will both be heterozygous is $\frac{1}{4} \times \frac{1}{4}$ or $\frac{1}{16}$. When this is the case, the chance that their child will be homozygous for this allele is $\frac{1}{4}$. The chance, then, that an offspring of a first-cousin marriage will be homozygous for a specific allele present in one of the grandparents is $\frac{1}{16} \times \frac{1}{4}$ or $\frac{1}{64}$. However, there are four alleles present in the two grandparents, each of which has this same probability of homozygosity. The total chance of homozygosity is therefore $4 \times \frac{1}{64}$ or $\frac{1}{16}$.

Disregarding the much smaller probability that homozygosity can also result from the fortuitous combination of two alleles of independent origin, the offspring of a first-cousin marriage then should be homozygous for a particular allele one sixteenth as often as the allele is found in the grandparents or, more generally, one sixteenth as often as the allele is found in the population at large. From the frequency of homozygotes from first-cousin marriages, it is possible to make a simple estimate of allele frequencies in the population: if PKU's were to be found among the offspring of first-cousin marriages with a frequency of one in 800, one would conclude that the frequency of the recessive allele in the population is $1/800 \times 16$ or $\frac{1}{50}$.

The argument need not be limited to one locus or to visible phenotypes. If the genes in question are recessive lethals, then any such lethal carried heterozygous in one of the grandparents will contribute $\frac{1}{16}$ or about 6 percent loss in the progeny of first-cousin marriages.

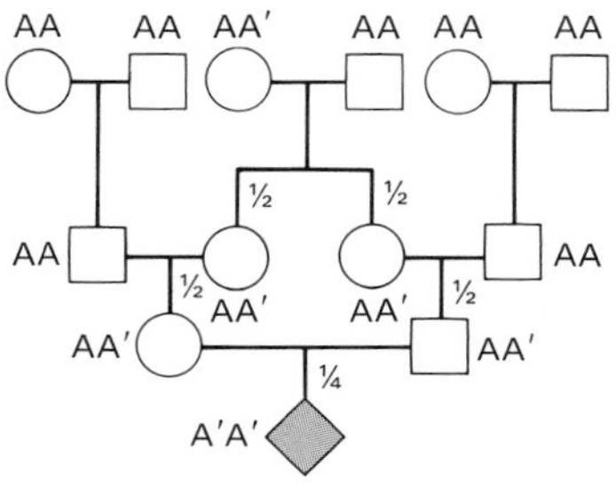

Figure 10-6 Pedigree showing the basic relationships in a first-cousin marriage along with the probabilities that a specific allele present in one of the common grandparents will appear homozygous in one of their children.

Table 10-4 Difference in mortality of progeny, when parents are unrelated or first cousins, for mortality at different periods. The numerical excess can be attributed to homozygosity for defective alleles. (By permission of N. Morton, *Progress in Medical Genetics*, Vol. 1:261–291, 1961. Grune & Stratton Inc., New York.)

| Type of Mortality | Authority | Relationship of Parents | | Difference in Mortality |
		Unrelated	First Cousins	
Miscarriages	Slatis et al. (1958)	.129	.145	.024
Stillbirths and neonatal deaths	Sutter and Tabah (1958)	.044	.111	.067
Postnatal deaths	Slatis et al. (1958)	.024	.081	.057
Infant and juvenile deaths	Sutter and Tabah (1958)	.089	.156	.067
Juvenile deaths	Bemiss (1958)	.160	.229	.069
Early deaths (Hiroshima)	Schull and Neel (1958)	.031	.050	.019

From Table 10-4 it can be seen that the mortality rate among the offspring of first-cousin marriages is greater than that for the general population, at all ages studied. If the prenatal deaths are in excess by about 6 percent and the postnatal deaths another 6 percent totaling 12 percent, then the data suggest that an average of two lethal genes per person would account for this greater mortality from first-cousin marriages.

It is not necessary to postulate completely lethal genes to account for the increased mortality observed, which might equally well be the result of four different genes, each of which when homozygous would cause mortality in half the offspring, as well as other combinations having comparable effects. To take such possibilities into account, the term *lethal-equivalent* has been coined, indicating any gene or group of genes which, acting separately or together, cause one death when homozygous.

The data available do not give information on very early losses of the embryo, a likely time for homozygous lethals to exert their effect, or, for that matter, on premature deaths later in life. When these are taken into account, several more lethal-equivalents appear to be part of the average person's make-up. An educated guess is that each person carries three or four lethal-equivalents.

Not only is viability affected by consanguinity, but also the frequency of other defects, both physical and mental, increases (Figure 10-7). One study in Chicago shows that, whereas the frequency of physical and mental defects runs at about 10 percent in the normal population, it goes as high as 16 percent in first-cousin marriages. Once again, such studies must be limited to differences that are clear and unmistakable. Although the above difference of 6 percent would correspond to an average of only one detrimental gene per person, some workers in the field suggest that the total number of lethal-equivalents plus detrimental genes carried by the average person is in the vicinity of ten. This number probably does not differ significantly from one population group to another. It is the totality of these detrimental genetic conditions that constitutes our *genetic load*.

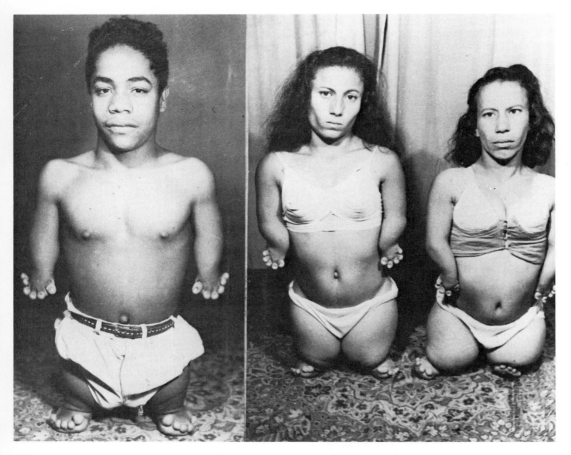

Figure 10-7 A recessive developmental defect found in a small village in Brazil; more than 60 percent of the marriages, which produced 68 such affected individuals, were consanguineous. (From A. Quelce-Salgado, A new type of dwarfism. *Acta Genet.*, S. Karger AG, Basel, **14**:63–65, 1964.)

References

BOSTIAN, C. H., M. WHITTINGHILL, W. S. POLLITZER, and L. MURO. 1969. Evidence of iris-earlobe linkage in a mosaic man. *J. Hered.*, **60**:3–9.

CROW, J. F. 1961. Mutation in man. *Prog. Med. Genet.*, **1**:1–26.

MORTON, N. 1961. Morbidity of children from consanguineous marriages. *Prog. Med. Genet.*, **1**:261–91.

NEEL, J. V. 1972. The detection of increased mutation rates in human populations. In H. E. Sutton and M. I. Harris, eds., *Mutagenic Effects of Environmental Contaminants.* New York: Academic Press.

PENROSE, L. S., ed. 1961. *Recent Advances in Human Genetics.* Edinburgh: Churchill Livingstone.

QUELCE-SALGADO, A. 1964. A new type of dwarfism with various bone aplasias and hypoplasias of the extremities. *Acta Genet. Stat. Med.,* **14**:63–6.

SCHULL, W. J., ed. 1968. *Mutations.* Ann Arbor: University of Michigan Press.

VOGEL, F. 1964. Mutations in man. Genetics today. *Proc. XIth Int. Cong. Genet.,* The Hague 1963. New York: Pergamon.

VOGEL, F., and G. RÖHRBORN, eds. 1970. *Chemical Mutagenesis in Mammals and Man.* New York: Springer-Verlag.

WITKOP, C. J., E. J. VAN SCOTT, and G. A. JACOBY. 1961. Evidence for two forms of autosomal recessive albinism in man. *Proc. IInd Int. Cong. Hum. Genet.,* Sept. 6–12 (Rome: Instituto "Gregorio Mendel"), 1064–65.

Questions

Useful terms: somatic mutation, phenocopy, penetrance, expressivity.

1. What is the important distinction between a somatic and a germinal mutation?
2. If somatic mutations were at all frequent, would we be likely to recognize them? How?
3. What difficulties are encountered in recognizing new mutational events when they occur?
4. Which is the easier to estimate, the mutation rate of a recessive or of a dominant gene? Explain.
5. How is the principle of equilibrium used in estimating recessive mutation rates?
6. What is the probability that a newborn child will carry a new mutation of spontaneous origin that was not present in either of the two parents? This rate is fairly high. Why do we not see more new mutations among children?
7. It may reasonably be argued that the distinction between a dominant and a recessive gene is a trivial one or that, on the contrary, it is of tremendous significance. Discuss these two different points of view.
8. What natural experiment makes it possible to estimate the number of deleterious genes that a human will, on the average, carry?
9. Discuss the relevance of the following conditions to the concepts discussed in this chapter: phocomelia, polydactyly, achondroplasia, albinism, muscular dystrophy, syndactyly, kwashiorkor.

11

Locating Genes in the Genome

Gregor Mendel was lucky. If he had continued his experiments with the garden pea, he would eventually have gotten results in disagreement with his second generalization, the law of independent assortment. He worked with seven pairs of alleles; the garden pea has only seven chromosome pairs. Actually several of the allelic pairs had loci on the same chromosomes, but his work was not extensive enough to include any pair of loci that might have given unusual results. If he had extended his work to include many more loci, he would have uncovered some cases where the law of independent assortment no longer applied. Thus, he would have discovered *linkage*, but this premature observation would not have fit into his conception of the rules of inheritance. As a result he might not have published his results, or, if he had, his presentation would have been much less convincing.

Chromosome Mapping

One of the aims of human genetics research is to find the positions of as many loci as possible within the chromosome complement. This involves either determining whether any two loci are actually located on the same chromosome (i.e., whether the loci are *syntenic*) or determining which of the 23 chromosomes carries the locus in question, and then finding its position along the chromosome length. In principle this task of finding the location of, or *mapping*, loci is simplest for certain loci for which the population is highly heterogeneous,

178

such as the blood-group antigens, histocompatibility antigens, and biochemical variations in the proteins found in the serum of the blood. Curiously, the inheritance of some common morphological traits such as hair and eye color is sufficiently confused by modifying factors that these characteristics are among the most difficult to work with. For polygenic traits such as height, and for those more diffuse polygenic traits such as behavioral characteristics in which heredity and environment play complicated interacting roles, the task becomes formidable and, in the long run, perhaps not even worthwhile, because of the indefinite and variable contributions of large numbers of loci under different environmental and developmental conditions. Many simple morphological traits, such as albinism, occur so rarely that the right genetic conditions for mapping have never occurred and the positions of such loci in the chromosome complement are unknown.

DETERMINING LINKAGE AND SYNTENY.   Assigning a locus to a specific chromosome is easy only in the case of sex-linked genes. It can be stated unequivocally that the loci for color-blindness and hemophilia are located on the X-chromosome; otherwise, these alleles would not show the typical sex-linked pattern of inheritance. Without further evidence we can say that the loci for these two defects are syntenic.

For autosomal loci, the classic test for linkage depends on the occurrence of two different alleles at each of two different loci nonrandomly in sibships. To understand this, we must again examine the course of the meiotic divisions.

In the simplest general case in which two syntenic loci are marked by different alleles on the two homologues (Figure 11-1), the homologues are separated at meiosis; the allelic positions are unchanged (and are found on the same chromosomes) in the gamete. If meiosis were as simple as this, there would be complete linkage of alleles (*A* with *B* and *A'* with *B'*), except for the rare case when mutation might change one allele into another. Therefore, from such a heterozygote *AB/A'B'* only two types of gametes would be produced, *AB* and *A'B'*.

However, either of the two chromatids of one homologue may be involved in exchanges of genetic material with either of the two chromatids of the other (Figure 11-2). These exchanges are seen cytologically as chiasmata. When the exchanges occur between two loci (as between the *A* and the *B* locus), they produce new chromosome types called *crossovers* or *recombinants*, *A* with *B'* and *A'* with *B*; these may be recognized in the progeny.

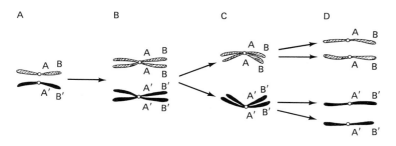

Figure 11-1   A pair of homologues heterozygous at two different loci (*A*), showing how they proceed through meiosis, replicating (*B*), followed by separation of the homologues at first anaphase (*C*) and separation of the chromatids at anaphase of the second division (*D*). The gametes receive only two gene combinations, *AB* and *A'B'*, those present in the parent.

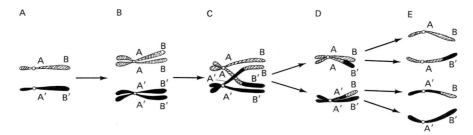

Figure 11-2 The sequence of events during the meiotic divisions showing how a chiasma occurring between two heterozygous loci gives rise to crossover or recombinant chromosomes. The first two steps, (A) and (B), are identical to those in Figure 11-1. (C) The occurrence of a chiasma, during prophase of the first division; (D) separation of homologous chromosomes (each consisting of two chromatids) at anaphase of the first division; (E) separation of the chromatids at anaphase of the second division, giving four different products, half of which will be recombinants AB' or A'B'.

Since chiasmata are more or less randomly distributed along the length of the chromatids, except for some regions like those near the centromere, and occur independently of the presence of heterozygous loci, a chiasma may or may not occur at some point between the loci, but clearly the probability that this will happen depends directly on the length of chromosome between them. When the loci are very far apart, the chance of at least one chiasma between them is excellent; when they are very close together, the probability is very low.

The frequencies of recombinants may be used directly to construct a map. If loci A and B show 5 percent crossing over—that is, 95 percent of the original combinations AB and A'B' and 5 percent A'B and AB'—then we can say that they are 5 units apart. Let us suppose, further, that loci B and C on the same chromosome show 15 percent crossing over. How much crossing over will then be between A and C? The previous results, A-B of 5 percent and B-C of 15 percent, might have happened in either of two ways (Figure 11-3). If B is the middle locus then we would expect to get 20 percent crossing over in an experiment involving A and C, but if A is the middle locus we would expect only 10 percent. Additional data involving the A and C locus will show which of the two alternatives is the correct one.

From experiments of this sort it can be concluded that the loci appear to be linearly arranged along the chromosome, a situation we may have suspected from our examination of the polytene chromosomes in Chapter 5. The proposal (based on crossover data) made by Sturtevant in 1913 that genes are linearly arranged along the length of the chromosome was severely criticized by other workers at that time on the grounds that, since the chromosome was clearly a three-dimensional sausage-shaped body, the representation of distances between three or more loci could more appropriately form a triangle or more complicated geometrical figure instead of a straight line!

The reason for the lack of any indication of linkage when two loci are far apart is, of course, the high probability of one or more chiasmata showing up as genetic exchange between the loci, making it appear that they are not linked. Referring back to Figure 11-2, we see that if a person (necessarily a female in this case) is heterozygous for hemophilia (represented by A and A') and sex-linked

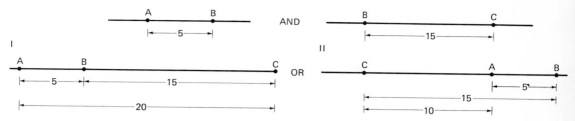

Figure 11-3 The principle of chromosome mapping, using crossover data, showing the additivity of crossover values. In one experiment A and B prove to be 5 units apart, and in another B and C are found to be 15 units apart. The quick conclusion is that A and C are 20 units apart (possibility I). However, there exists an equally likely possibility (II) that A, not B, is the middle locus. In that case, the expected distance between A and C would be 10 units. Additional information, such as data involving A and C specifically, would be needed to distinguish between these possibilities.

ichthyosis (B and B'), then, since the loci are quite distant from each other, the occurrence of exchanges between them will produce eggs of compositions AB, $A'B'$, AB', and $A'B$. Since the latter two will be very frequent, approaching if not equalling 50 percent, the loci will not be linked genetically. We would not know from the progeny of such doubly heterozygous women that the loci are syntenic, but the fact that both these defects are sex-linked tells us that their loci must both be on the X-chromosome.

Determining degrees of linkage, if any, between loci in experimental plants and animals is a relatively simple task. One parent is made heterozygous at two or more loci and is backcrossed to an individual who is homozygous recessive for those loci so that all combinations of the alleles produced by meiosis in the heterozygous parent are immediately detectable in the phenotypes of the progeny. The percentages of recombinant types then give an indication of the presence or absence of linkage. In humans, however, the problem is much more difficult.

DETERMINING LINKAGE FROM PEDIGREES. If we want to know whether some locus, say that for PKU (phenylketonuria), is linked with the locus for albinism, we would have to find a doubly heterozygous individual who has married another with both recessives. Matings of this sort are extremely rare, and in some cases impossible to detect. Loci for the blood groups are particularly valuable in making tests for linkage since different alleles have high frequencies of occurrence and can be detected in heterozygotes because of codominance. The chance of finding detectable heterozygotes in the population is quite large and matings occur frequently between all genotypes. Furthermore, the frequency of double heterozygotes, which must be quite small for rare diseases such as PKU and albinism, becomes moderately high in cases of loci with high allelic frequencies. It is not an accident that some of the earliest linkages discovered were those involving blood groups.

Another general problem is that it is difficult if not impossible to analyze data for linkage without complete information from the parents (and even the grandparents can be useful here) and they may not be available for examination. Ideally, family size should be as large as possible; this makes it possible to determine more accurately the ratio between parental allelic combinations and the recombinants. The trends toward very small family size viewed so enthusiastically by those concerned with the problems of overpopulation are a source of profound melancholy to those engaged in human genetics linkage studies.

Another problem is that loci on the same chromosome are often so far apart that the amount of observed recombination is very close to 50 percent. Detection of linkage under these conditions is virtually impossible until other loci are found in between those two. Nevertheless, human geneticists have made a heroic effort to determine linkages whenever possible, and have come up with some positive results. Usually these results are obtained by feeding pedigree data into a computer programmed to calculate the chance that the distribution of parental types and recombinants among the progeny is dependent on some degree of linkage compared to the chance that the distribution comes about by chance from independently segregating loci.

The beginner always imagines that the study of human genetics is impeded in some way by the number of chromosome pairs (23), large compared to the *Drosophila* number (4) or corn (10). That this is not the case can be seen by asking a simple question: how much more would we know about the human genome if it had only *one* chromosome? or *two*, one sex chromosome and one autosome? The answer is that we would know no more than we do now, and probably less. The chromosomes divide the genome into smaller discrete segments; allocation of loci to these parts serves as a convenient first step in the total task. Perhaps humans have close to the optimal number of chromosomes for the job of mapping.

Determination of Synteny by Chromosome Form

USE OF UNUSUAL CHROMOSOME VARIANTS.   A few years ago it was found that a medical student at Johns Hopkins University had an unusual chromosome 1 in his complement, and because it first appeared that part of chromosome 1 had uncoiled it was called "uncoiler 1" (Figure 11-4). This chromosome was shown to be *familial* (i.e., it tended to occur in families, in this case his) and a natural question was whether there were any other characteristics that were transmitted along with this peculiarity. As it turned out, different alleles for a blood group called Duffy were segregating in this kindred and it could be shown that its segregation followed along precisely with the abnormality in chromosome 1. In other words, the locus for the Duffy blood group was syntenic with the abnormality on chromosome 1. Clearly other loci segregating in such a family might be checked similarly, and if the linkage between any locus and the locus responsible for the uncoiling were sufficiently lower than 50 percent the probability of verifying this linkage could be very good.

Using a number of different techniques for staining chromosomes, it can be shown that other variant forms exist in the population; that is, they are *polymorphic*. Such different chromosomes segregate as distinct chromosome peculiarities in kindreds and have been used in assigning specific loci to chromosomes. A list that includes assignments made in this way is found at the end of this chapter in Figure 11-7.

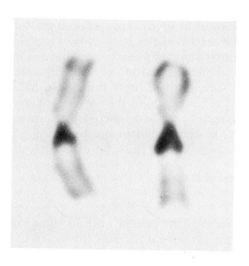

Figure 11-4 A comparison of a normal chromosome 1 and the variant known as "uncoiler 1." The staining method used shows an unusual amount of heterochromatin adjacent to the centromere. (Courtesy of P. Pearson, University of Leiden.)

USE OF TRISOMIES.   There are still other simple methods by which loci may be allocated to specific chromosomes. For instance, suppose that chromosome 21 carried the locus for the ABO blood groups. We would then expect the frequency of the phenotypes of the ABO blood groups in Down's syndrome patients to fall in a different pattern from that of normal individuals, because of the presence of three instead of two alleles. Thus, the gametic frequency of the 0 allele is about .68 and the frequency of homozygotes normally about .46 (= .68 × .68). We would then expect the frequency of 0 individuals with trisomy-21 to be 0.68 × 0.68 × 0.68, or about 0.31. By making similar calculations, taking alleles three at a time, we can show that the frequencies of the other groups would also be different in trisomies. Since the frequencies of the ABO phenotypes among Down's individuals are the same as for the normal population, we can conclude that the ABO locus and other blood group loci as well are not found on chromosome 21. This method, however, is limited to loci that allow a determination of gene dosages, three versus two, such as the blood group loci, or loci determining enzyme activity, which might be found at a higher level when there are three genes rather than two present. Although the procedure of determining enzyme levels is simple in theory, it is more complicated in practice because trisomic patients may show enhanced enzyme levels for secondary reasons related to a disturbed protein metabolism, rather than the presence of three (instead of two) genes responsible for the enzyme. Nevertheless, in one case, that of an enzyme called superoxide-dismutase, the activity is about 50 percent higher in Down's patients, in agreement with other information that the locus is on chromosome 21.

USE OF DEFICIENCIES.   Deficiencies, missing segments of chromosomes in one of the two homologues, can also be used for specifying the region in which a locus may, or may not, be found. If, for instance, an individual has a deficiency for the short arm of chromosome 9 on one of the two homologues and at the same time is of blood group AB (which one piece of evidence indicates to be on 9), then it is clear that the locus for AB must be on the long arm—otherwise the deficient person could not have both A and B alleles.

There seems to be no limit to the ingenuity shown by workers in the area of chromosome mapping in their efforts to pinpoint the location of loci within the human chromosome set, although, to be sure, most of these neat tricks have been borrowed from the area of classical genetics where they were first discovered and put into operation. A good example of this is a system for determining the position of the centromere relative to certain biochemical markers.

Determining the Genetic Location of the Centromere

A very common abnormality of development is the *benign ovarian tumor* of the female. When those cysts are tested for cytological markers for which the woman is known to be heterozygous, it is found that the markers are usually homozygous, indicating that the tumors originate, not from ordinary somatic cells, which would be heterozygous, but from oocytes which have gone through at least the first division of meiosis. On the other hand, when the cells are checked for biochemical markers, also heterozygous in the woman, the alleles in the cyst are sometimes heterozygous and sometimes homozygous.

These results have a simple explanation. If it is assumed that the cysts arise from single oocytes after the first meiotic division, then the cytological banding pattern present on one of the two homologues will be found in one of the two meiotic products and not the other (Figure 11-5*A*). In this way, a diploid line may then develop, homozygous for a chromosome that was heterozygous in the other somatic cells. Another possibility, however, develops when the heterozygous locus is located some distance from the centromere (Figure 11-5*B*). When an exchange occurs between that locus and the centromere, each of the two chromatids attached to the same centromere at the first anaphase division may carry different alleles and the developing cyst then may be heterozygous. On the other hand, if no exchange should occur between the locus and the centromere, then the cyst will be homozygous for one or the other alleles. In this way, the locus for the enzyme PGM3 (phosphoglucomutase) has been shown to have a position 17 units away from the centromere on chromosome 6.

These results are of unusual genetic interest, beyond the obvious relation to the manner of origin of cystic teratomas in women. According to this hypothesis, the chromosomal banding pattern becomes homozygous in cysts because no exchange occurs between the differentially staining region and the centromere; this is consistent with the usual cytological position of these regions close to the centromere. On the other hand, the heterozygosity for enzyme markers depends on exchange between the centromere and the more distant locus responsible for that enzyme (Figure 11-5*B*), with the probability of the occurrence of such an exchange being related directly to the distance between the locus and the centromere.

This is indeed a remarkable genetic experiment, for not only does the relative frequency of homozygous versus heterozygous cysts, from a heterozygous woman, indicate the amount of exchange, and therefore the relative position on the chromosome arm of the genetic marker being followed, but also the "map-

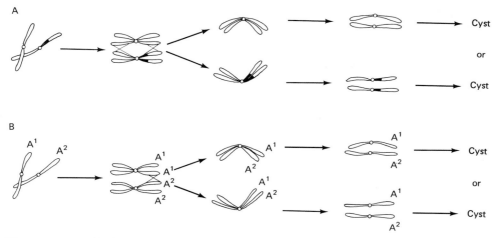

Figure 11-5 The hypothetical behavior of a heterozygous chromosome pair prior to the formation of an ovarian cyst. (*A*) Heterozygosity for a cytological difference at the centromere will produce two products after the first division, each homozygous for a distinctive chromosome type. (*B*) When the locus being followed is some distance from the centromere, a chiasma between that locus and the centromere will give heterozygous products after the first division. Since the chance that a chiasma will occur between the locus and the centromere is proportional to the distance between them, the relative proportion of heterozygous cysts serves as a measure of the genetic distance between the centromere and that locus.

ping" experiment is being achieved with only one heterozygous locus, where at least two are needed in standard linkage experiments. Furthermore, the distance so indicated is measured from the centromere, a chromosome organelle for which mutant variation is now undetectable and therefore with no possibility of mapping under ordinary circumstances.

Cell Hybridization

A most spectacular recent advance has been the development of somatic *cell hybridization*, a process by which somatic cells, grown in culture, are induced to combine with somatic cells from another source to produce a so-called hybrid. The properties of these hybrid cells can give valuable information about the nature of the parental types.

Like so many other important observations in genetics, cell hybridization was first noticed in the course of experiments set up to check on an entirely different cell property. In 1960, several researchers in Paris mixed together two different mouse cancer lines that differed not only in their morphology but also in their chromosome number. During these tests to see whether the two kinds of cancer cells had an effect on each other, a third type was found in the culture. Upon investigation, it turned out to be carrying the chromosomes of both the original types. These cells were hybrids that had arisen by the *fusion* of cells of the two "parental" types. It was found that these hybrid cells could be perpetuated as a distinct cell line, with some of the characteristics of both parents. Initially they had the total chromosome count of both parents together, but as mitosis continued over a period of time the cells gradually lost some of the chromosomes and developed into individual lines with chromosomes of varying number.

Ordinarily, cell fusion occurs spontaneously with a very low frequency, about one in a million. These particular hybrids were detectable because they had an advantage in rapid mitosis over the two parental types, eventually becoming a significant fraction of the total cell population. Since that time, a number of treatments have been discovered that will appreciably increase the frequency of hybridization. Perhaps the most common method is the addition of *Sendai virus*, related to those which cause influenza. After the inactivation of this virus by ultraviolet light or chemical treatment, it is added to a cell culture. It makes the cell surfaces "sticky," with the result that fusion occurs with a frequency from 100 to 1,000 times greater than with untreated cells.

SELECTIVE SYSTEMS.   Even with this high fusion frequency, it is necessary to have some kind of selective system that will automatically differentiate between the very large number of parental cells and the relatively few hybrids. In principle, this is a very simple thing to do (Figure 11-6). This scheme is an adaptation of a method perfected decades ago in genetic studies of molds and bacteria, for collecting rare combinations of new cell genotypes. One mutant cell line is unable to synthesize substance A, which is necessary for its survival and must be supplied as part of the basic medium in which the cells grow. Another strain is unable to manufacture substance B, and survives and multiplies only when B is provided in the medium. We shall call the first strain A^-B^+ and the second A^+B^-. These two cell types can be mixed together, and both will thrive as long as the medium contains both substances A and B. At this point a fusion-inducing

Figure 11-6 A simplified version of the procedure for obtaining cell hybrids. Two lines of cells, each with a synthetic disability requiring the addition of a substance to the medium, are mixed in the presence of a virus or other hybrid-promoting agent. The two extraneous growth requirements are removed, killing cells of the original lines and allowing the survival only of cells with both synthetic properties. Unless new cell types are produced by such rare phenomena as mutation, the survivors will be hybrids.

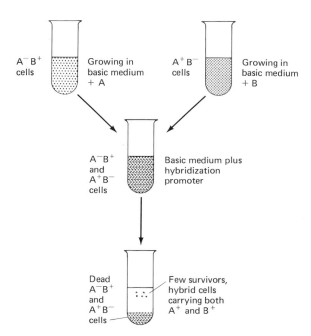

material such as Sendai virus is added, and, after an opportunity for fusion, the cells are removed from their former enriched medium and put into a new medium lacking both A and B. Under these conditions, cell types A⁻B⁺ and A⁺B⁻ no longer flourish and eventually die. A hybrid cell, however, can now manage perfectly well because it carries the normal alleles for the A deficiency from its A⁺B⁻ parent and the normal alleles for the B deficiency from its A⁻B⁺ parent, in a manner analogous to an F_1 heterozygote. In this way, even though the incidence of fusion may be very small, hybrid cells will be produced and detected very efficiently, and will be well differentiated from the parental stocks. In actual practice, the number of deficient cell types available from humans is quite limited; most genotypes characterized by this kind of deficiency would also be lethal to the organism and therefore not recoverable. One of the few enzyme-deficient types comes from individuals with the Lesch-Nyhan syndrome, a serious sex-linked defect characterized by mental and physical maldevelopment and very short life-span.

Another useful technique for the production of hybrids involves the use of one normal parent line and another lacking an enzyme necessary for growth, or with some other characteristic that can be readily selected against. If the normal parent line grows slowly, or is added in small numbers, with the enzyme-deficient parent subsequently eliminated by removing its necessary substrate from the medium, the culture will consist of a few normal parent cells and some hybrid cells. The latter can then be separated manually and their hybrid nature verified.

INTERSPECIES HYBRIDS.　It is part of the modern dogma of neo-Darwinism that as two strains become separated from each other during the early stages of divergence into species they become less able to intercross and produce fertile progeny. Generally speaking, after diverging, two related but distinct species will not mate and produce offspring. There are no known cases of hybrids produced between any of the upper primates. This restriction, however, does not apply to cell hybrids, and cells from quite unrelated organisms can be induced to fuse. Hybrids between cells of the mouse, the rat, the hamster, and humans are only a few of a large number which have been produced, and many other hybrid types will undoubtedly be possible when systems have been developed to select between the potential parent lines and their hybrids. In fact, the production of hybrid types among mammals is restricted only by the effort necessary to develop selection techniques versus the probability that the new hybrid will provide worthwhile information to the researcher.

It might be imagined that such an F_1 hybrid cell, once produced, would constitute a dead end, of minor value only as a scientific curiosity. There is, however, another aspect of cell hybrids of great importance when the parental lines derive from two different species. When the cells hybridize, their nuclear membranes also fuse so that all of the chromosomes of both species are found in a single nucleus. Thereafter the chromosomes become synchronous and progress together through the mitotic cycle, except that, for reasons that are not at all clear, the chromosomes of one species or the other may preferentially be lost.

In hybrids between mouse cells and human, the mouse chromosomes usually persist and the human are preferentially eliminated. This might suggest that mouse chromosomes have some inherent advantage over those of other species, but in the cross of mouse and Syrian hamster cells it is the mouse cells that are at a disadvantage. The loss of chromosomes makes it possible to perform a kind of segregation analysis or linkage study by isolating clones of cells with varying numbers of chromosomes still present from one of the species.

MAPPING THE LOCUS OF THYMIDINE KINASE.   Let us take the case of a deficiency of the enzyme thymidine kinase, which is known in mouse cell lines but has never been observed in humans. (This does not necessarily mean that the corresponding mutant allele has never occurred in humans. It could be, for instance, that individuals carrying this defect are lost very early during embryonic development.) From a mixed culture containing affected mouse cells (thymidine kinase-less, or TK⁻) and human cells, we obtain, using the procedures described above, an F_1 hybrid cell line. This hybrid line is normal with respect to the defect in question, so we have good reason for postulating that the human parental line has contributed genes which are capable of carrying out that function normally. As the hybrid cells undergo mitosis and the human chromosomes are preferentially lost, we can derive clones which have a small number of human chromosomes more or less randomly selected from the original 46. After determining precisely which human chromosomes are present in each clone, we test the clones for the original mouse enzyme defect. In the case in question we will discover that every clone containing chromosome 17 from the original human complement is capable of synthesizing the enzyme thymidine kinase (Table 11-1). In those cases in which chromosome 17 is missing, irrespective of which other chromosomes still remain from the human contribution, the clone is defective and thymidine kinase activity is absent. It is a reasonable conclusion that the normal allele responsible for the synthesis of that enzyme in humans is located on chromosome 17.

A little thought tells us that we have performed a most remarkable experiment; we have used the mouse cells, which can be manipulated experimentally quite freely, to study the genetics of humans. By way of interspecific cell hybridization we have made a significant contribution to our knowledge of the inheritance of various characteristics without the direct involvement of affected individuals. In the case of the thymidine kinase deficiency, we can reasonably postulate that if it should occur it would be a disease with a simple genetic basis, even though it has never been observed in humans. We have even been able to state with some assurance that the locus responsible is carried on chromosome 17, a conclusion that would be very difficult to come to even if this were a well-known characteristic in humans. The technique makes it possible, in principle, to draw on a store of mutant genes available in experimental animals, or to

Table 11-1   Tests of clones derived from mouse-human hybrids, in which the original mouse line was thymidine kinase deficient (TK⁻). Note that whenever the clone is TK⁺ it carries the human chromosome 17.

Human Chromosomes Persisting in Clone	TK?
5, 9, 12, 21	−
3, 4, 17, 21	+
5, 6, 14, 17, 22	+
3, 4, 9, 18, 22	−
1, 2, 6, 7, 20	−
1, 9, 17, 18, 20	+

induce new ones in those animals using radiation or other mutagenic agents—procedures that would be scientifically as well as ethically prohibitive in humans. Furthermore, we have been able to accomplish in a very short period of time an experiment that, were it not for the cell fusion technique, would have required many generations, hundreds of years of human time. In fact, although this technique has been utilized for less than a dozen years, it has been responsible for a large proportion of the information we have with respect to the allocation of loci to specific autosomes in humans.

LIMITATIONS AND PROSPECTS FOR CELL HYBRIDIZATION.   Of course, it should be emphasized that these techniques currently can be applied only to a small percentage of genetic changes in humans, those which are identifiable in cell culture. At present, this excludes a large number of important morphological characteristics, developmental and neurological anomalies, even certain traits that are characterized by the absence of enzymes (such as PKU, in which the appropriate enzyme occurs only in differentiated liver cells), as well as some common cell properties such as the erythrocyte surface antigens. However, as time progresses, there will become available for this sort of analysis a greater number of such biochemical markers, at the same time that techniques will be developed for studying loci not now detectable in cell culture.

It is possible, in fact, to do a kind of primitive chromosome mapping because the chromosomes do not always maintain themselves as units but may, on occasion, become involved in translocations or fragment into smaller chromosomes. In this way two loci that are ordinarily linked can become separated. The frequency with which a random break occurs between two syntenic loci will obviously depend in part on the distance between them. It is not possible to be specific about actual genetic distances in this sort of experiment, but it is reasonable to guess that two loci are relatively far apart when they are frequently separated from one another by random breakage. A good example of the effectiveness of this technique is the progress that has been made in mapping chromosome 1, on which there are now 21 loci placed, with varying degrees of certainty, along the chromosome length (Figure 11-7).

In these diverse ways, knowledge of the human genome has suddenly blossomed. The allocations shown in Figure 11-7 bear witness to the effectiveness of this technique, and such a listing will need updating with every passing month because of the information being generated in this fertile area of research.

References

BOONE, C. M., and F. H. RUDDLE. 1969. Interspecific hybridization between human and mouse somatic cells: enzyme and linkage studies. *Biochem. Genet.*, **3**:119–36.

EPHRUSSI, B., and M. C. WEISS. 1969. Hybrid somatic cells. *Sci. Am.*, **220**:2–11.

HARRIS, H., et al. 1965. Mitosis in hybrid cells derived from mouse and man. *Nature*, **207**:606–8.

MILLER, O. J., P. W. ALLDERDICE, and D. A. MILLER. 1971. Human thymidine kinase gene locus: assignment to chromosome 17 in a hybrid of man and mouse cells. *Science*, **173**:244–45.

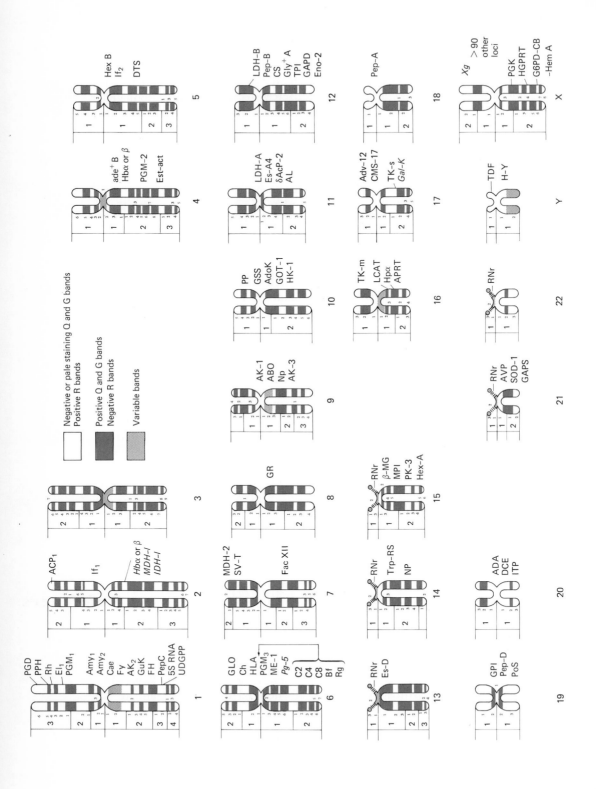

NABHOLZ, M., V. MIGGIANO, and W. BODMER. 1969. Genetic analysis with human-mouse somatic cell hybrids. *Nature,* **223**:358–63.

RENWICK, J. G. 1971. The mapping of human chromosomes. *Ann. Rev. Genet.,* **5**:81–120.

SCHWARTZ, A. G., P. R. COOK, and H. HARRIS. 1971. Correction of a genetic defect in a mammalian cell. *Nature (New Biol.),* **230**:5–8.

SHAW, R. F., and H. GERSHOWITZ. 1963. Blood group frequencies in Mongols. *Am. J. Hum. Genet.,* **15**:495–96.

WEISS, M. C., and H. GREEN. 1967. Human-mouse hybrid cell lines containing partial complements of human chromosomes and functioning human genes. *Proc. Natl. Acad. Sci. U.S.A.,* **58**:1104–11.

WESTERVELD, A., R. P. L. S. VISSER, P. MEERA KHAN, and S. BOOTSMA. 1971. Loss of human genetic markers in man-Chinese hamster somatic cell hybrids. *Nature (New Biol.),* **234**:20–24.

ZEPP, H. D., J. H. CONOVER, K. HIRSCHHORN, and H. L. HODES. 1971. Human-mosquito somatic cell hybrids induced by ultraviolet-inactivated Sendai-virus. *Nature (New Biol.),* **229**:119–21.

Questions

Useful terms: linkage, syntenic, crossover, recombinant, codominance, polymorphic.

1. Explain why Gregor Mendel might have run into severe problems if he had simply continued making crosses between more and more characters of the garden pea.
2. What is the distinction between linkage and synteny? Is it likely that in man there are more groups of genes that are linked than there are syntenic ones?
3. Is it likely that in a relatively short period of time man's genes will be mapped so completely that he will be understood physically, psychologically, and in all other important aspects? Explain your answer.
4. For what group of genes can the loci be allocated unambiguously to a specific chromosome directly from a simple pedigree?
5. Imagine that the loci for two independent characteristics are found at opposite ends of chromosome 18. Will the alleles inherited from each parent on specific strands always be found together in the gametes of their progeny? Or will there be a redistribution of the alleles?
6. Make a diagram showing the meiotic processes by which new gene combinations are made up.
7. A genetic test for linkage in an experimental animal is usually fairly simple. Why is it more difficult in humans?

Figure 11-7 The location of genes on the human chromosome map. Note that many of the locations are tentative and not confirmed, that the loci are usually those of enzymatic properties, and that many common defects, such as albinism or cystic fibrosis, are conspicuously absent, unless their chromosome allocation is unambiguous because of sex-linkage. (Courtesy of P. Meera Khan, University of Leiden.)

8. What general kinds of characteristics lend themselves more readily to determining linkage than others? Would a clear characteristic such as albinism be an excellent one to use for this purpose?

9. Is the allocation of loci to specific chromosomes hindered by the fact that man has so many chromosome pairs?

10. How can trisomies be used in allocating loci to chromosomes? heritable chromosome anomalies?

11. Can genetic distances ever be determined when only one heterozygous locus is under investigation?

12. Somatic cell hybridization is a powerful technique in allocating loci to specific chromosomes. Show how this works, using an actual illustration.

13. Would there be any point in trying to make a hybrid between the cells of man and some insect?

14. Discuss the relevance of the following conditions to the concepts discussed in this chapter: hemophilia, Duffy blood group, cystic teratoma, Lesch-Nyhan syndrome.

12

The Basis for Induced Genetic Change

Early Attempts to Induce Mutation

Just as the alchemist, in the early beginnings of chemistry, dreamed of the trans-mutation of one element, lead, into another, gold, so did the first geneticists recognize the importance of inducing new heritable changes, the exact nature of these being unknown at that time, but now recognized as gene mutations. And just as those early efforts of the alchemists were doomed to failure, not because of the incorrectness of the idea itself, but because knowledge essential to carrying out the experiment and detecting the resulting changes had to be developed systematically over a long period of time, those early geneticists were equally frustrated in their attempts to modify the genetic system of experimental plants and animals. Curiously, some of the agents tested unsuccessfully during the early days of experimental genetics, in the 1910's, such as high-energy radia-tion (from radium and x-rays) and certain chemicals, proved later to be among the most effective mutation producers, or *mutagens.* We now know that other physical agents (heat and ultraviolet light) and chemical compounds, running into the many hundreds, are mutagenic.

X-RAYS.   The first announcement of success in attempts to induce gene muta-tion came in 1927 when H. J. Muller reported that he had succeeded in pro-ducing mutations in *Drosophila* with a high frequency after x-ray treatment. This was followed by a similar report in 1928 by L. J. Stadler, who, treating barley with x-rays and radium, had found large numbers of new mutants present in the treated plants, and not in unirradiated control plants. In each case, the mutations appearing were random with respect to which locus was affected and the type of allele produced; that is, these first beginnings were far from approaching the goal (still unattained) of causing a specified kind of change in a particular gene.

ULTRAVIOLET LIGHT.   After this, it was simply a matter of time before other mutagenic agents were uncovered. Ultraviolet light, radiant energy like x-radiation but of lesser potency, also produces mutations, in those cases when it can reach the genetic material. Since *transmissible* mutations can be produced only if the agent reaches the nuclei of the germ cells in the gonads, ultraviolet light can be ruled out as a potential mutagenic agent of any genetic importance in humans. Ultraviolet light is absorbed in the first few millimeters of the epidermis of the skin, as every person who has suffered from severe sunburn can unhappily testify, and it is responsible for an increased frequency of skin cancer in persons who receive high exposures. It is undoubtedly capable of producing mutations in the somatic cells located near the surface of the body; these might be important to the individual in whom they appear, such as the increased frequency of skin cancer in persons who receive high exposures. Such doses of ultraviolet would not, of course, have any mutagenic importance to future generations.

CHEMICALS.   Although reports of positive results from treatment with various chemicals appeared sporadically during the 1930's, it was not until World War II that an unquestionable chemical mutagen was found. An accident in an industrial plant in Great Britain left several workers with severe burns from mustard gas, a poisonous gas not used during that war, but manufactured in great quantities nevertheless. Someone remarked on the similarity of the burns on these victims to the burns produced on the skin of a heavily x-rayed person; this led to an actual test with *Drosophila* to determine whether mustard gas was also mutagenic. Such proved to be the case. Subsequently more than a thousand compounds have been tested for mutagenicity; the number giving positive results is in the hundreds. Some of these will be discussed in a later chapter.

The Nature and Action of X-Rays

Of considerable interest and importance to humans at the present time is the high-energy radiation originating from x-ray machines, from cosmic rays, and from radioactive isotopes resulting from bomb blasts and production of nuclear power. Before considering the effects on human tissue, we must first understand the nature of this energy.

THE ELECTROMAGNETIC SPECTRUM.   The most obvious section of the radiation spectrum (Figure 12-1) is, of course, that which we see as visible light. A convenient way of classifying radiation is in terms of its wavelength, which for visible light is about a millionth of a meter, or 1 μm. Radiant energy comes in indivisible units called photons, which have very specific amounts of energy, called quanta (singular, quantum). For a photon of a specific wavelength there exists a corresponding quantum energy value. Because the product of the wavelength of a given part of the spectrum and its frequency is a universal constant, the speed

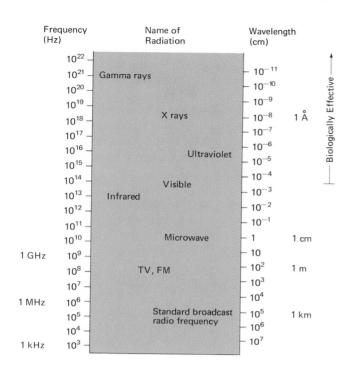

Figure 12-1 The spectrum of electromagnetic radiation.

of light, wavelength and frequency are inversely related. Radio waves, which have a long wavelength (measured in many feet), have a low energy per quantum, whereas x-rays, gamma rays, and cosmic rays have high values.

Because electromagnetic radiation is emitted by such a wide variety of sources, from radio stations to electric light bulbs to x-ray machines, it is important to understand the general principles that determine whether or not any given radiation type is biologically important. The effectiveness of radiation depends primarily on whether or not it can produce chemical reactions, and whether these occur will depend on the energy per quantum (not on the total amount of energy) in the exposure. We know that visible light must be capable of causing some reactions; otherwise we could not see, since vision depends on a chemical reaction in the retina of the eye. However, relatively few chemical reactions can be caused by energies lower than the quantum value for visible light; in the case of radio broadcasting or television waves, the energy per quantum is so small that there is no possibility that they can promote a chemical reaction by atomic absorption. Since the ability to produce chemical reactions is independent of the total number of quanta involved, a person standing near a radio station that produced a total power of hundreds of thousands of watts would not have to worry about the possible induction of mutations. Biological effects such as temporary male sterility that have been attributed to these longer wavelengths (from radio and radar transmitters) are produced by ordinary heating of the living tissue.

PRODUCTION OF X-RAYS. In principle, high-energy radiation may be generated very simply. All that is needed is an ordinary flask, with a piece of metal (target) at one end and a wire (filament) that can be heated by an electric current at the other. Then the air is removed from the flask, and a source of high voltage (tens of thousands of volts) is connected across the filament and the target (Figure 12-2), with the positive end connected to the target. When the filament is heated, it emits electrons, which, being negatively charged, are attracted to the target by its positive charge. Unimpeded in the vacuum, the electrons accelerate under the influence of the high voltage until they smash into the target. They may be stopped immediately, whereupon their kinetic energy is transformed into photons of radiant energy. An electron accelerated by 50,000 volts, and then suddenly stopped, would release 50,000 electron volts in a single photon. This much energy per photon puts it in the x-ray range.

Actually small amounts of energy—only 5 electron volts, for instance—can produce chemical reactions when delivered to the molecule itself, and the question immediately arises as to why ordinary household voltages, given the right physical setup, do not produce damaging radiation, and why it is necessary to apply tens or even hundreds of thousands of volts across an evacuated tube to produce biologically damaging radiation. The answer lies in the fact that, fortunately, different regions of the electromagnetic spectrum have different transmissibilities through ordinary air (see Figure 12-3). We know from everyday experience that radio waves and visible light are highly transmissible. Transmissibility falls off in the ultraviolet region and stays at zero until we reach the region of x-rays. An x-ray machine operating at 10,000 volts will produce "soft" x-rays that are mostly absorbed by the first few inches of air through which they pass.

The physical conditions ordinarily used to produce x-rays (an evacuated tube, a hot filament, a high voltage between the filament and another metallic electrode) are found in many household appliances—in fact, wherever vacuum tubes are found (radios, television sets, etc.). In most cases, one need not be concerned about x-ray production from these sources, because even if the voltage amounts to hundreds or thousands of volts the low-energy photons that are produced are largely absorbed by the glass wall of the tube and the surrounding air. Certain color television sets have been found to emit radiation; this phenomenon will be discussed later.

Unstable Isotopes

Another source of high-energy radiation is the disintegration of unstable configurations of neutrons and protons in certain atomic nuclei. This more potent radiation emanating from within the nuclei of atoms is called *gamma* radiation, to distinguish it from the x-rays produced by changes of electron position outside the atomic nucleus.

Any given element is characterized in its uncharged state by a specific number of negatively charged electrons surrounding it, and by a corresponding positive charge carried by an equal number of protons in its nucleus. In addition, the nuclei of all elements except ordinary hydrogen contain a number of uncharged particles called neutrons. Different classes of atoms of the same element may differ by the number of neutrons in their nuclei; they will still, however, be identical in their behavior in ordinary chemical reactions. Atoms of the same element differing from each other by the number of neutrons in their nuclei

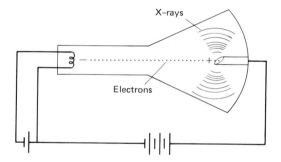

Figure 12-2 Schematic of a typical x-ray tube. The air is removed from a flask, and a hot filament is placed at one end and a solid target at the other. When a source of high voltage is connected across the two elements, x-rays are generated at the target.

are called *isotopes*. The isotope of hydrogen with two neutrons, tritium, is characterized by an instability of the nucleus—there seem to be too many neutrons present for the small positive charge of one proton and such a nucleus may suffer spontaneous degeneration by expelling a particle, producing helium.

THE HALF-LIFE OF UNSTABLE ISOTOPES.   What causes some isotopes of a given element to be stable and others unstable is not completely understood; it appears that the number of neutrons and protons must bear some proportion to each other for stability. If there are too many, or too few, neutrons or protons, the nucleus will spontaneously disintegrate and during this process will eject one or several of the types of particles making it up, along with gamma radiation, which has all of the properties of x-rays, except that it is more energetic. The particles ejected are also important; virtually all the biological damage to tissue caused by the disintegration of *tritium* and *plutonium* comes from high speed particles which are quite energetic but can travel only very short distances in solid matter.

Because the components of the nucleus change, there may be a change in the element that the nucleus represents, and this new nucleus may be either stable or unstable. The rate of spontaneous disintegration differs from one isotope to the next; the measure of the rate is the *half-life*, which is the time required for the disintegration of one half of the atoms in a quantity of a given isotope. Note that half of the remaining atoms will disintegrate during a second half-life interval, leaving 25 percent unaffected, and that during a third such period half of

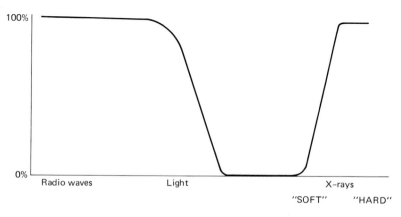

Figure 12-3   Relative transmissibility through air of different wavelengths of electromagnetic radiation. Some irregularities in this smooth curve, caused by specific absorption by certain molecules, have been omitted.

the 25 percent will disintegrate, leaving 12.5 percent unaffected. This type of reduction is called an exponential decay; regardless of how much time is involved, in units of half-lives, there will still remain a mathematically calculable amount of the original isotope unchanged. For biologically important elements, some radioactive (i.e., radiation-producing) isotopes and their half-lives are as follows:

Potassium 42 (^{42}K)	12 hr
Sodium 24 (^{24}Na)	15 hr
Phosphorus 32 (^{32}P)	14.2 days
Sulfur 35 (^{35}S)	87 days
Calcium 45 (^{45}Ca)	164 days
Strontium 90 (^{90}Sr)	28 yr
Plutonium 239 (^{239}Pu)	24,000 yr
Potassium 40 (^{40}K)	1,000,000,000 yr

The energy from the disintegration of radioactive isotopes is of biological importance in several ways. The explosion of an atomic bomb causes the formation of vast quantities of radioactive isotopes, which are blown into the atmosphere and released gradually over weeks, months, and years as fallout, exposing millions of humans to radiation. The production of energy by atomic fission in nuclear power plants creates large quantities of radioactive isotopes. While virtually all of these are ordinarily contained, their potential effects on life must be seriously reckoned with. Unstable isotopes are increasingly used in industry, medicine, and scientific research.

Chemical Effects of Radiation

When a photon collides with an electron, it will impart some or all of its energy to that electron, setting it in motion, or, if that electron was originally part of an atom, knocking it out, thereby *ionizing* the atom. It is this ionization that has important biological consequences. Suppose that two atoms form a molecule because they share an electron. Removing any electron from either atom will be the equivalent of removing the shared electron, so that the two atoms fall apart. In this manner, irradiation can cause compounds to disintegrate. For compounds of great length, consisting of a long chain of atoms hooked together, it can easily be imagined that this sort of event happening at any one of a large number of places can effectively destroy the molecule. This class of action is called *direct*, since it represents the immediate effect of the radiation on material.

INDIRECT ACTION.   There is still another way in which radiation acts. The most common compound in any biological system is, of course, water, and the absorption of the energy of x-ray quanta, with the addition and subtraction of electrons, can cause the hydrogen and oxygen of the water molecules to rearrange themselves into highly active agents. Hydrogen peroxide, H_2O_2, is one of the dozen products of such breakdown. These active products may persist in solution for some time until, by chance, they happen to come in contact with another molecule with which they can react. When this happens, both molecules are changed chemically. This type of reaction is called the *indirect* action of radiation.

Experiments on lower organisms suggest that almost half of the damaging effects of x-rays are due to direct action and the other half to indirect.

ABSORPTION OF HIGH-ENERGY RADIATION. The greatest hazard of x-radiation arises from its efficient penetration through solid matter. That it does penetrate readily is illustrated by the typical x-ray photograph of human tissue. The radiation that succeeds in penetrating the tissue exposes the photographic plate, and objects with different degrees of absorption (metallic objects, bone, cancerous tissue, etc.) give a characteristic image on the plate. The fact that energy is absorbed, with the molecular consequences described above, should forewarn us of the likely biological effects of such radiation.

If we select a certain thickness of material (metal, or living tissue, for instance) such that 50 percent of the radiation striking it succeeds in passing through it, then a second thickness placed next to the first will decrease the radiation passing through it to 25 percent of the original dosage. Clearly, as we add additional thicknesses we decrease the transmitted radiation by successive fractions of one half and a sufficient thickness of such shielding can be used to decrease the transmitted radiation to any desired amount. This exponential decrease obviously has serious implications in the field of radiation protection.

The amount of absorption is dependent on the atomic weight of the material—heavier elements being much more efficient absorbers than lighter ones. For this reason, lead, a heavy but relatively inexpensive element, is commonly used for confining high-energy radiation, i.e., for shielding. How do we decide how much shielding is required for various therapeutic and diagnostic applications? One criterion might depend on the amount of background radiation (from all natural sources) to which humans are ordinarily exposed—if the radiation to which a person is exposed, after shielding, is of the same intensity, roughly, as the natural background radiation, then the added dose might be acceptable. A more realistic approach might come from weighing the benefits resulting from the radiation exposure (or from the activities relating to that exposure) and the possible damage resulting from it. Much of the following discussion will be concerned with this kind of appraisal.

Measurement of Radiation

The appropriate measurement of radiation for biological purposes must be closely connected to the amount of ionization produced by that radiation in a given volume of tissue. The increase in mutation will be much the same, independent of the wavelength of the absorbed radiation or the time span over which it is absorbed. As long as we know the total amount of ionization produced during the exposure, we will have a good measure of the expected biological effect.

The original measurement of dose was the *roentgen*, or *r-unit*, named for the German physicist W. C. Roentgen, who discovered x-rays in 1895. In its original technical definition, it was the amount of x-radiation that produced 1 standard unit of charged particles in 1 cm^3 of air. For our purposes, we can consider it to be the amount of x-radiation that produces 1.6×10^{12} ion-pairs in 1 cm^3 of water. An approximately equivalent unit is the *rad*, which is based on the total amount of energy absorbed, and which therefore includes other biologically effective agents such as neutrons. In discussions of effects on humans, the unit commonly used is the *rem* (roentgen-equivalent-man). In any case, for x- and gamma radiations, these units are roughly equivalent, and may be used interchangeably for purposes of comparison without making any appreciable error. In most cases radiation applied to humans is considerably less than an r-unit, and instead a thousandth of an r-unit, a milliroentgen or millirem (*mr or mrem*), is the unit applied. Since the mr is only a thousandth of the r, it should be very carefully noted which one is being used in any statement of dosages.

To give a feeling for the meaning of the r-unit as it applies to doses human beings are exposed to, the doses for various purposes are indicated in Figure 12-4. Note that the scale is logarithmic, that each unit on the scale represents a change by a factor of 10. At one extreme are the massive doses applied in x-ray treatments to eradicate cancerous tissue, running into the tens of thousands of rems. At the other extreme are the rather small doses received as background radiation, from fallout, and from nuclear reactors.

Doses Lethal to the Organism

When the radiation dose is higher than several hundred r-units, the individual is killed because too many cells are destroyed for continued survival. This results from destruction of the intracellular components, those molecular structures necessary for normal functioning but not necessarily part of the genetic make-up. Thus a dose of 100 r will produce about 10,000 ion pairs in a typical cell with the result that more than 1 percent of the proteins and nucleic acids will be ionized. However important this damage may be to the immediate cells, our main concern will center on the changes induced in the genetic material, which will be propagated from an exposed cell to the next generation, and from an exposed individual to his or her progeny.

ACUTE VERSUS CHRONIC EXPOSURES. The biological impact of a given radiation dose may differ depending on whether it is applied in a short period of time (*acute*) or is spread out over a long period (*chronic*). Generally, those effects that result from cell destruction, effects that can be "repaired" by the recovery of the cell or by replacement of injured cells by unaffected ones, are more sensitive to acute exposures than to chronic ones. Thus a heavy radiation dose, sufficient to kill a human if given acutely in a few minutes, would have much less disastrous consequences if spread out chronically over a lifetime. On the other hand,

Figure 12-4 Logarithmic scale showing approximate x-ray doses used for therapeutic treatment of disease in specific organs and for diagnostic examinations, and the yearly whole-body dose from background radiation.

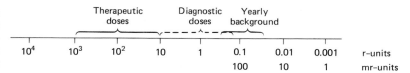

for other kinds of damage—mutation, for instance—from which there is no comparable recovery, acute and chronic doses may be about equivalent, although acute doses may be several times more effective than chronic. Generally there is no confusion as to which type of application is meant; unless it is specifically stated that a dose is chronic, it is considered to be acute, provided, of course, that the source of the dose does not itself imply a chronic exposure (e.g., fallout).

CELL LETHALITY.   If a person survives the first few weeks after exposure, there is a new hazard to be faced stemming from the destruction of certain blood cells, normally characterized by rapid mitosis, which the body requires in large quantities. These include the leukocytes, erythrocytes, and platelets. The first are essential for disease resistance, the second for oxygen transport, and the third for blood clotting. Thus, if any of these cell populations is injured so that normal mitosis is interrupted, the affected person may experience low resistance to disease, anemia, or excessive bleeding once the mature cells present in the circulatory system have run the course of their normal life-span and an inadequate number of new ones are formed to replace them (Figure 12-5). The main loss of such cells comes from the action of radiation in breaking chromosomes.

Radiation-Induced Chromosome Breaks

SINGLE CHROMOSOME BREAKS.   As the radiation traverses a cell, it may break chromosomes. At the point of a chromosome break, the two newly formed free ends behave as though they were "sticky," or unsaturated, in the sense that they can now rejoin with other broken ends. This is in contrast to normal chromosome ends, which are stable and do not join with other ends—if they did, the

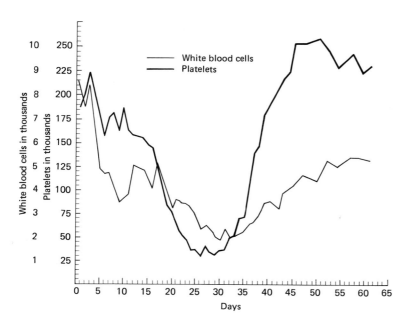

Figure 12-5 Counts of different cells of the blood system in five men exposed to 236 to 365 r total body radiation in an accident at Oak Ridge, Tennessee. (Courtesy of the Medical Division, Oak Ridge Institute of Nuclear Studies, Oak Ridge, Tennessee.)

fusion of two normal chromosomes, end-to-end, would give rise to serious problems in normal mitosis. In many cases, a broken chromosome may simply *restitute* to its original condition with no net damage, unless a gene located at or near the point of breakage has been damaged, in which case the chromosome may now carry a new mutation (Figure 12-6A).

At the next replication of the chromosome strand, newly broken ends of the sister chromatids may unite with each other in what has been called *sister strand union* (Figure 12-6B). As can be seen from the illustration, this gives rise to serious problems at the following anaphase, for we now have two abnormal chromosomes, one with no centromere, called *acentric*, and a second with two centromeres, or *dicentric*. An acentric fragment, having no centromere to direct it to the poles of the division, will be lost during succeeding mitoses, leaving the daughter cells with a large deficiency for that genetic material carried in the

Figure 12-6 Normal restitution of a chromosome break versus the production of acentrics and dicentrics resulting in an anaphase bridge.

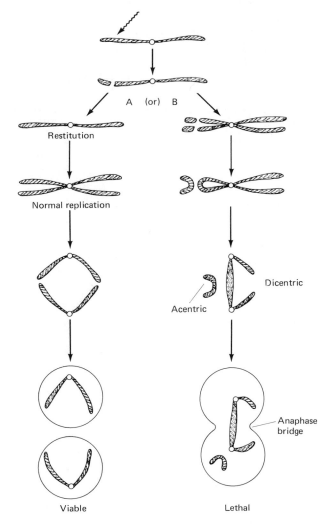

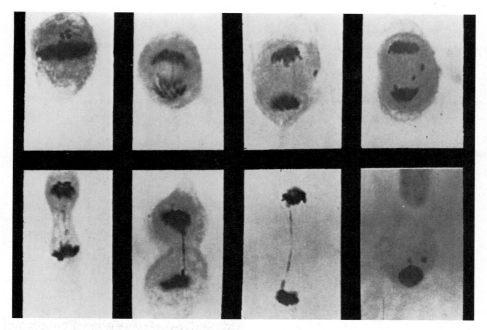

Figure 12-7 Cells of the mouse at telophase showing dicentric bridges formed at anaphase, along with acentric fragments. (Courtesy of T. H. Roderick, The Jackson Laboratory.)

acentric fragment. Figure 12-7 shows a mouse cell with a dicentric chromosome, along with acentric fragments.

The greater damage to the cell, however, may be done by the dicentric. At anaphase a *bridge* is formed, running between the two poles (Figure 12-6B). In many cases, this bridge itself is sufficient to prevent the normal development of the two daughter cells. If the bridge should break, with the two opposing centromeres carrying fragmented chromosomes to their respective poles, at the next replication the broken chromosomes will rejoin upon themselves to produce additional dicentrics. Eventually such cell lines must die out or produce abnormal cell lines in which there is a broken chromosome or in which the broken chromosomes have been eliminated after successive anaphases with dicentric bridges.

Kinds of Chromosome Rearrangements

TRANSLOCATIONS. If breaks occur in two chromosomes there are several possible consequences (besides restitution, which we shall disregard because it leads to nothing new). Figure 12-8A illustrates the formation of a *translocation* after two different chromosomes are broken and the acentric fragments of each unite with the centric fragment of the other chromosome to form a *reciprocal translocation*. Although the genetic material of one chromosome is moved to another chromosome and vice versa, it should be noted that the cell still contains the

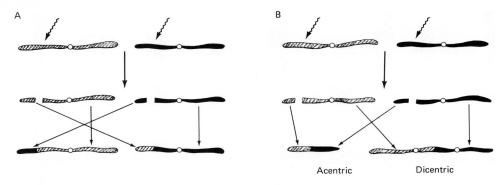

Figure 12-8 *A.* Breakage of two chromosomes resulting in a reciprocal translocation. *B.* Breakage of two chromosomes resulting in two possible kinds of translocations involving acentrics and dicentrics.

same total amount of genetic material, and that the chromosomes involved in reciprocal translocations are structurally normal in containing one and only one centromere. There is no reason why such a cell should not be able to proceed through mitosis without difficulty. Note, however, that when such a translocation is induced in a normal diploid cell, there still exists one normal homologue of each altered chromosome in that same cell. If the cell in question is a gonial cell, then problems may arise when the homologues attempt to pair at meiosis.

DICENTRICS AND ACENTRICS. The rejoining may occur in a different way, shown in Figure 12-8B. In this case, dicentric chromosomes and acentric fragments are produced; their behavior will be much like that of the dicentrics and acentrics produced by single breaks.

DEFICIENCIES. If both breaks occur in the same chromosome, the results depend on whether they both occur in one chromosome arm or one on either side of the centromere. If the first is the case (Figure 12-9A), and if the breaks do not restitute, then the piece between the two breaks may be lost so that a *deletion* or

Figure 12-9 Double breakage of a chromosome resulting in a deficiency (*A*) or a paracentric inversion (*B*).

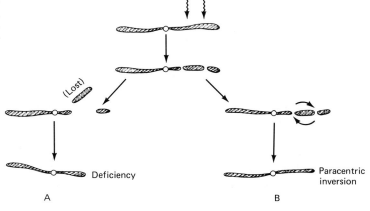

deficiency is produced. A chromosome deficiency that involves a small number of genes may not have any obvious effect, because of the presence of the unaffected homologue in the normal diploid cell. Larger deficiencies, however, are more often detrimental to normal development.

INVERSIONS. In some cases, the piece of chromosome is rotated and reunites in reversed order (Figure 12-9B) to form a *paracentric inversion*. The structure of the chromosome is now restored, and, in fact, this chromosome would not be detectable as having been altered under ordinary circumstances. Without special staining techniques it would appear normal and, with its normal genetic complement, would probably have no detectable genetic effect. Paracentric inversions might be expected to occur in the population as one of the more common types of abnormal chromosome structure, although their detection might be difficult.

Still another possibility exists when the two breaks occur in the same chromosome but with one on each side of the centromere (Figure 12-10). Obviously it is not possible to have a viable deletion of the middle segment since this would also remove the necessary centromere region. However, that segment could rotate, with the broken ends uniting to produce an inversion, in this case a *pericentric inversion* (Figure 12-10A and D). Depending on the relative distances of the breaks from the centromere, the new chromosome may have altered arm lengths, as indicated in the figure. However, the chromosome does have the same total number and kinds of genes that it had originally.

RINGS. Another very interesting anomaly that may arise from this type of breakage is the *ring* chromosome (Figure 12-10B and C), produced when the two acentric chromosome ends are lost and the two ends of the centric section join together. The capability of the two chromatids of a ring chromosome to separate at mitosis has always intrigued biologists. Any simple model that we might construct simulating the replication of a ring of great size (molecularly speaking) leads to the expectation that the daughter rings will be interlocked, and lost at anaphase. Human patients with ring chromosomes do tend to show a high degree of morphological variation of that chromosome; perhaps the more interesting question is why the loss does not always take place within a few divisions after formation.

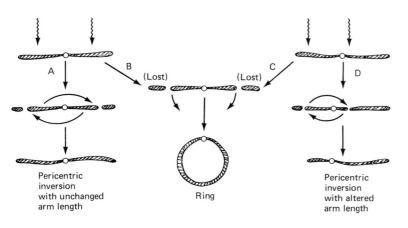

Figure 12-10 Double breaks in chromosomes resulting in pericentric inversions with unaltered (*A*) and altered arm lengths (*D*), or in ring chromosomes (*B* and *C*).

Relative Frequencies of Single and Multiple Breaks. From the standpoint of simple probability, a single break will be the most common type when the overall frequency of all types is low. In the most elementary form of this calculation, if the chance of a single break is small and is given as p, then the chance that a cell will suffer two breaks simultaneously is $p \times p$, or p^2, which must be even smaller. (For instance, if $p = 5$ percent or .05, then $p^2 = .05 \times .05 = .0025$, which is a quarter of 1 percent.) Three-break rearrangements would then be proportional to a frequency of p^3, a proportion which for our purposes is negligibly small and can be disregarded.

Most cells suffering unrestituted chromosome breaks will be eliminated within the first few mitoses after the breaks occur, because many of the breaks result in bridges or deficiencies that may themselves be lethal. This is a positive therapeutic benefit since it must prevent most damaged cells from becoming permanent features of the mature organism, or, in the germ line, from becoming gametes and giving rise to grossly defective zygotes. In any case, we may be sure that any rearrangements actually seen in a population of human cells represent a small fraction of the breaks that originally occurred in those cells.

Induced Changes in Human Cells

Chromosome Rearrangements. Cytological examinations of cells of irradiated individuals show a higher frequency of chromosome abnormalities than in control individuals. Studies on 456 survivors of the Japanese atomic blasts showed an increased frequency of translocations and pericentric inversions, with dicentrics and rings also present. Table 12-1 shows a comparison, made in 1966, of 77 heavily exposed (200 rads or more) persons from Hiroshima and Nagasaki who were all more than 3 km from the center of the blast, with 80 controls who were estimated to have received less than 1 rad of exposure. By far the most common type of rearrangement was the translocation; pericentric inversions also occurred with a frequency higher than in the controls. The appearance of dicentrics over a 20-year period may have stemmed from a high initial frequency and a slower rate of mitosis. Possibly some of these cells had been in a dormant phase and had not undergone mitosis at all. The fragments, on the other hand, appeared with the same frequency in control and exposed persons (9 vs. 12) and may have represented new spontaneous events, the radiation-induced acentric fragments having been lost as anticipated. Similarly, a study of the exposed Marshall Islanders showed chromosome aberrations in 23 of 43 exposed persons 10 years after the exposure. These included 13 translocations, 8 dicentrics, and 1 ring. Although there appeared to be a high frequency of acentrics (21 cases or 1 percent), there was an equally high frequency of acentrics in the controls. In another survey, 16 men given from 17 to 50 r-units were found to have a high frequency of rings and dicentrics in their cells. Individuals being treated with radioactive gold for rheumatoid arthritis showed 8.5 percent damaged cells, mostly with dicentric chromosomes, as compared to 0.48 percent for a control set.

Destruction of Cell Lines. Clearly, if enough primitive cells are destroyed in one of these ways, the body may lose an important function, such as its ability to fight disease if the leukocytes are affected. On the other hand, this suscepti-

Table 12-1 The numbers of chromosome aberrations in lymphocytes of heavily exposed Japanese, 21 years after the Hiroshima and Nagasaki bombings, compared to control populations. (From Bloom, Neriishi, Awa, Honda, and Archer, *Lancet*, **2:**7520, 1967.)

	Hiroshima		*Nagasaki*		*Total*	
	Control	*Exposed*	*Control*	*Exposed*	*Control*	*Exposed*
Number of survivors	58	55	22	80	80	77
Total cells examined	5,374	4,742	1,814	2,036	7,188	6,778
Percent of cells with 46 chromosomes	96.5	95.3	95.4	94.8	96.2	95.1
Number of survivors with one or more complex aberrations	10	35	3	12	13	47
Number of cells with						
a. Single chromatid gaps and breaks	145	168	66	44	211	212
b. Double chromatid gaps and breaks	44	29	31	10	75	39
c. Rings	0	2	0	1	0	3
d. Dicentrics	1	6	0	2	1	8
e. Fragments	8	7	1	5	9	12
f. Translocations	3	54	2	18	5	72
g. Pericentric inversions	3	7	0	4	3	11
h. Others	7	18	3	4	10	22

bility of cells in rapid mitosis is taken advantage of in the treatment of cancer. Here the cancerous cells are exposed to massive doses of radiation and, in many cases, can be completely eliminated. Of course, much depends on the skill of the radiologist, who must direct the radiation precisely at the site of the cancer, with minimum radiation to other organs and tissues that might be damaged. The massive doses applied to cancerous tissue, sometimes hundreds or even thousands of r-units, are not a reflection of any resistance of the tissue to radiation as much as they represent the necessity of killing all of the cancer cells, and the realization that the patient will almost certainly succumb to the disease if drastic measures are not taken.

CARCINOGENESIS. Ironically, another effect of radiation is its production of cancers or its *carcinogenic* effect. Adult cells that are completely differentiated and engaged in specialized tasks may be induced by radiation to revert to a primitive condition of rapid mitosis. How radiation induces cells to behave in this way is not known. Perhaps in some cases the radiation destroys an intracellular mechanism that suppresses or inhibits mitotic divisions. Perhaps in other cases it decreases the cell's normal resistance to invaders and allows a carcinogenic virus to enter the cell. More likely, different cancers may be produced by any one of several mechanisms, including the preceding. In any case, the probability that any one person will develop an induced cancer is very low even when that person has been given heavy radiation doses; otherwise, diagnostic or therapeutic exposures would not be acceptable.

The effect of radiation on the induction of cancer depends on the specific type of cancer and on the dose involved. Twenty-eight percent of painters of radium dials for watches, exposed over many years to the equivalent of 1,200 r-units, developed bone cancer, but those receiving less developed none. A study of 6,000 x-ray technicians from World War II has shown no higher incidence of cancer than in others in the population. On the other hand, 200 r of x-rays applied to the thyroid region of children has produced cancer of the thyroid in 3 percent of the recipients. In one survey of 19 children with thyroid nodules indicating an abnormal growth, 18 had been irradiated on the head, neck, or chest from 5 to 17 years previously. Of 67 exposed Marshall Islanders exposed after the H-bomb blast of 1955, 21 developed thyroid abnormalities, undoubtedly induced by the radioactive iodine 131 present in the fallout. Figure 12-11 shows one Marshallese boy at age 12 suffering from thyroid deficiency, along with his improvement after 3 years of treatment with thyroid hormone.

SPECIFIC RADIATION EFFECTS ON BLOOD TISSUE. The decrease in blood cells (Figure 12-5) seen after a minor whole-body radiation exposure comes from the destruction of the cells that have suffered chromosome breaks and are prevented from completing normal mitoses and differentiating into normal blood cells. As the mature leukocytes die off over a period of several weeks, and their normal replacements do not appear, the total count is diminished, giving rise to a decreased number of leukocytes, or *leukopenia*. It is at this time that danger from external invaders becomes most acute; death can result from infections that the immune system could take care of quite effectively under ordinary circumstances. However, with the passage of time, the unaffected leukoblasts, those cells giving rise to mature leukocytes, may proceed through sufficient normal mitoses to compensate for the deficiency, and the person's recovery, from this aspect at least, will be complete.

A long-range hazard comes from the induction of a blood cancer called *leukemia*. In this disease, the cells are induced to go through rapid uncontrolled mitosis, without subsequent differentiation, so that they cannot function normally. In contrast to leukopenia, in which the white cell count decreases, in leukemia it increases. Treatment follows the same pattern as in other malignancies; an attempt is made to kill the mitotically superactive cells. Leukemia differs from leukopenia also in its greater lethality and the greater delay between the exposure to an inducing agent such as radiation and its appearance in the human body. Observations of survivors from Hiroshima suggest that those who developed leukemia after receiving 500 r manifested the disease some 3 years later, whereas those who developed it after receiving 200 r did not manifest it until 5 years later.

In addition to the survivors of atomic bomb blasts, evidence that radiation produces leukemia and other types of cancer comes from several other sources. About 3,000 children in the 0–10 year age group in the United States die each year of all types of cancer, half of these from leukemia. Children have been exposed in utero during prenatal measurements of pregnant women's pelvic dimensions and other observations of the state of fetuses during the last trimester of

Figure 12-11 Marshallese boy at age 12 suffering from thyroid deficiency as a result of exposure to fallout from atomic tests; on the right is the same boy 3 years later, after medical treatment. (Courtesy of R. A. Conard, Brookhaven National Laboratory.)

the gestation period. When such exposed children have received a dose of 2 r-units, they are about twice as likely to die of a malignancy before their tenth birthday as other children; it is estimated that perhaps 400 deaths per year in this group can be attributed to prenatal x-ray examinations. The obstetrician making use of such an examination might counter with the argument that without the information provided by such examinations unexpected problems arising during childbirth might cost more than 400 lives per year. It is undoubtedly true, however, that this type of examination may not be required for all pregnancies in which it is made, and should not be used routinely. One estimate made several years ago was that such examinations were made five to ten times more often than necessary.

EMBRYONIC AND FETAL DAMAGE.   Another kind of somatic effect is that of interference with the normal development of an embryo or fetus. The factor giving rise to such malformations is referred to as a *teratogen*. If a woman is x-rayed during the first few months of pregnancy, there is some chance that vital tissue or organ development of the fetus will be stopped, or misdirected, so that the child may be born with severe defects. X-ray radiation is an example of a potent teratogen. In mice, doses of 200 r have produced abnormalities (defects of the vertebrae, ribs, skull, and limbs) in 100 percent of offspring, when the dose was applied during a critical period early in development. The possibility of teratogenic effects has led some states to prohibit x-raying pregnant females before the sixth month of pregnancy, unless immediate critical problems of the mother's health are involved.

Surveys of the Japanese survivors at Hiroshima and Nagasaki between the second and sixth months of intrauterine life at the time of the blast showed a reduced average head size and increased incidence of mental retardation. Rough calculations of the relationship between the degree of effect and the dose suggest that the frequency of mental retardation with reduced head size is 10 percent per 100 r. Thus the fetus in the uterus may be much more susceptible to the induction of mental defect during this period than to leukemia.

SHORTENING OF LIFE-SPAN.   Virtually every observation made on other mammals leads to the conclusion that exposure to high doses of radiation (100–300 r) decreases life-span. This is consistent with our views of the biologically destructive effects of radiation; we would expect general debilitation of a nonspecific sort to result from prolonged radiation doses accumulating to a sizable total dose (in the tens or hundreds of r). It is not possible to make an accurate guess of the extent of this effect since large-scale experiments on humans under controlled conditions are clearly out of the question. In mice it appears that the life-span shortening is of the order of 1 to 1.5 percent of total life-span for each 100r. Some guesses can be made from cases in which individuals have received appreciable doses over a period of time; it has been pointed out that radiologists of the past generation, for instance, had an average life expectancy 5 years shorter than that of their medical colleagues not involved with radiation. One suggestion is that there may be on the average a loss of 1 or 2 days of life for each r received per day beyond a 1.5 r/day dose.

Mutational Changes

INDUCED GENE MUTATION.   Prior to the discovery of the effects of radiation on gene mutation, its deleterious action on living systems was considered to be much the same as the deleterious action of chemical poisons. At very low concentrations, poisons will ordinarily have no effect, there being some minimum concentration (the *threshold*) necessary before any biological consequences can be noted. If this pharmacological concept were applicable to radiation damage, a person receiving a small dose day after day (as a careless x-ray technician might) would not have to be concerned because the threshold dose would not be reached on any one day. It was also assumed that if a person recovered from an exposure that had perceptible deleterious effects they would wear off over a period of time and he or she would be in the same condition as if the exposure had never been received.

With respect to gene mutation, neither of these assumptions is correct, and a quick review of the way in which radiation acts shows why. Radiation, when absorbed, manifests itself in ion production—roughly a million million in each cubic centimeter of tissue for each r-unit. Any one of these ions could be responsible for a genetic change, so that if we received a dose of only a millionth of an r-unit there would still be a million ions present to damage the molecules. Reducing the dose simply reduces the probability of a biological change. Consequently, any dose however small carries with it its proportional amount of mutational damage. This relationship, indicated in Figure 12-12, is commonly referred to as a *linear relationship* and implies the absence of any threshold below which genetic damage will not be produced.

Furthermore, genetic damage accumulates. If a mutation occurs during one year of early life, and a second mutation during another year, the individual will have the equivalent of the sum of the two events. The reason for this is that, even if many cell divisions should occur between the two events, the first mutation change will perpetuate itself by cell division just as the nonmutated cells do, and to the same extent, so that the frequency of a mutational change will not be altered by cell division. A second mutation whenever it occurs will simply add its proportional effect to whatever has happened before.

From this point of view, then, the total amount of radiation reaching the gonads during the individual's reproductive lifetime is important. Indeed, even doses applied to the mother during an individual's embryonic and fetal life should be added in as well. In the past, some pediatricians exposed infants to

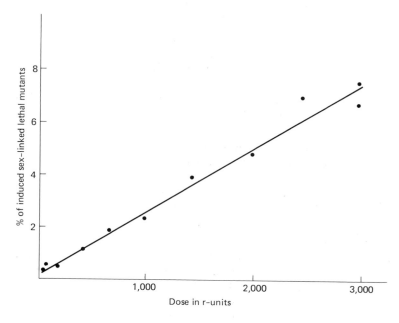

Figure 12-12 The linear relationship between radiation dosage and induced mutation rate, as suggested by numerous studies in *Drosophila*. Although it is not possible to make an accurate determination of the rate at very low doses, because the erratic spontaneous mutants are indistinguishable from the induced ones, there is every reason to believe that the line is straight, even at low doses, i.e., there is no threshold.

bimonthly fluoroscopic examinations for the first 2 years of life, as an aid in monitoring childhood development, with whole-body doses of more than 100 r-units. These should be viewed with some concern.

In summary, the *mutational* damage caused by radiation has two properties different from those of most other types of biological damage that are of utmost significance. The first of these is that the dose-response relationship is linear, without any threshold, and the second is that doses are strictly cumulative. These two points are of tremendous significance in estimating genetic damage to the population under present conditions. Because there is no threshold, each small dose makes its proportionate contribution to the pool of new mutations; thus a minute amount of radiation (e.g., from fallout) applied to the population of the world can result in significant numbers of mutations. Similarly, since these exposures, although very low-level, may be continuous throughout the life of an individual, the cumulative effect may add up to a significant exposure even in a single individual.

EFFECTS ON MUTATION RATE.   It is clearly not possible to design an experiment to measure the increase in mutation rate caused by radiation to humans. In fact, studies of the children of the survivors of Hiroshima and Nagasaki have failed to uncover any significant increase in new mutations. It can be stated further that there is little possibility of attributing any specific instance of mutation to radiation exposure since, regardless of the intensity of exposure, there can be no absolute assurance that any new mutation appearing (e.g., a rare dominant) resulted from that exposure and not from some spontaneous event.

The best information on a general increase in mutation rates after radiation comes from extensive studies on the house mouse. Here it is found that the mutation rate per locus per r-unit of chronic radiation is about 2.5 per 100 million (2.5×10^{-8}); we can assume that the mutation rate of humans is about the same. If we assume the number of loci in man to be about 20,000 (2×10^{4}) by multiplying these two numbers together ($2.5 \times 10^{-8} \times 2 \times 10^{4}$), we arrive at a total mutation rate for all loci, per r-unit, of 5 per 10,000 (5×10^{-4}). Since a zygote is formed from the combination of two gametes, each with the full number of loci, we would expect a total rate of 10×10^{-4} mutations per r, and if each person received on the average 10 r to the gonads during his lifetime up to the time the new progeny is conceived then we would expect 1 percent ($= 10 \times 10^{-4} \times 10$) of all zygotes to carry a new mutation induced by that radiation. It has been shown that acute doses are approximately four times more effective in inducing mutation than chronic doses; if the preceding calculations were based on acute rather than chronic doses, the net result would be four times as great, even though the total dose was the same in both cases.

THE DOUBLING DOSE.   In an equally simple fashion we can make an educated guess about the amount of radiation that it would take to produce the same number of mutations as are produced spontaneously. Spontaneous mutation rates have different values for different loci, but a good overall average can be taken as 1×10^{-6}. With an induced mutation rate of 2.5×10^{-8} per r, it would take 40 r to bring the number of induced mutations up to the number that occur spontaneously ($40 \times 2.5 \times 10^{-8} = 1.0 \times 10^{-6}$). For this reason, we can refer to 40 r as the *doubling dose*. Of course, such a calculation is based on figures which themselves are rough estimates, and therefore the values calculated here should not be taken too literally. Each investigator who attempts a calculation of this sort, based on rates that are themselves open to question, will come up with a final answer different from others. Most estimates generally agree, however, in placing the doubling dose somewhere between 20 and 200 r-units; some more recent ones suggest an even higher doubling dose—in the several hundreds of r-units.

Two aspects of induced mutation should be kept in mind. First, whereas beneficial mutations must form only a minute fraction of all spontaneous mutational changes, they must be even rarer among radiation-induced mutations, since radiation breaks chromosomes, with attendant damage to genes, loss of chromosome segments, and so on. On the other hand, in most cases these mutations will be recessive and so will be expressed only in the homozygous state, produced by chance some generations after the initial induction. When they do become homozygous, they will often be lethal to the embryo at an early stage so that society will not suffer from the deleterious effects of those particular mutations. Others, however, will be viable and will be responsible for gross physical defects, a reduction in physical vigor, premature death, or other "semilethal" conditions.

References

ARENA, V. 1971. *Ionizing Radiation and Life.* St. Louis: Mosby.

BLOOM, A. D. 1972. Induced chromosomal aberrations in man. In H. Harris and K. Hirschhorn, eds., *Advances in Human Genetics.* New York: Plenum.

EVANS, H. J., W. M. COURT BROWN, and A. S. MCLEAN, eds. 1967. *Human Radiation Cytogenetics.* Amsterdam: North-Holland.

MULLER, H. J. 1927. Artificial transmutation of the gene. *Science,* **66**:84–7.

MULLER, H. J. 1954. The manner of production of mutations by radiation. In A. Hollaender, ed., *Radiation Biology,* Vol. I. New York: McGraw-Hill.

RUSSELL, W. L. 1964. Evidence from mice concerning the nature of the mutation process. In S. J. Geerts, ed., *Genetics Today.* Oxford: Pergamon. (Proceedings of the Eleventh International Congress of Genetics.)

SANKARANARAYANAN, K. 1974. Recent advances in the assessment of genetic hazards of ionizing radiation. *Atomic Energy Rev.* **12**:47–74.

SHILLING, C. W., ed. 1964. *Atomic Energy Encyclopedia in the Life Sciences.* Philadelphia: Saunders.

STERN, C. 1973. *Principles of Human Genetics.* San Francisco: Freeman.

UPHOFF, D. E., and C. STERN. 1949. The genetic effects of low intensity irradiation. *Science,* **109**:609–10.

Questions

Useful terms: mutagen, quantum, gamma radiation, neutron, tritium, half-life, fission, ionization, electromagnetic spectrum.

1. Why did such a long period of time elapse between the first observations on the genetics of experimental organisms and the discovery that certain agents are mutagenic?

2. Ultraviolet light and x-rays are both powerful mutagenic agents when bacteria are exposed to them. Why is ultraviolet much less mutagenic in its action on mammals?

3. Describe the electromagnetic spectrum and indicate the point at which the energy per quantum becomes great enough to exert a mutagenic action.

4. An ordinary radio involves circuits with high voltages, sufficient to produce radiation, but it is not ordinarily considered to be dangerous. Can you explain why? Is this also the case for television sets?

5. How does gamma radiation differ from x-radiation?

6. What is the difference between the direct and the indirect action of high-energy radiation?

7. Is there any difference in the end result if a radiation dose is delivered all at once or if it is spread out over a longer period of time?

8. If a person is given a heavy dose of radiation, what are the physiological consequences resulting from the destruction of the cells in his body?

9. If a chromosome strand is broken by irradiation, what are the different possible end results?

10. What different kinds of chromosome rearrangements are produced by radiation? Make a drawing of each type.

11. Under what circumstances might the destruction of cells by radiation have a beneficial effect on the person?

12. What are the two important ways in which the mutagenic action of radiation is basically different from the action of poisonous chemicals?

13. What is a "doubling dose" and how is it used to indicate the magnitude of radiation effects?

14. Discuss the relevance of the following conditions to the concepts discussed in this chapter: anemia, leukopenia, leukemia.

13

Exposures of Humans to Radiation

Although the deleterious effects of high-energy radiation on living tissue have been recognized since the turn of the century, its dangers have not been appreciated by the population at large. For one thing, none of our sensory organs responds to high-energy radiation. Its damage may appear weeks or months after the exposure, blurring any obvious cause-and-effect relationship. Even when severe damage occurs, it may be easily misdiagnosed as coming from some other cause. Certain damage, such as that to our genetic material, can result from very low doses and accumulate over a long period of time.

For these reasons it is essential that external authority such as legislative bodies prescribe the conditions under which sources of radiation may be used, as well as maximum doses to which individuals may be exposed for other than medical reasons. Exposures to humans under any set of circumstances must be kept as low as possible.

During this discussion, the distinction between *acute exposures*, received within a short period of time (i.e., minutes or hours), and *chronic exposures*, in which the total dose is spread out over weeks, months, or years, must be kept in mind. High acute doses may cause immediate death, whereas chronic exposure to the same total amount of radiation will allow some degree of recovery of the organic systems and individual cells, during the period of time in which the chronic dose is applied. Therefore, although both types of exposure may eventually be lethal, the chronic type may have a delayed effect and may require a greater total dose to produce the same physiological damage as the acute.

215

A Brief History

Just 1 year after Roentgen identified x-rays in 1895, one of the first casualties from x-ray burn was an assistant of Thomas Edison, one Clarence Dally, who suffered from an inflamed and ulcerated scalp and a loss of hair after prolonged exposure to an unshielded x-ray tube. Some 8 years later he died, with cancerous growths on his arms and legs.

"Radiation sickness" was described in 1897, and the first successful medical use of x-rays came in 1900 when a small tumor on the nose of a patient was destroyed. In 1901 it was shown that heavy doses of x-rays were lethal to guinea pigs, even though no outward manifestations of damage appeared. A year later, skin cancer in humans was reported after x-ray exposure. The early workers with radium, A. H. Becquerel and Pierre Curie, deliberately exposed themselves to their experimental material to determine its effects. It has been estimated that more than a hundred persons died during the early decades of x-ray development. However, this is undoubtedly a gross underestimate, since many individuals may have died from exposures received years earlier, without the true cause of their death being known.

Doses of Radiation to Humans

LETHAL DOSES.   When a human receives an acute dose of many thousands of r-units to the entire body, death comes quickly (within hours or days), with obvious damage to the central nervous system as evidenced by convulsions, lack of muscle control, and erratic behavior. Exposures of close to 1,000 r will cause death in 30 to 60 days, and damage to the blood cells and the gastrointestinal tract will be evident.

Information on the effects of high radiation doses to the whole body comes from (1) industrial and laboratory accidents, (2) the accidental exposure to fallout during atomic bomb tests in the Pacific in 1954, (3) the bomb explosions in Hiroshima and Nagasaki, and (4) the medical exposure for cancer therapy or similar reasons. Data on the first two of these four categories are considered here in detail (Table 13-1) because they provide the most precise information available on high radiation doses to humans. In addition, they may emphasize the point that, to the present time at least, accidental death by exposure to radiation is an extremely rare event.

When the atom bombs were dropped on Hiroshima and Nagasaki, about 100,000 were killed and another 60,000 were injured. Most of the fatalities resulted from the explosions themselves—from the blast, heat, and shock. Only a small fraction of the immediate deaths can be attributed to the high radiation exposure.

In 1946, an accident involving an atomic experiment at Los Alamos resulted in the irradiation of one of the workers, Dr. Sloton, with an estimated total of 880 r. He survived only 9 days.

In 1958, a worker at Los Alamos accidentally transferred about 34 kg of material with a high concentration of the isotope ^{235}U from storage tanks to a waste tank that had dimensions such that the solution went "critical." There was a miniature explosion, of no great consequence in its blast effect but producing

Table 13-1 List of all accidental radiation exposures up to 1974. The type of accident is classified as A when a mass of isotope achieved critical dimensions, as B when the radiation was produced by an accelerator or other x-ray emitter, or as C when the source was a radioactive isotope. (Courtesy of C. Lushbaugh and V. Bond, Brookhaven National Laboratory.)

Date	Place	Exposed	Fatal	Serious Injury	Type
6/6/45	Los Alamos, New Mexico	3	0	0	A
8/8/45	Los Alamos, New Mexico	2	1	0	A
5/21/46	Los Alamos, New Mexico	8	1	1	A
5/14/48	Eniwetok	4	0	4	C
9/7/48	Los Alamos, New Mexico	1	0	1	C
6/2/52	Lemont, Illinois	4	0	0	A
?/53	U.S.S.R.	2	0	2	A
3/1/54	Bikini (Japanese)	23	1	22	C
3/1/54	Bikini (Marshallese)	267	0	110	C
6/19/58	Oak Ridge, Tennessee	5	0	5	A
10/15/58	Vinca, Yugoslavia	6	1	5	A
3/8/60	Lockport, New York	9	0	2	B
6/8/60	Moscow, U.S.S.R. (suicide)	1	1	0	B
11/9/60	U.S.S.R. (ingestion)	1	0	1	C
1/3/61	Idaho Falls, Idaho	3	3	0	A
?/61	Madison, Wisconsin	1	0	1	B
4/1/62- 7/22/62	Mexico City, Mexico	5	4	1	B
4/7/62	Richland, Washington	3	0	0	A
7/24/62	Mayaguez, Puerto Rico	7	0	0	B
7/24/64	Wood River Junction, Rhode Island	3	1	0	A
2/18/65	Chicago, Illinois	1	0	1	B
12/30/65	Mol, Belgium	1	0	1	A
5/3–4/68	La Plata, Argentina	18	0	1	B
8/28/68	Chicago, Illinois	1	1	0	C
2/4/71	Oak Ridge, Tennessee (UT-AEC)	1	0	1	B

a high level of radiation. It is estimated that the worker received approximately 12,000 r to his upper extremities. He became semiconscious and incoherent, and died 35 hr later with clearly manifested brain damage.

SUBLETHAL DOSES.   Of seven survivors of the Los Alamos accident in 1946, four received exposures estimated at between 100 and 400 r, and another three workers received from 30 to 55 r.

In 1958, six workers in a Yugoslav nuclear energy plant received an estimated 200 to 450 r of whole-body radiation. They were immediately rushed to Paris, where they received intensive medical treatment and all but one survived.

During the U.S. atomic tests of 1954 at Bikini Island in the South Pacific, some 300 persons received an "unpleasantly large" dose. Of 236 Marshall Islanders caught in the fallout when the wind shifted unexpectedly, 64 had radiation burns (Figure 13-1) and suffered from the usual symptoms of nausea, vomiting, and diarrhea, and in 39 cases loss of hair was reported. It is estimated that those receiving the highest doses got about 175 r. Twenty-three crew members of a Japanese fishing boat, *The Lucky Dragon*, were also caught in the local fallout and suffered from burns, fever, and swelling. One died several weeks later, but it is not clear whether this death is attributable to the exposure. Their total radiation dose was probably between 150 and 175 r.

From these unplanned "experiments," we learn that an exposure of several hundred r-units will cause nausea, vomiting, bleeding, and fatigue; other symptoms include a reduction in lymphocyte and platelet counts and a high body temperature. Subsequently the victim may experience loss of hair and temporary sterility. If the dose is closer to 1,000 r, the person will usually die, if it is closer to 100 r, he will probably survive. The probability of survival is, of course, in-

Figure 13-1 A young boy suffering from radiation burns received during the U.S. atomic tests at Bikini Island in the South Pacific, when an unexpected shift in wind direction exposed people on the Marshall Islands, about 500 miles away. (Courtesy of R. A. Conard, Brookhaven National Laboratory.)

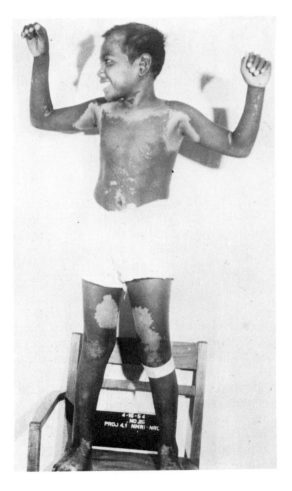

creased appreciably if he is provided with continuous medical monitoring so that no serious consequences can arise from a minor infection at a time when the white blood cell count is low.

THE MEAN LETHAL DOSE.   Since it is difficult to determine the dose at which a killing effect is just perceptible, or the dose that is completely lethal, investigators in the field have adopted the convention of a 50 percent lethal dose, or LD_{50}, defined as the dose that will kill about 50 percent of a sample of exposed animals in a specified period of time, such as thirty days after treatment ($LD_{50}/30$). Clearly, the most elementary ethical considerations forbid any attempt to derive an LD_{50} for humans, and in those cases where humans have accidentally received massive doses, there is considerable uncertainty surrounding the actual doses they received. It is therefore necessary to use figures from experimental animals, such as the mouse and rat, along with some fragmentary data on survival or death of humans after accidental exposures of the sort we have already discussed. From these it could appear that a whole-body dose of about 450 r would probably be lethal to half of a sample of humans. The *mean lethal doses* for other organisms are given in Table 13-2.

Doses from Medical Technology

THERAPEUTIC DOSES.   Treatment for cancer and other pathological conditions may demand the use of many hundreds and even thousands of r-units, directed specifically at the tissue involved; whole-body doses of this magnitude would

Organism	LD_{50} in r-units
Guinea pig	250 (175–400)
Pig	350–400
Dog	335
Goat	350
Monkey	about 600
Man	400–450
Mouse	550–665
Rat	665 (590–970)
Rabbit	750–825
Hamster	610
Chicken	600–800
Goldfish	670
Frog	700
Tortoise	1,500
Snail	8,000–20,000
Yeast	30,000
Amoeba	100,000
Bacillus mesentericus	150,000
Paramecium	300,000–350,000

Table 13-2 LD_{50} for x-ray irradiation of a wide variety of plant and animal species. (Adapted from C. W. Schilling, *Atomic Energy Encyclopedia in the Life Sciences*, Saunders, Philadelphia, 1964.)

more often than not be lethal. To minimize destruction of normal tissue, the patient (or the source of radiation) may be rotated during treatment so that the skin exposure will be spread out as much as possible. These massive doses are justifiable when the alternative (no treatment) is likely to lead to the early death of the patient. Such doses given to large numbers of people would be expected to have a significant effect on the overall production of new mutations in the population, but, as a rule, patients with malignancies tend to be older and generally have completed their families by the time of treatment. In any case, if a decision must be made between the present welfare of a medical patient and the hypothetical future difficulties that may be caused by new mutational events, it is invariably made in favor of the welfare of the patient.

DIAGNOSTIC DOSES.   Doses from several to about a hundred r-units are needed for fluoroscopy when the nature of the examination requires that the physician view an organ in action, as in gastrointestinal examination. The high dose is required in order to cause the sensitive fluorescent (zinc cadmium sulfide) screen to emit visible light as the x-rays strike it. Some typical figures for fluoroscopic examination are as follows: lower gastrointestinal examination, 27 r; upper gastrointestinal examination, 35 r; average x-ray abdominal examination during first half of pregnancy, 56 r. Another set of figures for the doses delivered during different types of medical examinations is found in Table 13-3. Children receiving x-ray examinations under carefully monitored conditions receive doses as in Table 13-4. It should be noted that the doses to patients are much less when x-ray film is exposed photographically than when a fluoroscopic screen is used, because of the much greater sensitivity of the film.

In discussions of radiation doses, the distinction should always be made between exposure to the whole body and exposure to the organ involved, in particular the gonads, since damage to the latter would be of concern to future generations. In the case of dental x-rays, for instance, when the dose to the jaw is about 100 mr for a single exposure, the dose to the gonads, in either male or female, is less than one ten-thousandth as much. On the other hand, an x-ray photograph of the pelvis which may require 220 mr will deliver more than a third as much to the gonads of either sex. From these concerns comes the concept of the *genetically significant dose*, which is defined as the total gonadal dose from a given source with possible genetic effects, divided by the total number in the population, to give an average dose per person. This calculation must take into account the number of persons irradiated, the type of examinations, and the exposures received, also the calculated accumulated dose to the gonads as well as the anticipated number of future children of the irradiated individuals. In 1964 the genetically significant dose in the United States from medical x-rays was 55 mr per person per year (compared to only 1 percent as much from dental x-rays).

Because of the greater appreciation by radiologists of the dangers of the dose to the reproductive organs, greater care is being taken now than in the past to shield the gonads from unnecessary exposure. On the other hand, many unnecessary x-ray photographs are being taken routinely when the probability of revealing any new medically significant condition is nearly zero. One of the reasons for this is that such photographs can be subsequently used as evidence

Table 13-3 Dose from x-ray examinations of adults showing the difference between the dose applied to the skin, and the gonadal dose, for different methods of examination. (Adapted from D. Antoku and W. J. Russell. *Radiology,* **101(3):**672–75, 1971.)

Region of Body Examined	Method Used	Dose at Skin (mr)	Gonadal Dose (mr)	
			Male	Female
Skull	Photograph	111	0.93	0.11
Thoracic spine	Photograph	249	0.19	0.26
Lumbar spine	Photograph	221	14.7	70.4
Pelvis	Photograph	219	83.0	79.0
Arm, forearm, wrist	Photograph	—	<0.01	<0.01
Upper gastrointestinal series	Fluoroscopy (conv.)*	1,220/min	12.4/min	81.1/min
	Fluoroscopy (image)*	328/min	0.348/min	12.8/min
Lower intestinal series	Fluoroscopy (conv.)	1,800/min	37/min	320/min
	Fluoroscopy (image)	400/min	5.7/min	77/min
Gallbladder series	Fluoroscopy (conv.)	1,200/min	0.2/min	1.8/min
	Fluoroscopy (image)	400/min	0.1/min	0.6/min
Barium enema	Fluoroscopy (conv.)	2,310/min	62.0/min	562/min
	Fluoroscopy (image)	480/min	11.0/min	140/min

*See text for explanation of difference between image and conventional fluoroscopy.

Table 13-4 Representative radiation dosages to children during x-ray examinations. (Adapted from S. Campbell, Pediatric Low Dosage Medical Radiography, 1969, Indiana State Board of Health, Indianapolis.)

Age	Number of Films	Dose at Skin	Total Female Gonadal Dose	Total Male Gonadal Dose
Upper Gastrointestinal Examinations				
3 mo	7	19,620 mr	4,592 mr	184 mr
13 yr	14	55,888 mr	223 mr	21 mr
Lower Gastrointestinal Examinations				
11 yr	8	9,420 mr	3,224 mr	111 mr
Heart Fluoroscopy Examinations				
4 mo	4	1,936 mr	450 mr	18 mr
12 yr	4	3,493 mr	23 mr	2 mr

Figure 13-2 Simple measures drastically change the exposure of the patient shown in (*A*). Collimation of the beam and repositioning of the patient (*B*) reduce the gonadal dose to about one fifth of the value in (*A*). (Courtesy of S. Campbell, *Pediatric Low Dosage Medical Radiography*, 1969, and Indiana State Board of Health.)

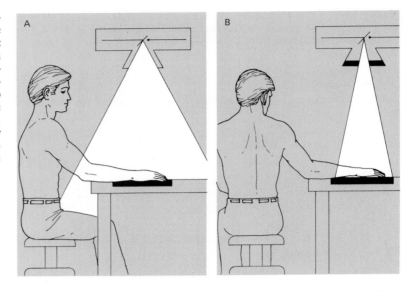

in any legal action (i.e., a malpractice suit brought by a patient against the doctor) to show that the physician in fact provided adequate medical care. However, it might be argued that the opposite is true, that excessive unnecessary x-raying of patients is equally reprehensible.

Several simple procedures can reduce the dose appreciably. Figure 13-2 shows two of these. The beam is cut down in size, or "*collimated*," so that only as large a beam is used as is needed to take the picture of the relevant organ. At the same time the individual is so positioned that his gonads are not exposed to the direct diagnostic dose. Another simple precaution emphasized these days by radiologists is the perfection of techniques to reduce the percentage of retakes necessary. A simple aluminum filter will remove a good part of the "softer" and more readily absorbed radiation, leaving the "hard" fraction for delivery to the organ being irradiated. Such simple precautions may reduce the exposure to the gonads by a sizable amount; the genetically significant dose from medical x-rays dropped from 55 mr in 1964 to 36 mr in 1970 in spite of an increase in examinations from 173 million to 210 million during that time. The decrease is attributed primarily to improved collimation.

Technical improvements continue to decrease the dose necessary by sizable fractions. The use of an "image intensifier" amplifies the image that would be produced by a light x-ray exposure and so reduces the necessary dose from one half to one fifth of that needed by ordinary fluoroscopy. Not only may that image, intensified electronically, be viewed immediately on a modified TV apparatus, but also a record can be kept of the examination on video tape.

Radioactive isotopes are used quite frequently in the diagnosis and treatment of certain diseases. In the United States the estimated average gonadal dose from the use of radioisotopes applied from outside the body is calculated to be 12 mr/year; the contribution from internal use is much less, not amounting to as much as 1 percent of the total of all medical exposure. Iodine 131 is a standard treatment for cancer of the thyroid, highly effective because of the accumulation

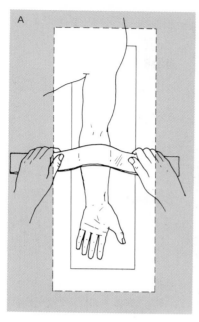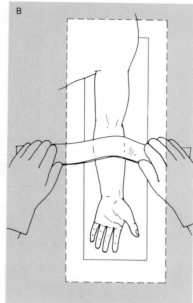

Figure 13-3 Immobilization of the hands of an infant during an x-ray examination. The simple procedure of wearing lead gloves on the right reduces the dose to the technician's hands to one five-hundredth of the dose otherwise received. (Courtesy of S. Campbell, *Pediatric Low Dosage Medical Radiography*, 1969, and Indiana State Board of Health.)

of iodine in that gland. In one typical case in which a female, age 16, was treated on two occasions, it could be shown that the total gonadal dose was probably between 80 and 160 r. At the last report, she was apparently cured.

DOSES TO TECHNICIANS. One large group of persons has traditionally been regularly exposed to unusually high doses—the radiologists and technicians who constantly live with high-energy radiation in their work. Although regulations provide for mandatory precautions that would prevent excessive over-exposure, there is no doubt that in many cases these rules are not observed, partly because of the lack of obvious immediate damage caused by the x-rays, discussed earlier, and partly because of the slight inconvenience that such precautions necessitate. Studies made in the 1950's showed at that time that x-ray technicians received an average of 100 to 300 mr/wk, that they had a frequency of leukemia about nine times greater than that of the rest of the population, and that the life-span of radiologists was reduced by about 5 years. Approximately 20 percent of radiologists checked at that time were receiving more than the "acceptable" dose of 300 mr (or .3 r) per week. More recent figures show that medical x-ray workers now receive a *mean annual dose* of 320 mr, dental x-ray workers 125 mr, radioisotope workers 262 mr, and radium workers 540 mr. Clearly, there is progressively less exposure of medical personnel.

Sometimes very simple procedures can be unusually effective in reducing exposures. Figure 13-3 illustrates the two ways of immobilizing an arm during an x-ray examination. The technician in *A* will receive 5.5 mr to the hands; in *B* only one five-hundredth as much (.01 mr) will be received because the technician is wearing lead gloves. Figure 13-4 illustrates two procedures, bad and good. In the first case, the unshielded operator receives .3 mr (even though she is leaning away from the machine) while the patient receives 890 mr to the gonads.

Figure 13-4 Excessive exposure of both technician and patient during x-ray photography (*A*). The collimation of the beam and shielding of both technician and patient (*B*) reduce the gonadal radiation exposures of both to a few percent of the previous value. (Courtesy of S. Campbell, *Pediatric Low Dosage Medical Radiography*, 1969, and Indiana State Board of Health.)

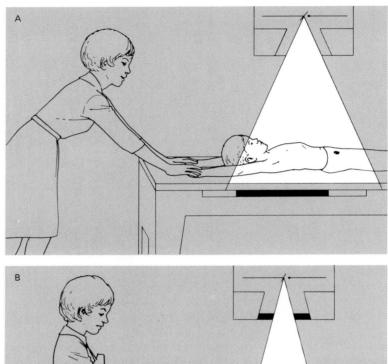

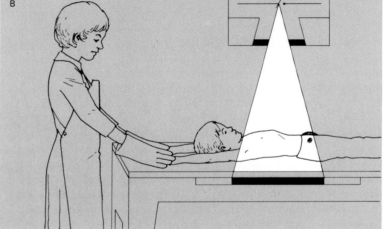

When both are shielded, as in *B*, the operator receives less than .01 mr and the patient's gonadal dose is 3.4 mr.

When children are being examined, the parent may be asked to hold the child still on the reasonable assumption that it may be better for the parent to receive a single small dose than for the nurse or technician to receive the same small dose to the hands day after day. At one time, dentists would hold film in position in the patient's mouth during dental x-raying; it is now becoming common practice for them to ask the patient to hold the film in place himself.

The service performed for the benefit of mankind by the medical profession in applying x-ray technology for diagnostic and therapeutic purposes cannot be questioned. If, in the course of their work, radiologists and their technicians should be subject to small but inevitable doses of radiation slightly above background level, this is unfortunate, but can be considered a sacrifice that they are willing to make in return for an obvious benefit to society. Although some young women are deterred from entering this vitally important diagnostic area because of the fear of genetic or other damage to their future progeny from the inevitable exposures incurred during such work, it is to be hoped that the figures showing the relative doses from various situations, along with the very low probability of their producing any biological effect, will serve to reassure prospective technicians, provided that care is taken to adopt standard safety precautions.

Miscellaneous Exposures

Occasionally persons will receive small exposures from miscellaneous sources. The radium dial on a watch will give .7 r/yr to the wrist and 10 mr/yr to the whole body. Pilots of large commercial aircraft, with their panels of radium-illuminated dials, receive radiation at the rate of 1,200 mr/yr (this value would, of course, have to be adjusted to that proportion of a year during which they are at the controls); and for each transcontinental round trip that a pilot makes he and his passengers will accumulate an additional 2 to 4 mr, because of the higher cosmic ray exposure at high altitudes. A person spending a full year in Grand Central Station would receive an additional 500 mr from the granite in the building.

In other cases, exposures of high doses seem to be completely unjustified. Not too many years ago shoe stores used fluoroscopes to reveal the bones of the foot within a shoe. It is questionable whether any real advantage in shoe-fitting resulted; the greatest effectiveness of the machine may have been a promotional one. In any case, the dose delivered to the foot was about 50 to 150 r/min, and 1 to 10 r/min to the whole body. In retrospect, this appears to have been a most incredible misuse of high-energy radiation. The machines were poorly shielded, without knowledgeable operating personnel in attendance (but exposing clerks, salesmen, and bystanders in large numbers), with no time restrictions on exposure and readily available to small children, for whom the machines may have served as a useful entertainment and baby-sitting device. These machines are now generally prohibited by state laws.

Before 1958, some 200,000 children in the United States had been irradiated for ringworm of the scalp, some more than once. After the radiation many experienced severe headache, nausea, and vomiting, an understandable reaction since they received 450 to 850 r to the scalp (Figure 13-5), with an average brain dose of 98 to 140 r. More recently, about 20,000 children were similarly treated in Israel.

It was discovered during the 1930's that radiation could overcome problems of sterility when it was caused by a failure of ovulation in the female. The dose necessary to induce ovulation is about 70 r-units. Since the rays must be focused specifically on the female germ cells to accomplish the purpose, gynecologists were severely criticized for this practice, and it can be assumed to have been stopped. For one thing, a greater sensitivity on the part of the medical profession to the genetic hazards of radiation has led to greater restraint in its use in exposing the germ cells. The possibility of an occasional defective infant attributable to the radiation (although, obviously, cause and effect in such a case could never be proved) definitely makes the procedure seem less attractive. In any case, the

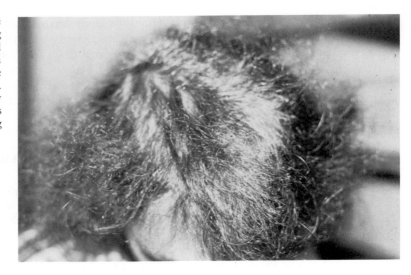

Figure 13-5 Premature graying of hair in a young man who had been treated during childhood with a massive x-ray dose to the scalp to cure ringworm. The radiation apparently incapacitated the cells responsible for producing normal hair pigment.

availability of "fertility" hormones now makes it possible for gynecologists to solve, quite easily, many problems of fertility resulting from failure of ovulation. Unfortunately, in many such cases, multiple ovulation results, and the consequences of the ensuing multiple pregnancy and premature birth may be fatal to the newborn.

Doses from Television Receivers

It was explained earlier how television picture tubes as well as other high-voltage tubes in the set can generate x-rays. The energy of the radiation depends on the voltage applied across the elements of the tube; Figure 13-6 shows the x-ray intensity, in millionths of an r-unit per hour, produced by a color television set, operating at unusually high voltages supplied by a circuit specially designed for that purpose. Black and white sets operate at voltages well below those shown on the graph; their x-ray production can be considered to be zero. In the late 1960's, measurements showed that the radiation produced by color TV sets was quite variable, but in some instances excessively high, ranging up to 12 mr/hr, and that 20 percent (in one sample) were emitting excessive doses of radiation. In some instances, poor shielding of the internal components was responsible; in others, excessive voltages. Some repairmen increase internal voltages in sets to brighten and sharpen the picture and unknowingly convert a safe set into a weak x-ray emitter. In other cases, fluctuations in the line voltage coming into the set have been multiplied by the power transformer of the set to excessively high voltages on the set components. Many brands have been recalled for alterations to reduce the external radiation.

This dose is reduced considerably simply by maximizing the distance from the set. Figure 13-7 shows the relative reduction in radiation level as the distance increases. In theory, radiation from a point source decreases in intensity as the

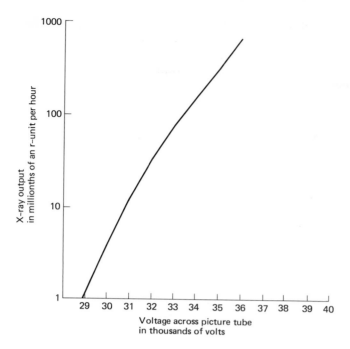

Figure 13-6 X-ray doses in millionths of an r-unit per hour after atypically high voltages have been applied to the picture tube of a color television set. Note that a typical color television set would not be capable of delivering these high voltages and that improvements in shielding, circuit design, and component construction further decrease the potential hazards of more recent color TV. (Courtesy of S. P. Wang, S. Sario, and H. Hersch, Zenith Radio Corporation.)

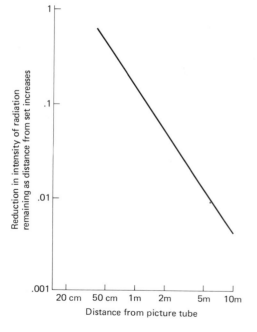

Figure 13-7 The relative reduction in radiation level from a radiating color television set as distance from the set increases. Note that at about 3 ft (or 1 m) the intensity is only 10 percent of the level at the set and that with greater distances the level drops off exponentially. (Courtesy C. P. Braestrup, Conference on Detection and Measurement of X-radiation from Color Television Receivers, 1968.)

square of the distance from that source. For distances near the set, the reduction is not as great as theory would predict since the emitting element is not a point, but at about 1 m the square law becomes effective. In any case, the simplest way to reduce one's potential exposure (other than shutting off the equipment) is to put as much distance as possible between the person and the set.

By taking into account the number of hours that the average person watches television in the United States (3 hr per day for children under 14 and 2.6 hrs for older persons), the average distance from the television receiver (5 ft 5 inches for children and 8 ft 2 inches for older persons), the absorption by body tissue, and so on, it is possible to calculate the dose to various organs (Table 13-5) if the set is emitting at the maximum rate of .5 mr/hr. In fact, from studies in Washington, D.C., it would appear that the average set emits less than one-tenth of this amount, about .043 mr/hr, so that the values in Table 13-5 should be divided by 10. Taking into account that many persons in the United States do not watch color TV (70 percent) or even do not watch TV at all (6 percent), the genetically significant dose to the U.S. population is approximately 0.5 mr/yr, or about one half of 1 percent of the average dose from natural background radiation.

All sets manufactured after June 1, 1970, have been designed in such a way that, even with maximum voltages, a new *maximum permissible dose* would not be exceeded. This maximum permissible dose amounts to .5 mr/hr at a distance of 5 cm in any direction from the set.

Background Sources of Radiation

As our discussion progresses from the heavy doses to the body to the lighter ones, we eventually reach the low, constant dosages originating in the natural environment. This is radiation from which we cannot escape, coming from outer space as cosmic rays, from the buildings we live in, from the ground we walk on, and from the elements making up our own bodies, as radioactive isotopes. Because of the small doses involved, the unit will invariably be thousandths of an r (mr) per year.

COSMIC RAYS.   Cosmic "rays" consist of high-velocity charged particles, originating from our sun and outer space. As they enter the earth's atmosphere, they collide with atoms and yield a variety of additional high-velocity particles,

Table 13-5   Estimated average dose from the front face of color television tubes radiating at the rate of .5 mr/hr. Note that most sets radiate at a rate less than a tenth as great. (From R. H. Neill, H. D. Youmans, and J. L. Wyatt, *Radiological Health Data and Reports*, **12[1]**:3, 1971.)

Organ	Children Less Than 15 Years Old (mrem/yr)	People 15 Years of Age and Older (mrem/yr)
Skin	42	18
Thyroid	30	11
Male gonads	18	8
Female gonads	5	2
Lens of the eye	34	15

as well as gamma rays, and, cascading toward the earth, they produce additional charged particles and radiation. At the same time, the particles may be slowed and part of the radiation may be absorbed during the passage through the atmosphere. For this reason, the intensity of the penetrating radiation is strongly dependent on altitude. Cosmic rays are about twice as intense at Denver, slightly more than 5,000 feet in altitude, as they are at sea level. Also, the paths of the charged particles are curved by the earth's magnetic field and so are not uniformly distributed around the earth but differ depending on the latitude. Table 13-6 shows the annual cosmic ray doses for ten American cities, along with the excess dose being received per year because of the altitude and latitude of each.

RADIATION FROM ROCKS AND SOIL. The elements contributing most to the natural radioactivity of the ground are the products of disintegration of uranium 238 and thorium 232, which give an average figure of 50 mr/yr. This average figure, however, is subject to wide variations because of the great variation in the concentration of these elements from one place to another; the range around this mean value extends from 25 to 75 mr/yr.

In some areas, the soil is unusually rich in radioactive elements. About 50,000 people live over radioactive deposits in two states of Brazil—they receive from 500 to 1,000 mr/yr. In a village in another state of Brazil, a few hundred people receive as much as 12,000 mr/yr (12 r/yr). In an area of India occupied by about 10,000 people, the exposure from radioactive soil ranges from 200 to 2,600 mr/yr. Some 1,800 residents of Grand Junction, Colorado, live in houses built on "tailings," rock and soil removed from uranium mines. They may receive as much as 107 mr/yr additional radiation and the total number of persons in the state of Colorado similarly exposed may be 10 to 20 times as great as the number of Grand Junction residents given above. Uranium miners are an unusually highly exposed occupational group—they may receive doses as high as 6,600 mr/yr to the skeleton and 500 mr/yr to the gonads.

Table 13-6 Approximate cosmic ray doses for a number of American cities. (From V. L. Sailor, Population Exposure to Radiation: Natural and Man-Made, Sixth Berkeley Symposium, 1971, Berkeley.)

Location	Elevation (ft)	Dose Rate (mr/yr)	Difference from Sea Level (mr/yr)
Denver	5,280	67	39
Albuquerque	4,958	60	32
Salt Lake City	4,260	54	26
Oklahoma City	1,207	34	6
Phoenix	1,090	33	5
Atlanta	1,050	33	5
Kansas City	750	32	4
Chicago	579	31	3
San Francisco	65	28.5	.5
Baltimore	20	28	0

RADIATION FROM BUILDINGS. A person inside a building may receive less radiation of cosmic or soil origin, but this may be matched or even greatly exceeded by radiation coming from the building itself. Thus, while an ordinary wooden house may reduce exposure to about 70 percent of the level outside, brick, stone, or concrete buildings may contribute 50 to 100 mr/yr more than the exposures received out-of-doors in the same area. Table 13-7 shows the radiation exposure inside various buildings. Although no correction has been made for the cosmic rays present at the same time, it is clear that the doses are far above the expectation for cosmic rays alone.

INTERNAL EXPOSURE. Since all the air, water, and food that enters the human body contains naturally occurring radioactive elements, the body itself accumulates a small amount of radioactivity. Most of this comes from potassium 40 (20 mr/yr); some from carbon 14, radium 222, and so on (3 mr/yr). This total is not very great; from another point of view, however, it may appear more significant since it is responsible for about 500,000 atomic disintegrations per minute within the human body.

DOSE FROM FALLOUT. At one time, the ever-increasing increment to natural radiation from the man-made product of atomic explosions seemed to present a threat of unlimited dimensions. The agreement of the major powers to halt aboveground nuclear testing has largely removed this danger, although the continued testing by certain countries (China, France, India) does add a small amount of additional radiation. In spite of these small additions, the *radiation from fallout* is presently decreasing. Among the major components of fallout listed in Table 13-8, iodine 131, niobium 95, and zirconium 95, with their very short half-lives of a few months or less, clearly must be less important than the others. The exposure to humans depends on the degree to which the isotopes are retained in the stratosphere before deposition on the ground as well as their

Table 13-7 Gamma exposure rates inside various buildings, cosmic ray component included. (From V. L. Sailor, Population Exposure to Radiation: Natural and Man-Made, Sixth Berkeley Symposium, Berkeley, 1971.)

Location	Structure	Exposure Rate (mr/yr) Typical	Extreme
East Germany	Brick	106	1200
New York City	Brick	79–118	
Grand Central Station	New York granite		500
United States	Wood	60	
	Concrete	130	
Aberdeen, Scotland	Granite	81	110
Cornwall, England	Granite	145	
Sweden	Wood	48–57	
	Brick	99–112	
	Concrete	158–202	

Radioactive Isotope	Half-Life
Zirconium 95	30 days
Niobium 95	35 days
Cesium 137	30 yr
Strontium 90·	28 yr
Carbon 14	5,730 yr
Iodine 131	8 days

Table 13-8 The most common radioactive isotopes found in fallout, along with their respective half-lives. (M. S. Terpilak, C. L. Weaver, and S. Wieder, *Radiological Health Data and Reports*, 1971.)

fate afterward. Strontium, in particular, poses a special threat because of its involvement in the food chain, where it competes with calcium for incorporation into living tissue (see Chapter 14).

The cesium 137 and strontium 90 produced by nuclear tests in the 1950's and 1960's will have spent out one full half-life by the year 2000 and so will be only about one-half as effective in their biological effects starting at the turn of the next century. On the other hand, carbon 14, with its half-life of about 6,000 years, will continue to expose humans to a total of 13 to 16 mr every 30 years for an extremely long period of time. One way of viewing this problem is to

Table 13-9 Fifty-year dose commitments in mr from fallout in the northern hemisphere from atomic tests carried out before 1968, classified according to the main isotopes involved and the tissues affected. Except for the dose from ^{14}C, which is uniform throughout the world, the doses in the southern hemisphere are less than a quarter of those shown. The relative magnitude of the internal (1) and external (2) radiation from fallout can be assessed by comparing the doses from fallout with those from natural background radiation (3). (L. D. Hamilton, Am. Phys. Soc., 1971.)

Source	Whole Body	Gonads	Cells Lining Bone Surfaces	Bone Marrow
1. Internal radiation from fallout:				
Cesium 137	21	21	21	21
Carbon 14	13	13	16	13
Strontium 90	—	—	130	64
Strontium 89	—	—	<1	<1
2. External (total) radiation from fallout:	50–100			
Short-lived	—	36	36	36
Cesium 137	—	36	36	36
3. Natural background radiation:				
External	4,000	4,000	4,000	4,000
Internal	1,050	950	1,200	1,000

estimate the total dose from fallout that a person will have received from the start of extensive nuclear testing to the year 2000, at which time, barring unhappy incidents, the dose from fallout will reach a very low level. This dose commitment to the year 2000 for various tissues is given in Table 13-9; as an average it can be taken to be 200 mr for the U.S. population, or the equivalent of 2 years of natural background radiation.

References

ANTOKU, D., and W. J. RUSSELL. 1971. Dose to the active bone marrow, gonads, and skin from roentgenography and fluoroscopy. *Radiology,* **101:**669–78.

BLOOM, A. D., S. NERIISHI, and P. G. ARCHER. 1968. Cytogenetics of the in-utero exposed of Hiroshima and Nagasaki. *Lancet,* **2:**7558.

BOND, V. P., T. M. FLIEDNER, and J. O. ARCHAMBEAU. 1965. *Mammalian Radiation Lethality: A Disturbance in Cellular Kinetics.* New York: Academic Press.

CONARD, R. A., B. M. DOBYNS, and W. W. SUTOW. 1970. Thyroid neoplasia as late effect of exposure to radioactive iodine in fallout. *JAMA,* **214:**316–24.

HERSKOWITZ, I. H. 1957. Damage to offspring of irradiated women. *Prog. Gynecol.* **3:**374–84.

LOUTIT, J. F. 1962. *Irradiation of Mice and Man.* Chicago: University of Chicago Press.

MULLER, H. J. 1973. *Man's Future Birthright.* Albany: State University of New York Press.

NEEL, J. V. 1963. *Changing Perspectives on the Genetic Effects of Radiation.* Springfield, Ill.: Thomas.

RUSSELL, L. B. 1957. Genetic considerations in the practice of ovarian irradiation for the treatment of sterility. In *The Year Book of Obstetrics and Gynecology,* pp. 537–42. Chicago: Year Book Publishers.

SAILOR, V. L. 1971. Population exposure to radiation: natural and man-made. Pollution Conference of the Biology-Health Section of the Sixth Berkeley Symposium.

SAVIC, S. D. 1970. Results of a study of color television set assembly and repair workers x-ray safety. *Radiological Health Data and Reports,* **11:**519–25.

SOBELS, F. H. 1969. Estimation of the genetic risk resulting from the treatment of women with [131]iodine. *Strahlentherapie,* **138:**172–77.

TERPILAK, M. S., C. L. WEAVER, and S. WIEDER. 1971. Dose assessment of ionizing radiation exposure to the population. *Radiological Health Data and Reports,* **12:**171–88.

Questions

Useful terms: acute exposures, chronic exposures, mean lethal dose, therapeutic dose, diagnostic dose, fluoroscopy, genetically significant dose, collimation, mean annual dose, cosmic ray, maximum permissible dose, fallout dose.

1. Why is it important to make a distinction between acute and chronic exposures?
2. Under what circumstances have humans received lethal doses of radiation?
3. Why is it difficult to determine with reasonable accuracy the value of the mean lethal dose for humans?
4. Do different organisms agree in having roughly the same sensitivity to radiation?
5. Why is the genetically significant dose particularly important from the standpoint of the overall population?
6. Would you say that the medical profession uses x-rays for diagnosis more or less often than would be desirable?
7. In the past the most regularly exposed group of individuals has been involved in medical treatment and diagnosis. Can you explain why this happened and whether this is still true?
8. List a number of uses of x-rays that seem, in retrospect, to have been completely unwarranted.
9. What hazards do color television sets present, particularly to young children?
10. What is the origin of the various types of natural radiation to which every individual is exposed? How much radiation comes from each of these sources?
11. What is the annual dose received by people in the United States from fallout produced by atomic explosions in earlier years? How does this amount compare with the amount from natural radiation?

14

The Nuclear Energy Controversy

In recent years there has been an extensive debate between those individuals favoring the development of nuclear power sources and those actively opposed. The majority of people on both sides of the issue are motivated by a desire for at least a continuation and perhaps even an improvement of their present mode of life, and its extension to others they consider to be less fortunate. However, most do not have the scientific background for independent judgment in this area and must rely on secondhand information from experts in the field. This information gap has accentuated the controversy, and in some places the issue has been placed before the voters for their decision.

The important issues are primarily biological and genetic ones. How much damage will be wrought upon the existing world's population if widespread radiation induces cancers, particularly leukemia, or other ill health? To what extent do we jeopardize future generations by exposing germ cells to mutational change that will be revealed only generations hence? These are important questions for which we now do have some quantitative estimates; in this chapter we shall consider in some detail the probable and known human exposures from different phases of nuclear power production, and the biological consequences of such exposures.

Our discussion will be limited to the biological problems posed by nuclear power generation. Other aspects, such as the question of engineering problems, and aesthetic considerations, important as they may be in their own right, are properly not part of this treatment, nor is any discussion of the argument forwarded by some that the solution to the world's energy problems may rest in a greatly decreased power consumption rate per capita, and a decrease, or at least a stabilization, of the world's population size. Also, the feasibility of other energy sources—wind, solar power, nuclear fusion—to supplement our present reliance on fossil fuels is a matter for discussion elsewhere.

The Chain Reaction

At the outset, the nightmare of atomic holocaust with its characteristic mushroom cloud originating in an explosion from a nuclear power plant must be disposed of. To understand the difference between nuclear energy in a reactor and that in atomic explosions, it is necessary to consider the nature of the *chain reaction* itself (Figure 14-1*A*). The basic material of an atomic chain reaction is an unstable isotope of uranium, plutonium, or a comparable element. A free neutron, colliding with the nucleus of such an atom, will cause it to disintegrate into smaller pieces called fission fragments, at the same time emitting several neutrons and releasing energy. Each of these neutrons can now cause the disintegration of several more isotope atoms, with the release of still more neutrons and more energy.

If the neutrons are used effectively, with few escaping to the outside, an uncontrolled fission reaction may occur. In order for this to happen, a certain minimum amount of fissionable material called the *critical mass* must be used;

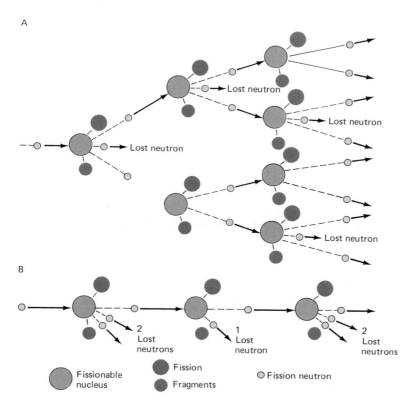

Figure 14-1 *A*. A series of nuclear fissions in which one neutron causes a fission reaction that, in turn, yields several uncontrolled neutrons. The reaction cascades almost instantaneously into an explosion, *B*. A controlled fission reaction in which an average of only one of the released neutrons takes part in a subsequent fission.

A

Lost neutron

Lost neutron

Lost neutron

Lost neutron

Lost neutron

B

2 Lost neutrons

1 Lost neutron

2 Lost neutrons

Fissionable nucleus

Fission Fragments

Fission neutron

the atomic bombs of 1945 contained about 5 kg, or 10 lb. This mass must conform to the proper geometrical configuration: a sphere will allow fewer neutrons to escape than any other shape because it has the minimum surface area for a given volume. Finally, the right configuration and mass must be achieved by bringing together, very rapidly, smaller noncritical masses in other shapes. For example, a conventional explosion that forced together two hemispheres of uranium 235 within a few millionths of a second would bring about an uncontrolled fission reaction. To accomplish such atomic explosions requires considerable engineering design. Newspaper reports that a probably functional design for an A-bomb has been worked out by high school or college students with only published documents to aid them are more a commentary on the precocity and sophistication of the younger generation than any reflection on the simplicity of A-bomb construction.

The Nuclear Reactor

If, on the average, one neutron triggers an event that produces only one active neutron, any others being lost (Figure 14-1*B*), then the fission reaction will be self-sustaining but *not uncontrolled.* This control is achieved in nuclear reactors by choosing a suitable geometric design for the isotope mass, the proper concentration of fissionable elements with other materials, and a suitable damper material such as graphite to moderate the reaction.

In the reactor core, an isotope with a long half-life is induced to undergo fission by the capture of a neutron. Thus ^{235}U, with a half-life of 10^{17} years, will undergo fission immediately (within 10^{-21} sec) upon the introduction of a free neutron to its nucleus. When the uranium atom undergoes fission, it transmutes to two elements of lesser atomic weight and releases, on the average, between two and three neutrons. As an example:

Uranium 235 + 1 neutron $\longrightarrow$ barium 139 + krypton 95 + 2 neutrons

These fission products separate with high velocities and about 90 percent of the energy released is in the form of kinetic energy, or heat. The amount of energy released from such disintegration is so great that 1 lb of ^{235}U undergoing fission will produce enough heat to bring 100 million gallons of water from room temperature to the boiling point. Water heated in this fashion can be used to run steam turbines to produce electricity. Of the remaining energy released during fission, another 6 percent is in the form of x-rays, either immediately or from the fission fragments, which, if unstable, may disintegrate later.

In a nuclear reactor, the essential requirements for a bomb—purity of fissionable material and a precise timing device—are missing. In addition, the neutrons present in reactors would prematurely ignite any small amounts of fissionable material and prevent a critical mass from being formed. The serious reactor accident that might happen is the "meltdown," caused by runaway reaction releasing excessive heat and melting the fuel elements. Most reactors have an airtight hemispherical dome that holds all the nuclear components; the dome is engineered to contain the radioactive gases released in case of a meltdown (Figure 14-2). Whether, in fact, the dome is capable of doing this in all accidents is unknown, since no such accidents have occurred and, for obvious reasons, an actual test has not been made.

Figure 14-2 The San Onofre Nuclear Generating Station, located near San Clemente, California. The hemispherical structure is the containment shell. (Courtesy of Southern California Edison Company.)

HAZARDS TO EMPLOYEES.   Those persons receiving the largest doses of radiation from reactors are obviously the personnel employed within the reactor housing itself, and within the confines of the installation. The doses they receive are closely observed and regulated, and efforts are made to minimize exposure. Table 14-1 shows the amount of radiation received by employees of six nuclear power plants. In the first column is the total dose received by all employees. The last column gives the average dose per employee. The unit *man-r* is introduced in this table; it is calculated by summing the doses received by all individuals. Thus, if 100 persons receive 1 r and 50 receive 3 r, the total is $(100 \times 1) \times (50 \times 3)$, or 250 man-r. This unit is useful in calculating the total exposure to the population. If one reactor has a high emission rate but is located in a sparsely inhabited area so that 10,000 persons receive 0.1 mr, totaling 1 man-r ($10^4 \times 10^{-4}$), the mutational effect may be less than one with a lower rate, 0.01, but exposing 1,000,000 persons for a total of $10^6 \times 10^{-5}$ or 10 man-r. It is also worth noting that the total exposure of all employees in each of these plants, averaging about several hundred r-units, would not be considered excessive if given by a medical doctor to one patient during therapeutic treatment.

237

Table 14-1 Average employee exposures at six nuclear plants during the year 1972. The six selected out of a total of 15 plants include the three with highest exposures and the three with lowest. (Courtesy of M. I. Goldman, Environmental Safeguards Division, NUS Corporation.)

Name of Plant	Total Plant Dose Man-r	Number of Workers	Average Man-r/ Person/Year
The three plants with highest employee exposures.			
Dresden	728	239	3.05
Millstone	596	232	2.57
Humboldt	254	129	1.97
The three plants with lowest employee exposures.			
Yankee Rowe	255	769	.33
Quad Cities	56	173	.32
Pilgrim	16	57	.27

A possibility for higher radiation exposure of workers occurs during reactor accidents. Table 13-1 lists such accidents.

HAZARDS TO THE GENERAL PUBLIC.   Although we may dismiss the likelihood of a bona fide nuclear explosion occurring in an atomic power plant, we must nevertheless assume that other types of accidents (meltdowns, for instance) will occur at such installations and that radioactive material will be released to the outside air, and therefore to the general public.

One such accident occurred at Windscale, England, in 1957. The graphite serving as the moderator in the reactor core became too hot, swelled, and obstructed the air flow intended to keep the reactor cool. Sections of the reactor overheated, and the graphite and uranium burned. Because this reactor was an experimental model, it lacked a containment dome and radioactive material was carried up the stacks and dispersed over the immediate countryside. The workers receiving the highest exposures during the accident received only 4.5 r in two cases, 3.3 r in one case, and slightly more than 2 r in four others. Two days later, milk samples from cows in the neighborhood proved to be contaminated with iodine 131, which had settled on the grass and was eaten by cows. In 4 days, milk produced in an area approximately 8 miles wide and 30 miles long had to be destroyed. Because the half-life of iodine 131 is 8 days, the concentration soon diminished. Concentrations of other radioactive isotopes never reached levels that were considered hazardous.

Assessing Biological Damage

Uranium disintegrates into a wide variety of elements, some stable and some unstable. It is the group of unstable, radioactive elements produced by fission that are primarily responsible for biologically damaging radiation from reactors. Radiation from radioactive isotopes is measured in a unit called the *Curie*, defined as the amount of radiation produced by a gram of radium. The unit used in describing the total radioactivity on earth for a given isotope is usually thousands of curies (kCi) or millions of curies (MCi), whereas the unit used with reference to human exposures is millionths of a curie (μCi).

THE RETENTION OF ISOTOPES IN THE BODY.   When a small amount of radioactive isotope is introduced into the body (by inhalation, by swallowing, or medically by way of intravenous injection), the amount of radiation the body receives will depend on the half-life of the isotope ingested and its relative accumulation and retention in various organs. Some reactor by-products, such as the noble gases xenon and krypton, are not chemically reactive and so are not retained within the body, but other elements may be metabolized and stored for long periods. The time required for the body to eliminate one-half of the dose of any substance is called the *biological half-life* of that substance, and this value differs from one organ to another.

The effect on the body of a radioactive isotope will be the product of its radioactive and biological half-lives; the effectiveness of the isotope in producing biological damage must depend on the amount not excreted from the tissue during a given period, and the natural radioactivity of the amount remaining. Iodine 131 has a biological half-life of 138 days in the thyroid and 7 days in the kidneys, liver, spleen, and testis. Thus, 1 μCi of ^{131}I will give only 0.5 to 3.5 mr to the whole body, 2 to 3 mr to the gonads, but 1,000 mr (1 to 2 r) to the thyroid gland, where it preferentially accumulates (Figure 14-3). In the Windscale incident, the concentration of iodine 131 in milk depending on the location of the pasture downwind from the reactor, but some samples gave readings as high as .8 μCi/liter, exceeding the .1 μCi/liter considered the highest allowable concenration for children.

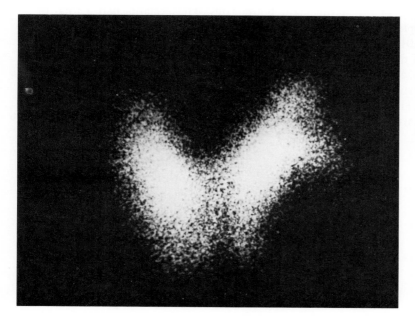

Figure 14-3 Autoradiograph showing the incorporation of radioactive iodine into the thyroid gland. The exposure of the film is brought about by the radioactive disintegrations within the thyroid itself. (Courtesy of R. A. Conard, Brookhaven National Laboratory.)

ACCUMULATION OF STRONTIUM 90. Biological half-life becomes particularly important in considering strontium 90. This element has many of the properties of calcium and may be metabolized in body processes that ordinarily use calcium. If the body is given equal amounts of calcium and strontium, from 25 to 50 percent as much strontium will be used as calcium. A cow feeding on equal amounts of the two elements will produce milk in which 11 percent of the calcium is replaced by strontium.

A newborn child will possess half the strontium/calcium ratio of the mother. The discrimination in favor of calcium is not sufficient to prevent considerable accumulation of strontium in the bones, and the half-lives (radioactive 10,000 days, biological 18,000 days) combine to give an overall effective half-life of ^{90}Sr in the body of 6,400 days. Since humans may accumulate strontium by way of contamination of the food chain (Figure 14-4), it assumes an unusual importance.

Radiation Release of Nuclear Energy Plants

It can be predicted that if nuclear reactors become prevalent we shall, from time to time, become aware of reactor accidents that involve the release of radioactive isotopes and necessitate further precautionary measures to minimize their impact on humans. A clear distinction should be kept in mind, however, between the amount of radioactivity released in this kind of accident and that which would result from an atomic bomb explosion. Although some incidents have been reported involving research and experimental reactors, of over 300 civilian and military reactors in the United States, operating for a total of 2,000 reactor-years up to 1975, there is no record of any malfunction of those reactors leading to the release of a significant amount of radioactivity to the environment.

The amounts of radiation released by 13 nuclear energy plants for the year 1969 averaged 14 mr at the plant boundaries. These plants had a total of 34 million people living within a 50-mile radius, and the average yearly dose per person was .0145 mr. The grand total in man-r amounts to 493, a dose will within the range of that sometimes applied to one person for therapeutic reasons. Two of the 13 reactors are of an early design with greater release of radioactive elements with short half-lives; if these are excluded, the average annual dose at the plant boundary drops to .9 mr-year and the combined total of all the doses to all persons within a 50-mile radius is 24 r-units.

Figure 14-4 The accumulation of strontium 90 as it passes through the food chain.

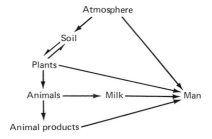

A

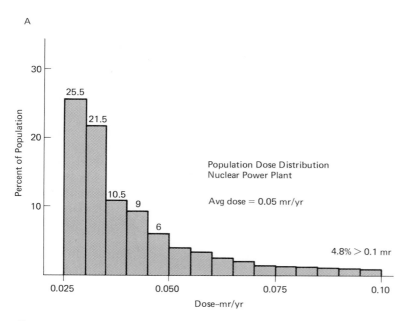

Figure 14-5 *A.* The distribution of radiation doses to individuals living within a 50-mile radius of a nuclear plant with a boundary dose of 5 mr/yr. *B.* The known distribution of natural radiation to the U.S. population. (Courtesy of A. P. Hull, Brookhaven National Laboratory.)

B

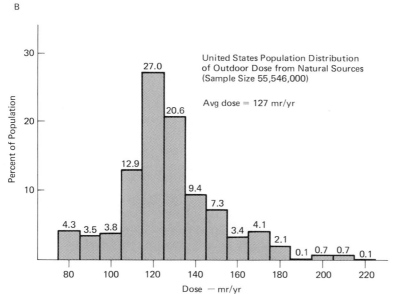

When the dose at the boundary of a plant is 5 mr, the doses to the population within a 50-mile radius will be distributed as in Figure 14-5*A*. This can be compared to the distribution of doses received by the U.S. population from natural sources (Figure 14-5*B*).

TRITIUM AND KRYPTON EFFLUENT.   A problem specific to atomic reactors is the outflow of unstable isotopes, especially ^{3}H, or tritium, which has a half-life of 12 years. The amount of tritium released from a large pressurized water reactor might be as much as 10,000 Ci per year; from a boiling water reactor, 100 Ci/yr. Although this sounds like a lot, it is estimated that cosmic rays produce between 4 and 8 million curies of tritium each year worldwide.

Residents of cities and towns along the shores of Long Island Sound have been understandably apprehensive about the future of that body of water if nuclear plants should proliferate along its edges. Engineers favoring such construction have calculated that if ten nuclear reactors were to use the Long Island Sound as a discharge basin, and they each contributed 3,000 Ci of tritium per year, a person eating nothing but animals and plants from the Sound would receive a dose of .07 mr/yr. If it is assumed that nuclear plants proliferate at a rate suggested by those who foresee a need for a thousand such plants (along with 15 fuel reprocessing plants) by the year 2000, the body tissue dose rate will increase as in Table 14-2. The effect of the accumulation of krypton 85 at that time is also shown.

PROBLEMS OF WASTE DISPOSAL.   When a nuclear power plant is refueled, the spent elements being removed dissipate energy at the rate of 1 percent of the reactor operating power. At first, these wastes are stored in "ponds" to allow the isotopes with short half-lives to lose most of their radioactivity. Iodine 131, with a half-life of 8 days, is one such. After 5 or 6 months the remaining isotopes are concentrated and stored in a more permanent form. From this point on, the long-lived elements such as strontium and cesium become the more important components; it will take 20 years for the energy to drop to 2 percent of that at the time of concentration. As this energy dissipates, the concentration heats, and its temperature must be monitored and controlled to prevent the waste from destroying its containers. The leakages reported in the press in recent years have been from ordinary steel tanks at Hanford, Washington; in other installations in this country and in England, France, and Russia, storage is in tanks made of stainless steel, which is more resistant to the highly corrosive radioactive sludge. Waste tanks are generally located so that when and if leaks do occur the rate of release to the environment by seepage into a stream or underground water table is minimized.

When such leakages occur, a conflict arises, opposite stands being taken by the engineers on the one hand, who calculate the additional exposure of the public to the released radiation and find it "negligible," and those on the other hand who argue that any increment, however small, should not be tolerated.

When tank storage has exhausted the shorter half-life isotopes in the reactor waste the material is further processed, and those elements of some commercial value are "harvested." The rest is solidified, and if it is then imbedded in glass or ceramic material it may be stored for an indefinite period. Precautions must be taken to prevent water from infiltrating the storage area lest some of the ceramic material be dissolved, releasing its proportionate amount of radioactivity to the environment. There has as yet been no practical scheme devised for the unsupervised, permanent disposal of radioactive waste. Plans to get rid of these by-products by permanent storage in abandoned salt mines have met with considerable vocal opposition from the residents of those areas. Other plans, fanciful or not, include burial of the solid waste deep in the ice on the continental shelf of Antarctica and, perhaps the most imaginative suggestion, the loading of concentrated amounts of long-lived radioactive isotopes into rockets to be shot into the sun, a feat that would require considerably less skill than is now regularly demonstrated by both the U.S. and U.S.S.R. space programs. This could be carried out only if the probability of rocket malfunction were much less than it is today, if the spread of radiation after such a malfunction could be shown to be minimal, and if the dollar (or ruble) costs of such a project were within reason.

TRANSPORTATION OF SPENT FUEL.   Although there has not been a recorded case of an accident that has killed, injured, or overexposed people during the transportation of radioactive fuel materials, such accidents will probably occur in the future as more and more fuel must be transported. From 1949 to 1970 there were 160 accidents involving vehicles transporting radioactive materials; in four cases (2.5 percent) there was dispersal of some radioactivity. The effect of such a release would depend on several factors—meteorological conditions, amount released, population density, and so on. According to one set of calculations made by nuclear engineers, taking into account the increased population density and nuclear plants for the state of California in the year 2000, there will be, on the average, less than one incident per year in which one person will get as much as 300 r to the thyroid from iodine 131, the most important radioactive element in such waste, or 30 r to the whole body from krypton 85, the second most potent element.

The Fast Breeder Reactor

A good deal of the discussion of the safety of nuclear reactors centers around a type called the fast breeder reactor. In reactors of the type now in commercial use, uranium consisting of 97 percent ^{238}U and 3 percent ^{235}U, the latter having been enriched four times from an original concentration in ore of about .7 percent, provides the power for the reactor. In the "breeder" reactors, plutonium 239, one of the by-products of ordinary reactors, is mixed with the very abundant and ordinarily useless ^{238}U. As the plutonium disintegrates, it releases neutrons that are captured by ^{238}U, converting the ^{238}U into plutonium, so that more plutonium is created than is used up. Since ^{238}U is present in the earth in virtually unlimited quantities, the breeder reactor could provide an almost inexhaustible source of power. Thus to provide power for a city of a million inhabitants for a year, an ordinary reactor might require 150 tons of uranium; a breeder reactor could provide the same output on only $1\frac{1}{2}$ tons of uranium. It is obvious why such power generation would appeal to an engineer.

Unfortunately, there are several problems that prevent the immediate widespread acceptance of the breeder reactor. It operates at high temperatures with liquid sodium as the heat exchange agent, but liquid sodium is highly corrosive and presents many engineering problems in that part of the reactor where heat is transferred from the fuel unit to the steam generator. But the most serious problem is the high concentration of plutonium 239 that such reactors would necessitate.

BIOLOGICAL DANGERS OF PLUTONIUM.   Plutonium 239 is a particularly hazardous isotope. With a half-life of 25,000 years, 1 kg (or 2.2 lb) has a radioactivity output of 64 Ci, but only 16 billionths of a curie within the human lungs will deliver a yearly dose of about 15 r. Unlike strontium, plutonium does not enter the food chain to any appreciable extent because of its reluctance to pass through cell membranes and is very poorly absorbed by the digestive tract when ingested. However, it can enter the body through wounds or abrasions and accumulate in the bones and liver and may become lodged in the lungs if particles carrying it are breathed.

Great concern has been expressed by those who do not believe present-day society can safeguard the large amounts of plutonium being processed (50,000 tons per year by 1990) without allowing lethal quantities to escape into the environment. A frequently voiced worry is that terrorist groups might steal the small quantity (10–20 lb) of plutonium needed to construct an A-bomb to be used as a blackmail weapon to achieve their goals. It is to be hoped that the intelligence needed to design a workable bomb would be matched by the realization that the unusual dangers of plutonium would almost certainly result in the early deaths of those involved in the illicit effort. Governments can, of course, afford the elaborate precautions necessary in the processing and fabrication of plutonium; as the number of countries grows with the means for refining plutonium, a convenient but not necessary auxiliary to nuclear plants, the probability of nuclear confrontation increases enormously. Finally, there is the additional concern that the decisions being made now with respect to the fast breeder reactors and the quantities of plutonium surrounding its use will commit future generations of the human race to safeguard vast amounts of radioactivity for an indefinite period into the future.

The "Maximum Permissible Dose"

Although the danger of excessive radiation exposure was recognized as early as 1902, statements specifying an exact amount permissible were not made until the 1920's, when a "safe" dose was expressed as a fraction of the dose giving a skin burn, or *erythema*. In 1932 it was suggested that the "tolerance dose" might be defined as that which "experience has shown to produce no permanent physiological changes in the average individual." In 1934 a group now known as the International Commission on Radiological Protection (ICRP) recommended a tolerance dose of .1 r/day. In 1950, the tolerance dose was brought down to .05 r/day.

A report by the ICRP in 1958 on the Biological Effects of Atomic Radiation suggested the following guidelines: that the maximum dose (excluding natural and medical exposures) be 5 r/yr for those involved occupationally with atomic energy, that this be reduced to a tenth, or .5 r/yr, as the individual dose for any person not occupationally involved, and that this latter figure be cut by one third to .17 r/yr as the average dose for the population at large. This last value amounts to about 5 r/30 yr, or 5 r per generation. These values have been referred to earlier as the "maximum permissible dose." Essentially similar guidelines have been proposed by the Federal Radiation Council, the National Academy of Sciences, and the National Committee on Radiation Protection.

These limits, when applied to nuclear energy plants, have been interpreted to mean that the general population should not receive an exposure of more than 170 mr. However, this "maximum permissible dose" would be received only if a person camped at the plant boundary for 24 hr a day, 365 days a year. Figure 14-5 shows how sharply any initial dose of radiation drops off with distance from the boundary. To expose the entire U.S. population to an average of .17 r would require boundary doses of many hundreds of r-units, clearly an unacceptable level, particularly from the standpoint of the plant workers, who would not survive very long within the plant at such high exposures. In addition, the radiation at existing plant boundaries is quite low; for 11 such plants, the average emission at the boundary is less than 1 mr/yr and reaches higher levels only for plants of an early obsolete design. Responding to pressure from concerned citizen groups, the AEC has recently reduced the maximum allowable dose at the boundaries of nuclear installations from 170 mr to 10 mr.

A comparison of the doses from the major sources of radiation for 1960 to the year 2000 is given in Table 14-2.

The Mutagenic Effects of Radiation Exposures

All in all, these doses, present and anticipated, are not excessive when compared to other radiation exposures. What, then, is the biological basis for the controversy? It stems from the two basic principles put forward earlier: (1) that mutation production is linearly related to the dose and (2) that the mutational effects of the dose are cumulative over the lifetime of an individual. As an illustration of these points, let us calculate the number of mutations expected if each person in the population should receive the rather high dose of 170 mr each year. Over a 30-year period (one generation), the accumulated dose will amount to 5 r. If 20 r is the doubling dose, then the radiation will be responsible for $5/_{20}$ or $1/_4$ as many mutants as occur spontaneously. On the other hand, if 200 r is the doubling dose, the additional number of mutants will amount to only $1/_{40}$ of the spontaneous number.

Table 14-2 Estimated total annual average whole-body doses in the United States (mr/person). (A. W. Klement, Jr., et al., U.S. Environmental Protection Agency, 1972.)

Year	Natural Radiation	Fallout*	Gaseous Effluent	Fuel Reprocessing	Tritium	^{85}Kr	Medical Diagnosis‡
1960	130	13.0†	.0001	—	.02	.0001	35
1970	130	4.0	.002	.0008	.04	.0004	35
1980	130	4.4	.026	.02	.03	.003	35
1990	130	4.6	.082	.09	.02	.01	35
2000	130	4.9	.17	.2	.03§	.04	35

*By the year 1970, figures increase slightly because of accumulation of Strontium 90.
†For the year 1963.
‡No predictions are made for therapeutic use.
§This figure, which had been decreasing because of the disappearance of tritium produced by bomb tests, begins to increase at a slow rate as the projected number of nuclear reactors go into operation.

RELATIVE FREQUENCIES OF NEW MUTATIONS. The calculations for relative frequencies of new mutations are summarized in Table 14-3. For autosomal dominant traits, there are now about 10,000 cases per million live births, of which about 2,000 are new mutants (the other 8,000 being born of affected parents). If the doubling dose is 20 r, the effect of 5 r will be to produce 500 new cases in the first generation and, as these new mutants also produce progeny over time, 2,500 cases per generation. On the other hand, if the doubling dose is 200 r, there will appear 50 new cases per generation, which would eventually make a total of 250 cases (again, per million live births) attributable to the radiation.

A somewhat similar calculation can be made for X-chromosome traits where from $\frac{1}{4}$ to $\frac{1}{40}$ of the current number of new mutants will be attributable to radiation. On the other hand, for recessive traits, the newly mutated alleles will not appear until, by chance, they become homozygous, so that it is not possible to make any estimate of the number appearing within the near future. Calculations based on the known incidence of congenital anomalies, aneuploidy, and abortions are also given in this table.

Malignancies Induced by Radiation

A more immediate effect of radiation is the induction of malignant cell growths, particularly leukemia. Unlike the situation for gene mutations, where there can be reasonably assumed to be a linear dose-response relationship so that any dose, however small, will have its proportional effect, and any mutations so produced

Table 14-3 Estimated effect of 5 r per generation on a population of 1 million live births, compared with the current incidence. When two estimates are given, the first assumes a doubling dose of 200 r, the second a doubling dose of 20 r. (Data from the Report of the Advisory Committee on the Biological Effects of Ionizing Radiation, National Academy of Science.)

Type of Defect	Current Incidence per Million Live Births	Effect of 5 r per Generation	
		Immediate	Eventual
Specific genetic defects			
Autosomal dominants	10,000	50–500	250–2,500
x-chromosome traits	400	0–15	10–100
Recessive traits	1,500	Very few	Very slow increase
Congenital anomalies			
Chromosome rearrangements	1,000	60	75
Aneuploidy	4,000	5	5
Recognized abortions			
Aneuploidy and polyploidy	35,000	55	55
X0	9,000	15	15
Chromosome rearrangements	11,000	360	450
Total	71,900	215–1,010	860–3,200

at different times will persist to give a cumulative effect, for the induction of carcinomas the situation is not so clear. Do very low doses of radiation produce the proportion of cancers that would be expected on a linear dose-effect curve?

Much controversy has been generated by this question. Unfortunately, carefully controlled experiments on lower animals do not give us conclusive data. In one type of mouse leukemia, the simultaneous combination of an infection by a virus and a dose of radiation is required to cause leukemia, each of these being ineffective by itself, and neither the amounts of virus nor the radiation exposure shows a simple mathematical relationship to the probability of leukemia induction. Large-scale carefully controlled experiments with lower mammals fail to show a linear relation of incidence to dose at the lower doses, but neither do they exclude the possibility. For these reasons, many workers in this field have been reluctant to come to firm conclusions about the shape of the curve relating the incidence to dose at very low doses; a large number of them are convinced that a threshold for the induction of leukemia exists: that is, that very small doses of the order of fractions of an mr may be ineffective in inducing malignancies.

CANCER AND LEUKEMIA INDUCTION IN HUMANS. Innumerable studies on humans receiving high radiation doses show an increasing frequency of induced cancers with increasing dose. The difficulties in deciding whether there is a threshold, below which small doses of radiation produce no effect, or whether there is a square relationship with dose, implying a disproportionately small effect of low doses, or whether there is a simple linear relationship are graphically depicted in Figure 14-6. Two heavily irradiated populations are shown, persons

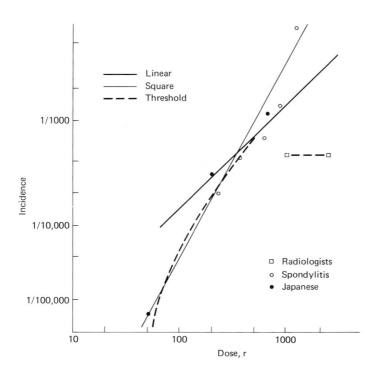

Figure 14-6 The incidence of cancer as a function of the dose received by radiologists, A-bombed Japanese, and people suffering from a disease of the spinal column. The three different lines show the expectations based on the assumption of a linear relationship, the existence of a threshold, or a square relationship. The similarity of the three lines and the scattering of the points illustrate the basic difficulty involved in coming to a firm conclusion on the basis of human data. (Courtesy of A. M. Brues. *Low-Level Irradiation*, Publ. No. 59, p. 84. copyright © 1959 by the American Association for the Advancement of Science.)

exposed therapeutically for a debilitating disease of the spinal column (spondylitis) and Japanese exposed at the time of the A-bomb explosions. These points have been compared with the various expectations based on the three assumptions given above. It can readily be seen that those data seem to fit all of the possibilities.

THE JAPANESE STUDIES.   Several such studies involve the survivors of the atomic explosions of Hiroshima and Nagasaki. Figure 14-7 shows the relationship between the doses (estimated) that the bomb survivors received and the numbers subsequently stricken with leukemia. There is considerable variation at the lower doses, particularly in the Nagasaki data, which resulted from an explosion with a ratio of destructive components (predominantly gamma rays, relatively few neutrons) more like the radiation we are commonly exposed to, as compared to the Hiroshima blast, which had a higher proportion of neutron release. If the low values from the Nagasaki data are representative, then there may exist a threshold dose for the induction of leukemia.

MALIGNANCIES FROM NUCLEAR ENERGY.   The problem of malignancies caused by nuclear energy is clearly most important. If the relationship is that shown in Figure 14-6A, then the linearity implies, as in gene mutation discussed earlier, that however small the dose, a corresponding effect occurs. In any case, for our purposes we do not wish to underestimate the potential disastrous effects on the population; we therefore will assume that a linear relationship holds. One articulate opponent of nuclear energy has proposed that the increased risk of cancer plus leukemia is 1 percent of the normal spontaneous rate for each r-unit received. Using this figure, we can readily estimate the number of cases of cancer per year that might be attributed to nuclear energy during the year 2000.

Figure 14-7 The incidence of leukemia per 100,000 of the exposed survivors of Hiroshima and Nagasaki, plotted as a function of the dose estimated to have been received by those survivors. (T. Ishimaru et al. Leukemia in atomic bomb survivors, Hiroshima and Nagasaki. *Radiat. Res.*, **45**:216–33, 1972.)

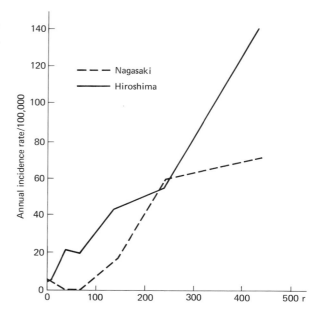

The spontaneous cancer incidence in the United States is about 280/100,000 (280×10^{-5}). If the population at that time is 300 million (300×10^6), the number of cases should be the product of those two figures, 840,000 cases. If, as the Environmental Protection Agency suggests, each person might receive on the average 0.5 mr/yr, he would receive 15 mr over a thirty-year period. If 1 r increases the incidence by 1 percent, .015 r (15 mr) will increase the incidence by .015 percent (1.5×10^{-4}). Multiplying the figure for the spontaneous cases by 1.5×10^{-4} gives an estimate of 126 cases of cancer that would, by this calculation, be attributed to radiation from nuclear energy plants.

MALIGNANCIES FROM MEDICAL RADIATION.   A similar calculation can be made for medical x-rays. If the average person in the United States receives 35 mr/yr from this source, and the rate of increase over the spontaneous rate is assumed to be 1 percent per r-unit, then .035 r over a 30-year period would give 1.05 percent increase or about three new cases per 100,000 inhabitants. This amounts to 9,000 cases attributable to medical x-rays in a U.S. population. The assumption that the increased number caused by radiation is strictly proportional to the additional dose makes it relatively easy to calculate the additional incidence of cancers so caused by any amount of additional radiation from any other source.

In the last two chapters the radiation exposures to humans from a wide variety of sources have been evaluated and, in particular, those doses originating from nuclear energy plants. Based on the data now at hand it is evident that these man-made sources are responsible for a small but steady exposure of humans to radiation. If those who maintain that nuclear plants are completely safe are in error, so are those who feel that the human race is now in jeopardy because of the high frequency of genetic diseases and leukemia being produced by atomic radiation.

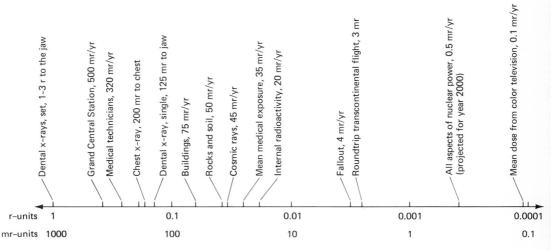

Figure 14-8   The relative exposures of the U.S. population (or the group specified) to various radiation sources, natural and man-made. Note that the scale of exposures is made logarithmic in order to include both relatively high and relatively low doses on the same plot.

At the present time, the total exposure of the population from atomic sources is much less than 1 per cent of the radiation either from natural sources, which is unavoidable, or from medical diagnosis and treatment, which may be somewhat excessive. The relative exposures from a wide variety of sources are summarized in Figure 14-8. In the next chapter we will discuss another potent source of biological damage, the chemicals to which we expose ourselves.

References

BRUES, A. M., ed. 1959. Low-Level Irradiation. Publ. No. 59, American Association for the Advancement of Science.

CLELLAND, W. 1974. Is clean power possible? *New Sci.*, **63**:786–89.

GLOWING, M., and L. ARNOLD. 1974. The early politics of nuclear safety. *New Sci.*, **64**:741–43.

HAMMOND, R. P. 1974. Nuclear power risks. *Am. Sci.*, **62**:155–60.

HULL, A. P. 1971. Some comparisons of the environmental risks from nuclear and fossil fueled power plants. *Nuclear Safety*, 12.

ISHIMARU, T., et al. 1971. Leukemia in atomic bomb survivors, Hiroshima and Nagasaki, 1 October 1950–30 September 1966. *Radiat. Res.*, **45**:216–33.

KLEMENT, A. W., JR., C. R. MILLER, R. P. MINX, and B. SHLEIEN. 1972. Estimates of Ionizing Radiation Doses in the United States—1960–2000. Rockville, Md.: U.S. Environmental Protection Agency.

LEAR, J. 1972. Radioactive ashes in the Kansas salt cellar. *Saturday Rev.*, February, pp. 39–42.

LOVINS, A. 1974. Fast reactors: the US debate goes critical. *New Sci.*, **61**:693–95.

SAGAN, L. A. 1972. Human costs of nuclear power. *Science*, **177**:487–93.

SEABORG, G. T., and W. R. CORLISS. 1971. *Man and Atom.* New York: Dutton.

STARR, C., M. A. GREENFIELD, and D. F. HAUSKNECHT. 1972. A comparison of public health risks: nuclear vs. oil-fired power plants. *Nuclear News*, October, pp. 37–45.

TAMPLIN, A. R. 1969. Fetal and infant mortality and the environment. *Bull. Atomic Sci.*, December, pp. 23–34.

WALLACE, B. 1974. Commentary: radioactive wastes and damage to marine communities. *BioScience*, **24**:164–67.

WEINBERG, A. 1972. Social institutions and nuclear energy. *Science*, **177**:27–34.

WEINBERG, A. 1973. Technology and ecology—is there a need for confrontation? *BioScience*, **23**:41–6.

Questions

Useful terms: chain reaction, critical mass, man-r, curie, maximum permissible dose, doubling dose.

1. How does the radiation dose received by workers in a nuclear plant compare to that received by those who work with radiation in the medical profession?

2. Are nuclear plants and A-bombs similar in their blast potential? What kind of serious accidents are nuclear reactors subject to and what are the consequences both to the workers and to the general public when this happens, judging from past experience?
3. What is the difference between the physical half-life of a substance and the biological half-life of that same substance?
4. Why are the radioactive isotopes of iodine and strontium particularly hazardous?
5. Of all the possible hazards attendant on nuclear power generation, what aspect presents the most perplexing problems?
6. Why is the breeder reactor subjected to greater criticism than other types of reactors?
7. What is the "maximum permissible dose" concept and who is responsible for determining such safety recommendations?
8. It has been argued that the concept of the maximum permissible dose allows the radiation from nuclear energy plants to reach a level such that the average dose received in the population will be 170 mr. Discuss both sides of this argument.
9. If we assume the doubling dose to be 100 r, how many new autosomal dominant genes would be expected per 10 million births (a) immediately and (b) over a long period of time? How does this compare with the number that appear spontaneously?
10. Why is there so much disagreement about the relationship between the incidence of cancer and the dose of radiation at very low doses?
11. To what extent are medical x-rays a hazard in producing new malignancies?

15

Chemical Compounds Causing Genetic Damage

Carcinogenesis, Teratogenesis, and Mutagenesis

The adverse effects of agents on cells and their constituents, particularly on the genetic material, have been described as *carcinogenic* (producing cancer), *teratogenic* (producing developmental abnormalities in utero), and *mutagenic* (producing mutations).

The first evidence of a carcinogenic effect appeared as early as the first part of the eighteenth century, when the high incidence of cancer of the scrotum in English chimney sweeps was attributed to their exposure to coal tars. At the present time, more than 500 chemical carcinogens are known, most of them relatively rare synthetic organic compounds. Teratogenesis has been discussed in Chapter 3.

DISCOVERY OF CHEMICAL MUTAGENS. Following the reports of the increased frequency of mutation after experimental x-ray treatments by H. J. Muller in 1927 (in *Drosophila*) and by L. J. Stadler in 1928 (in barley), a large number of investigators made serious attempts to discover mutagenic chemicals on the assumption that information about the composition of the gene might be obtained if it were known what kinds of chemicals could specifically alter gene structure. Most of the chemicals tested were those previously known to be carcinogenic in mammals. In general, the tests gave negative or rather uninteresting, slightly positive indications of an effect.

It was not until 1941 that a powerful chemical mutagen was found. This happened during the course of work in England with mustard gas, a chemical highly poisonous to humans that was being manufactured secretly in the event that circumstances forced its use as a weapon. It was noted that individuals who were accidentally exposed to mustard gas during its manufacture reacted with inflammation, blistering, and burning, consequences somewhat similar to those produced by x-rays. The similarities interested a group of pharmacologists, who, with the aid of a fruitfly geneticist, Dr. Charlotte Auerbach, showed that mustard gas, like x-rays, was indeed a powerful mutagen. This information was not reported to the scientific world until after the conclusion of World War II.

In retrospect, it should be noted that, prior to the war, observations had been made that colchicine could produce changes in chromosome number in plants and that nitrous acid could produce mutations in the mold aspergillus, commonly found on bread, cheese, and so on. We also learned that during the war the Germans and the Russians had discovered, in animal and plant experiments, that other compounds such as urethane and formaldehyde increased the mutation rate somewhat.

CHROMOSOME BREAKAGE AS AN INDEX. It is usually difficult to determine the activity of any potentially dangerous agent—a newly synthesized organic compound, for instance—by testing it on humans, because of the understandable reluctance to deliberately expose humans to large quantities of a new chemical. Of the three major categories of deleterious change in man, the possible mutagenicity of a new agent is the most difficult to determine. In the first place, most mutational changes are recessive and cannot be detected until homozygotes are produced by chance, many generations later. Secondly, even for rare dominants that would require (in theory) only the production of progeny to ascertain the effectiveness, the spontaneous rate is so low that even a tenfold increase appearing sporadically across the world would be difficult to detect and to relate to a specific cause.

Breaking of chromosomes by exposure to some test substance can, however, be observed very quickly both in the body (in vivo) and in the test tube (in vitro). Since agents that cause chromosome damage may also be responsible for one or more of the three major categories of biological effects, its occurrence is a valuable, but not infallible, index of potency. However, the reverse is not the case. Many substances now known to be mutagenic and carcinogenic do not break chromosomes. When an agent can produce chromosome breaks, it is said to be *clastogenic.*

When tests are made in vivo, it becomes necessary to resort to organisms other than man for these biological tests. One of the hazards in making deductions concerning the biological potency of a chemical comes from the inconsistency that is shown from one organism to the next. A chemical that shows strong mutagenicity in bacteria may show none whatsoever in the fruitfly, and one that is mutagenic in both of those may not show any effects in experimental mammals. Further, there appears to be no correlation among the kinds of effects a chemical may produce. Some agents, such as radiation, are known to be mutagenic, carcinogenic, teratogenic, and clastogenic. Thalidomide is teratogenic and not known to have any of the other properties. However, all chemical carcinogens seem to be mutagenic but not necessarily teratogenic.

Not only may mammals differ in their sensitivity to these agents, as, for instance, from mice to man to hamsters (Table 15-1), but also in some instances compounds shown to be biologically active in one strain of a given species prove not to be effective in another strain of the same species. Nevertheless, these organisms must provide the experimental material for specific information about possible dangers to humans, particularly when prior knowledge is required before releasing a compound for general public consumption. It is no wonder that the field is highly controversial, with quite contradictory claims being made by different investigators whose objectivity and impartiality are not open to question.

THE INFLUENCE OF THE METABOLISM. The most important reason for these conflicting results is that the metabolism of the organism may change the compound into closely related ones, which may or may not be biologically active. In one class of mutagens, the compound itself is inactive but its metabolic products are extremely potent. In another class, the compound is active but its metabolic products are not. Therefore, in each case, the mutagenic activity would depend on the exact metabolic pathways present in that species, and it would not be surprising to find that the same compound is mutagenic in one species and not in another. It has been shown in the guinea pig, for instance, that one compound (acetoaminofluorene or AAF) is mutagenic in one strain and not in another, and that the second strain lacks an enzyme present in the first that normally converts nonmutagenic AAF into a highly mutagenic product.

These effects may be further complicated by the fact that two compounds together may produce a result not found from each individually. Thus, simultaneous treatment of lymphocytes in vitro with a clastogenic agent an an antioxidant (vitamins C and E, for instance) gave a 30 to 60 percent reduction in the number of induced chromosome breaks, compared to a control with the clastogen only. It has been postulated that the declining incidence of stomach cancer in the United States may be an accidental result of the inclusion of antioxidants in foods as nutritional additives (vitamins) or as preservatives. Such a beneficial

Table 15-1 Effectiveness with which thalidomide produces birth defects. Note that not only is its teratogenicity quite variable from one species to the next but also a rather small dose may be effective in some cases (first column) whereas a much larger one may be ineffective in others (second column). (From H. Kalter, *Teratology of the Central Nervous System*, University of Chicago Press, Chicago, 1964.)

Species	Smallest Dose Producing Defects	Largest Dose Producing No Defects
Man	0.5–1.0	?
Baboon	5	—
Monkey	10	—
Rabbit	30	50
Mouse	31	4,000
Rat	50	4,000
Armadillo	100	—
Dog	100	200
Hamster	100	8,000
Cat	—	500

effect would be a refreshing departure from the almost universally valid generalization that preservatives and other food additives may be good for the food, but bad for the human body.

Clearly, the general rule to be followed in judgments concerning public health made before a definitive scientific conclusion is available is that whenever there is reason to suspect a biological effect in humans this information takes precedence over any evidence of a negative effect.

Substances Found in Food

SODIUM NITRITE. Sodium nitrite has been very widely used as both a preservative and a color enhancer in meat and fish products, such as frankfurters, bacon, smoked meat and fish, and vienna sausage. It is considered to be a hazard because it is converted into nitrosamines in the human stomach, and that class of compounds has been shown to be highly mutagenic in bacteria, viruses, molds, and other organisms. Some believe that its use in foods for humans should be drastically reduced, if not eliminated entirely.

SODIUM CYCLAMATE. For more than 20 years, but particularly in the last 10, sodium cyclamate has found widespread popularity as a sugar substitute in diet foods and soft drinks. Because of their universal use, these have been subjected to extensive tests by both manufacturers and governmental agencies.

Sodium cyclamate has been shown to have a carcinogenic effect in the bladders of mice in which pellets containing it were implanted. In one experiment, the proportions of mice developing carcinomas were 61 percent of those exposed and only 12 percent of control mice exposed only to the cholesterol used as a carrier for the cyclamate. The mouse bladder appears to be very sensitive to cyclamate; this organ was examined in individuals given cyclamate orally and its was found that in such cases carcinomas were also produced. On the other hand, rats fed diets containing as much as from 1 to 5 percent of sodium cyclamate (and also saccharine) for 2 years appeared to suffer no effect at all at the lower dose and no toxic effect at the higher dose.

It seems likely that the technique of implanting pellets into the bladders of mammals may not be a valid procedure for testing carcinogenicity of compounds that are ordinarily taken by mouth. There is no evidence that the use of cyclamate has been responsible for any congenital abnormalities in children, any carcinomas, or any other biological effect except possibly for a rare skin hypersensitivity. Nevertheless, in accordance with the rule that any suspicion of a deleterious effect should be sufficient reason for the removal of the product from general consumption, cyclamate was banned from general use. It was banned despite the fact that the mice that developed bladder tumors after feeding had consumed a daily amount per body weight equivalent to the human consumption of 250 bottles (each 16 oz) of a typical diet drink daily. Another report shows that in certain rat cells the metabolic breakdown product of cyclamate, cyclohexalamine, is clastogenic. A person would have to consume about a pound of cyclamate a day for life, however, to bring the exposure up to the level given the rats in this experiment.

CAFFEINE. One of the most widely ingested suspects is caffeine, found not only in coffee and tea but also in chocolate and in some soft drinks. It is a common component of a very large number of headache remedies and pain-relievers that have aspirin as their principal ingredient. In 1948 it was reported that caffeine produces mutations in bacteria as well as in fungi, and a year later that it produces chromosomal breaks in the root tips of onions. At extremely high concentrations it has been shown to be weakly mutagenic and clastogenic in *Drosophila* and clastogenic in cultured human cells. However, there is evidence that the action of caffeine is not to initiate the primary break but rather to inhibit the normal repair of breaks that have already occurred for some other reason. For this reason, caffeine may be most effective when applied in combination with some other agent that produces chromosome breaks.

On the other hand, large-scale studies on mice fed amounts of caffeine in excess of that ordinarily taken by humans did not reveal any increase in the mutation rate. When heavy doses are given to male mice, litter size is not reduced, indicating that chromosome breakage in the male mouse germ cells is not a result of high caffeine treatment. In another experiment involving female mice fed caffeine, the birth rate dropped from the customary five to seven per litter to one to two. The reason for the difference between the male and female sensitivities is not known. It is a reasonable deduction, however, for other chemical agents, that the decrease in litter size after treatment of the male parent can be attributed to chromosome breakage in the sperm exposed prior to the mating, but a decrease after treatment of the female can have several causes, including chromosome breakage as well as the loss of embryos in an environment rendered unpleasant by the drug.

Because of its universal consumption and its known ability to penetrate to the human germinal tissue and through the placental barrier to the fetus, where it is found in the same concentrations as in the maternal bloodstream, caffeine is considered by some to be one of the most potentially dangerous compounds now widely used by humans. One estimate of its effectiveness is that the average coffee drinker will, during his lifetime, be exposed to a total mutagenic action equivalent to at least 50 r-units.

FOOD DYES. Amaranth, the most common red food coloring in the United States and Great Britain, used in jam, ice cream, lipstick, candy, sausages, fish, and even dog food, has been suspected of being both carcinogenic and teratogenic. Its use has been banned in Russia and West Germany, and in 1972 the World Health Organization recommended strict limits on its use. How difficult the interpretation of experimental data can be is shown by the fact that whereas one group concluded that there was no evidence of danger to the fetus, another group, on the basis of the same evidence, concluded that a pregnant woman might exceed the safe level if she consumed as much as a third of a can of cherry soda a day.

YOGURT. When rats were fed nothing but yogurt, they grew at a normal rate and behaved normally in all respects, except that in every case tested they developed cataracts. Cataracts appeared in adult rats after 4 to 6 months from the start of the diet and in young rats of ages 2 to 3 months. It is likely that this developmental defect results from a very high concentration of the sugar galactose in yogurt, because it is known that pure galactose produces similar cataracts.

IRRADIATED FOODS.   As we have seen earlier, radiation causes the formation of a large number of highly reactive compounds such as peroxides. Studies on the feeding of heavily irradiated foods to rats and dogs show that there may be an effect on the number of offspring they produce, and on the body weight and life-span of the experimental animals. The reasons are unknown, but it is likely that a family of active compounds (such as the peroxides) produced by radiation are responsible.

AFLATOXIN.   Certain strains of the common mold aspergillus, found on beans, nuts, corn, grains, and other agricultural crops, produce a very potent mutagen known as aflatoxin. This has been shown to be highly effective in producing mutations in fungi and wasps and is clastogenic in human cell cultures. In pregnant mice it has been shown to increase the number of fetal deaths by tenfold, probably by the gross breakage of chromosomes. Buyers and processors of susceptible foods, particularly peanuts, have established rigid controls to prevent the dissemination of products contaminated by this mold.

Activity of Drugs

LSD.   Few compounds have received such widespread attention recently as the hallucinogenic drug lysergic acid diethylamide (LSD). Over the past decade, more than 100 studies have been published related to the adverse biological effects of this compound. Some of the early studies involved subjecting leukocytes in vitro to varying concentrations of LSD, and a number of workers have reported increased chromosome breakage, although there appears to be no simple relationship between the dosage and the extent of the clastogenic effect.

In evaluating such results it should be kept in mind that such tests involve cells which have been artificially stimulated to divide rapidly, a condition that does not exist in the body, and that therefore some chromosome breaks may show up in cultured cells which might not be present in a comparable in vivo test. In addition, the concentration or exposure used is usually greater than the human dose and the amount of breakage is not outside of the range induced by many common agents such as antibiotics, caffeine, and aspirin. In addition, in the in vitro tests the effects may be exaggerated over those found in the body where excretory and detoxification mechanisms might serve to reduce the response of the cells to the chemical.

When studies are made of the chromosome breaks of individuals who have taken LSD, different results are found, depending on whether the individual is in a medically controlled situation in which the purity of the LSD is predetermined or is exposed to illicit LSD, in which case not only is the purity not assured (having been "cut" with such chemicals as belladonna, arsenic, and cyanide) but also the substance taken may not even be LSD. Of 126 subjects given pure LSD, 18 showed a frequency of chromosome breaks higher than that of the controls, whereas 90 of 184 persons taking illicit LSD showed an increased frequency of aberrations. This latter frequency may be high for several additional reasons. Users of LSD may also regularly take other drugs. Furthermore, there is a high frequency of viral disease among drug abusers resulting from infection; viruses are known to produce chromosome damage. It is possible that some of the positive effects attributed to LSD in the early investigations may result from the general conditions of drug abuse.

Several reports have suggested that LSD may be carcinogenic. These reports are based on the observed breakage of chromosomes in the early studies reported above and the apparent similarity to chromosomes known to be broken in certain kinds of leukemia and other carcinomas. Although a few cases have been found in which leukemia has developed in individuals treated with pure LSD, at present there does not appear to be any strong evidence that LSD is a carcinogen. It is generally true that cancerous tissue shows a high level of chromosomal abnormality, but in most cases this is likely to be a consequence of the rapid mitosis and differenatiation of those cells, rather than the cause.

Experiments with *Drosophila* and fungi indicate that treatment with relatively massive doses of LSD does increase the mutation rate slightly. It is not known whether LSD has any mutagenic effect in humans.

A large number of studies have been made with rats, mice, and hamsters on the effects on their progeny, and it has been shown that there appears to be a wide variation depending on the species used and particularly on the time at which the LSD was applied during pregnancy. No effect has been reported when the exposure occurs late in the pregnancy in those animals.

In one instance in which LSD was injected into pregnant rhesus monkeys, one female delivered a normal infant, two were stillborn with abnormalities of the face, and the fourth died after 1 month of life. Two control animals produced normal progeny. The lowest dose used in these experiments was 100 times greater than the usual experimental dose in humans. In some studies in humans it was found that the ingestion of illicit LSD by pregnant women resulted in larger numbers of chromosomal breaks in their offspring but that the children were in apparent good health without any ill effects at the time of birth and immediately thereafter. On the other hand, a number of cases have been reported of malformed infants born to women who used illicit LSD before or during pregnancy, the defects generally involving the arms and legs. However, in most of the cases it is known that these women had also ingested other kinds of drugs, such as marijuana, barbiturates, and amphetamines. There is no evidence that pure LSD is teratogenic at moderate doses in humans.

Several investigations have questioned the medical significance of breaks reported in some of the studies with LSD. First of all, untreated people usually show some small frequency of lymphocytes with chromosome breaks, of the order of 2 to 5 percent; second, after certain diseases, especially those involving viruses, such as measles, smallpox, chickenpox, mumps, and hepatitis, and possibly including the common cold, the observed frequency of chromosome breakage may go up severalfold.

MARIJUANA.   Marijuana has been implicated in chromosome breakage. In one study, 49 marijuana users showed an average of 3.4 percent cells with broken chromosomes, whereas there were only 1.2 percent in the 20 controls. It has been suggested that the chromosome breaks found in individuals who have been exposed to LSD may in fact be the result of the marijuana very often taken by the same groups of persons who take LSD. In addition, recent observations indicate that cultures of leukocytes taken from marijuana smokers have an increased frequency of fragmented nuclei that do not have a full complement of chromosomes and result from abnormal mitoses. Studies show the rate of replication of DNA in lymphoblasts of marijuana smokers to be considerably depressed compared to that of normal controls, and at a level corresponding to that found in patients suffering from cancer or with a medically suppressed immune reaction after a transplant. Other more recent studies deny these effects.

Following a lead suggested after it was reported that several young male marijuana users showed unusual breast development, one team of workers found that the male sex hormone, testosterone, was on the average 44 percent lower in marijuana smokers than in controls and that in several cases the sperm count was so low that sterility resulted. In addition to the possibility that the drug may alter the course of normal masculine development in young males, it has been questioned whether it might not have a similar effect on male prenatal development when the mother uses marijuana heavily during the early months of gestation.

On the other hand, it should be noted that in many cultures the use of marijuana or closely related drugs in concentrations many times higher than those found in an ordinary marijuana cigarette has been a socially acceptable activity for centuries with no ill effects to those populations. So the controversy will continue. The daily press will continue to feature articles describing biological damage caused by marijuana alternately with articles denying any such effects. It is clear that the final judgment will depend on more information than is now available.

SMOKING. Because of the centuries-old exposure of humans to tobacco, and the absence of any obvious increased mutation rate in smokers, it is not very likely that nicotine is an effective mutagen in humans, although there have been reports that it is clastogenic in the dividing cells of some plants. However, one of the products of smoking is benzopyrene, which is formed after burning of tobacco or other organic materials. The number of fetal deaths found in the litters of female mice mated to males treated with this substance is more than 10 times greater than that of females mated to untreated males. The statistical evidence that cigarette smoking causes cancer of the lung has been sufficient to convince the U.S. government to require a warning to be placed on all packages of cigarettes. In recent years, a marked increase in the number of women dying of lung cancer has been noted; this has been attributed to the increase in women who took up smoking after World War II, followed by a 20- to 30-year lag period before the effects appeared.

ALCOHOL. As in the case of tobacco, the almost universal use of alcohol over many centuries makes it unlikely that any pronounced mutagenic or carcinogenic effect would have gone unnoticed. However, it has been widely reported as a cause of teratogenesis in pregnant women, and it has been reported that men who consume excessive amounts (four or more drinks per day) may also have abnormal offspring more frequently than expected.

ANTIMALARIAL DRUGS.   The antimalarial drug quinacrine (also called atebrine) was used on a very extensive scale during World War II in the South Pacific, where as many as 7 million persons may have been exposed to large quantities of the drug. This compound, which has subsequently been found to be very useful for staining chromosomes, is mutagenic in bacteria and viruses as well as fruitflies. No reports have yet been made of any similar adverse effect on humans.

ANTICANCER DRUGS.   Highly suspect are anticancer drugs, because of the ways in which they destroy living cells. In some cases, their effectiveness depends on chromosome breakage in mitotically active cells, producing anaphase bridges with subsequent loss of such cells. In other cases, they may be effective in non-dividing cells, producing their effect in some other way, as enzyme inactivation. In any case, any anticancer drug stands a reasonable chance of being either mutagenic, teratogenic, carcinogenic, or clastogenic, or any combination of these. However, since most individuals being treated with such drugs are critically ill, the question of the potential hazard becomes secondary to the possibility of destroying the cancer and saving the life of the patient. In addition, it can be argued that such individuals are usually beyond the childbearing years and so will not pass mutations on to subsequent generations even if such should be induced. On the other hand, youngsters should not be subjected to such drugs unless it is a matter of life and death. In particular, the drugs should not be used on young people for diseases that are not serious or can be treated otherwise. For instance, the anticancer agent methotrexate, which has been shown to be clastogenic, is used on occasion to treat extreme cases of the nonmalignant skin disease psoriasis. Such chemical treatments, like those involving radiation, obviously should be avoided.

Hazards of Technology

PESTICIDES.   Compounds used for combatting insect pests include one group that sterilizes (TEPA and TMA) and another group with direct lethal action (captans, DDT, and DDVP or Vapona). Since the former compounds sterilize insects by causing an increased frequency of chromosome breakage in the germ cells, which then causes the zygote to die, it is clear why there would be concern in exposing humans to such compounds. Both of them increase the frequency of fetal death in the mouse by a factor of 30 when the male parent has been exposed.

Much controversy surrounds the recent widespread use of DDT as a pesticide. Claims have been made that this compound is both carcinogenic and mutagenic. Part of the apprehension concerning DDT stems from its relative indestructibility and its tendency to accumulate in the food chain, where its concentration may reach values hundreds of times greater than initially. Nevertheless, its use in eliminating malaria throughout the world, where it may have been ingested by tens of millions of people, sometimes in considerable quantities, does not support any substantial carcinogenic effect in humans.

Although some studies made on mice and rats have suggested such a carcinogenic effect of DDT, other studies have not. These contradictions may be caused by metabolic differences of the sort described earlier; it is known that DDT may be metabolized by at least three different pathways. In one legal examination of the situation, the conclusion of the hearing examiner was that on the basis of the evidence available there appeared to be no evidence that DDT was either mutagenic, carcinogenic, or teratogenic in the amounts to which humans may ordinarily be exposed. This may be a fair statement; or, on the other hand, lack of evidence on these points at the present time does not preclude the possibility that future work may show such effects. Since as long a period as 20 to 30 years can separate the exposure and the appearance of a cancerous growth, it may take some time before any ill effects appear.

SOLVENTS.   Some solvents such as those used in dissolving glue have been found to be clastogenic. These substances are found in a very wide variety of mixtures in commercial products so that the specific active agent may be difficult to identify, although benzene has been named as a contributor to chromosome breakage. A study of 25 persons in Italy who had been exposed to benzene poisoning revealed that they suffered from a greatly increased frequency of chromosome breaks.

Late in 1973 it was reported that the solvents in spray adhesives might be responsible for both birth defects and chromosomal damage. The data of concern consisted of the cases of two infants with differing multiple congenital malformations whose parents had used the spray prior to their births and the apparent increase in the frequency of chromosomal breaks in peripheral blood lymphocytes from the parents and from other exposed persons. For a period of time, these adhesives were removed from the market; subsequent investigations failed to confirm this connection between the adhesives and the chromosomal damage, after which they were returned to the market. Vinyl chloride, which has appeared in a wide variety of plastic articles and aerosol sprays, has been correlated with a high frequency of cancer of the liver in persons exposed to high concentrations. It has been shown to be mutagenic in *Drosophila* at concentrations as low as 50 parts per million of air, whereas workers in some plants have been exposed to 20,000 parts per million. Its use has been banned in applications where people may receive high exposures, or low exposures over a prolonged period of time. It is no longer used in aerosol sprays.

OVERALL IMPACT OF MUTAGENESIS.   The frequency with which chemical agents produce transmissible genetic changes in humans is an open question at the present time, nor is there any simple means of arriving at a reasonable estimate, unlike the case for radiation effects, about which we can make some rough guesses. The problems encountered in analyzing the effects of new chemical compounds seem almost insurmountable; we do not even know whether the dose-effect relationship is linear at very low doses for any chemical in humans. The analogy with physiological poisons, which invariably have a threshold, is difficult to dismiss, even though some compounds appear to behave linearly at low doses. Clearly this is the most important consideration when hundreds of millions of humans may be ingesting small amounts of a mutagenic compound such as caffeine.

When other methods are developed permitting a more precise assay of mutagenicity, we shall undoubtedly learn that some of our apprehensions have been unfounded and, unfortunately, that some compounds now considered innocuous are having more influence than now realized. If anything seems clear, it is that no new compound should be put into widespread use without a thorough testing on a variety of animals in the most scientifically controlled fashion possible. With the synthesis of tens of thousands of new compounds each year and with the commercial production of thousands of them, new hazards are constantly being added to the environment with no means available at present for testing each one prior to its introduction to the population. It can be stated as a general rule that, although the evidence that specific chemicals are biologically deleterious can be confused, contradictory, and highly inconclusive, it would be well for humans, given freedom of choice, to ingest as few new synthetic or suspected chemicals as possible, particularly women in the early stages of pregnancy.

It has been estimated that during the year 1967, of all of the deaths of Americans attributable to chemical causes, the majority resulted from cigarette smoking, a variable frequency from 0 to 20 percent from poor diet, 3 percent from excess indulgence in alcohol, 3 to 8 percent from unknown chemicals that act as carcinogens, 4 percent from reactions to medication, .6 percent from drug addiction, .5 percent from occupational airborne particles, .25 percent from suicides involving chemicals, and .1 percent from accidents with chemicals.

Figure 15-1 The photo of man's worst enemy indulging in his exposure to countless chemicals of unknown danger. (Courtesy of A. H. Wiebenga, University of Amsterdam Medical School.)

From this it would appear that the greatest immediate damage to the well-being of the population comes from self-imposed treatments (Figure 15-1), rather than from accidental exposures to obnoxious chemicals in the environment.

To summarize the information on the wide variety of chemical compounds discussed in this chapter, we should note that there are four different ways by which biological damage may be manifest (carcinogenic, teratogenic, mutagenic, and clastogenic), and that species and strain differences make it difficult to extrapolate experimental data from laboratory experiments to man. Compounds that are suspect include sodium nitrate, sodium cyclamate, caffeine, food dyes, aflatoxin, pesticides, and some solvents. The data on certain drugs, i.e., LSD and marijuana, are contradictory, whereas the evidence that other more socially acceptable drugs (alcohol, tobacco) cause biological damage is less controversial.

References

AUERBACH, C. 1967. The chemical production of mutations. *Science,* **158:** 1141–48.

BADR, F. M., and R. S. BADER. 1975. Induction of dominant lethal mutations in male mice by ethyl alcohol. *Nature,* **253:**134–36.

COHLAN, S. Q. 1964. Fetal and neonatal hazards from drugs administered during pregnancy. *N.Y. State J. Med.,* **64:**493–99.

DRAKE, J. W., et al. Environmental mutagenic hazards. 1975. *Science,* **187:**503–14.

LITTLE, J. B., A. R. KENNEDY, and R. B. McGANDY. 1975. Lung cancer induced in hamsters by low doses of alpha radiation from polonium-210. *Science,* **188:**737–38.

MAUGH, T. H. 1974. Marihuana (II): does it damage the brain? *Science,* **185:** 775–76.

MAUGH, T. H. 1974. Vitamin A: potential protection from carcinogens. *Science,* **186:**1198.

NAHAS, G. G., et al. 1974. Inhibition of cellular mediated immunity in marihuana smokers. *Science,* **183:**419–20.

SHAW, M. W. 1970. Chromosome damage by chemical agents. *Ann. Rev. Med.,* **21:**409–32.

SMITH, D. J. 1975. Increased neonatal mortality in offspring of male rats treated with methadone or morphine before mating. *Nature,* **253:**202–3.

VOGEL, F., and G. RÖHRBORN, eds. 1970. *Chemical Mutagenesis in Mammals and Man.* New York: Springer.

WHITE, S. C., S. C. BRIN, and B. W. JANICKI. 1975. Mitogen-induced blastogenic responses of lymphocytes from marihuana smokers. *Science,* **188:**71–2.

Questions

Useful terms: carcinogenic, teratogenic, mutagenic, clastogenic.

1. What are the three major categories of action of chemical compounds on biological systems?

2. Briefly describe the history of the events leading to the discovery of chemical mutagenesis.
3. What are some of the problems involved in determining whether a chemical compound has some deleterious effect on humans?
4. How may metabolism determine the activity or inactivity of a compound?
5. Make a list of substances commonly ingested by humans that have been suspected of having a deleterious effect in the categories mentioned above.
6. Make a list of drugs that have been investigated for their possible mutagenic effects over the past several years, and indicate whether the evidence supports or refutes the possibility that they are biologically active.
7. Why is it not a simple matter to test each new chemical compound or drug for deleterious action prior to its being put on the market?
8. How would you compare the relative danger of damage from injurious chemicals compared to radiation on the one hand and accidents of civilization (automobile, etc.) on the other.

16

Twinning

One of the commonest criticisms directed at the science of genetics by social scientists is that the biological basis of human attributes is grossly oversimplified by geneticists. This criticism is, to a large extent, justified. During the early days of genetics, such characteristics as honesty, virtue, and criminality were commonly depicted in pedigrees as the consequence of single dominant or recessive genes and, even today, there is much discussion in the daily press of the spectacular "advances" to be made in the very near future as biochemists and geneticists learn to tinker with single genes in the human complement. Nevertheless, it seems self-evident that the important attributes of humanness, to whatever extent they are genetic, are not determined by single genes. Certainly the most obvious physical characteristics of height, weight, and bodily structure, as they appear as variants in the normal population, are not controlled by simple genetic factors, and one might imagine the genetic basis to be equally complex for behavior, intelligence, and other more important mental and emotional qualities. In fact, there are those who would argue the differences among individuals stem predominantly if not entirely from environmental, rather than genetic, variables. Where evidence for some genetic contribution to characteristics of this sort exists, it appears that alleles at many loci make contributions to the phenotype; that is, the genetic system is *polygenic.*

Because genetic analysis of any characteristic is difficult when there are large numbers of different alleles at different loci making variable contributions to the phenotype, it is desirable, whenever possible, to make the genetic background uniform, except for the locus, or loci, under investigation.

In the lower organisms and particularly in agriculturally important plants and animals, it is possible, by successive generations of inbreeding, to produce strains in which individuals are homozygous at almost all loci, with all members of the strain having essentially the same genetic composition; that is, they are *isogenic.* Variations that then appear among such isogenic individuals can be attributed to environmental causes. The progress of experimental genetics would be seriously hampered if genetic variation could not be kept under control in some way. However, it is obviously not possible to produce isogenic strains of humans. This problem can be circumvented to some extent by studying twins.

Use of Twins

Each person has a unique genetic constitution, not only different from every other living person, but also probably different from that of any other person who has ever existed. The exception to this rule is the case of identical twins, two different persons who are both the product of a single fertilized egg and therefore have the same genetic content. Consequently, twins provide a natural experiment that yields data not available from any other source.

GALTON'S CONTRIBUTION. The possibility of using twins in assessing the influence of inheritance was suggested by Francis Galton, a cousin of Charles Darwin. Galton was a pioneer in the application of mathematics to biology. His basic philosophy seemed to be that virtually anything could, and should, be measured. He attempted to ascertain the existence of a Supreme Being by comparing frequencies of shipwrecks of vessels with and without missionaries aboard, on the assumption that the latter might be shipwrecked more frequently. They were not. Along a similar line, he checked on the effectiveness of prayer by examining the longevity of royalty, who might be expected to live longer since their subjects prayed for their health and safety. They did not.

In his researches on twins, Galton obtained most of his information by correspondence with friends and the friends of friends who were twins, or who had twins in their families, 35 of whom were probably identical and 20 probably fraternal. Under these circumstances, the accuracy of his observations might be questioned. Nevertheless, he recognized in a general way the existence of the two types of twins, now known as *identical* and *fraternal,* and properly assessed their potential value in distinguishing between genetic and environmental contributions to specific traits. He came to the conclusion that twins who were very much alike in childhood remained so throughout their lifetimes, whereas those who were quite different were no more or less different in later life than ordinary sibs. He also noted that in many cases twins developed the same kind of mental illness.

Characteristics of Twinning

TYPES OF TWINNING. There are two fundamentally different kinds of twins: those who arise from two zygotes, *dizygotic (DZ)* or *fraternal twins,* and those who arise from a single zygote, *monozygotic (MZ)* or *identical* twins. Fraternal twins are of no greater genetic relationship to each other than sibs born at different times; about half the time they are of the same sex. Identical twins, on the other hand, come from a single fertilized egg that has divided during the course of early development into two cells or two cell masses, each forming a new individual. Such twins are genetically identical to each other, with rare exceptions to be noted later.

INCIDENCE OF TWIN BIRTHS.   The frequency of twinning in the White population of the United States is slightly more than one twin pair per hundred births, with about twice as many fraternal twin births as identical. Older mothers tend to bear more twins, and the recent decline in the frequencies of twin offspring may be the consequence of a reduced reproduction rate of older women. However, in other population groups in which older women still produce children, the twin rate is high. Thus in Ireland, where women tend to marry at an older age, the incidence of twinning is the highest in Europe, 14 per thousand births.

    There are also differences between population groups. In the Black population of the United States, the rate is about one twin pair per 73 births, or about 14 per thousand. In one African tribe, the Yorubas, the frequency has been reported to be as high as one pair per 30 births. On the other hand, among Asians, Japanese and Chinese have a very low rate, with an average of only one twin pair per 160 births. It is interesting that these groups differ only in frequencies of fraternal twinning; identical pairs appear to occur with the same frequency in all populations (Table 16-1). The ratio of fraternals to identicals is about 8:1 among the Yorubas, but the direction is reversed among the Asiatics, where the identicals are twice as frequent as the fraternals.

THE EMBRYOLOGICAL ASPECTS OF TWINNING.   At the time of a twin birth, the obstetrician usually makes a preliminary determination of the type based on the distribution of fetal and maternal tissues that make up the afterbirth. These include the *amnion, chorion,* and *placenta.*

    Early in development of the largely undifferentiated mass of cells, an amniotic cavity is formed within which the embryo will develop. The membrane surrounding the cavity becomes the amnion, which will be the innermost environment of the birth membranes (Figure 16-1). Part of the outermost layer of the cell mass will form a chorion, which as development proceeds becomes thinner and forms a second cell layer around the fetus. The placenta is of dual origin, formed at the juncture of the fetal chorion and the maternal tissue. Figure 16-2 shows

	MZ	*DZ*
Indians of Arizona, New Mexico, and Oklahoma	2.0	5.0
Oriental	4.5	2.5
U.S. White	3.8	7.4
U.S. Black	3.9	11.8
Johannesburg Black	4.9	22.3
Nigerian Black	5.0	39.9

Table 16-1   Twinning rates per thousand births by race of the mothers. (From tables by N. E. Morton, C. S. Chung, and M. P. Mi. *Genetics of Interracial Crosses in Hawaii,* S. Karger, AG, Basel, 1967; D. Hewitt and H. Stewart, *Acta. Genet. Med. Gemellol.,* **19:**84, 1970.)

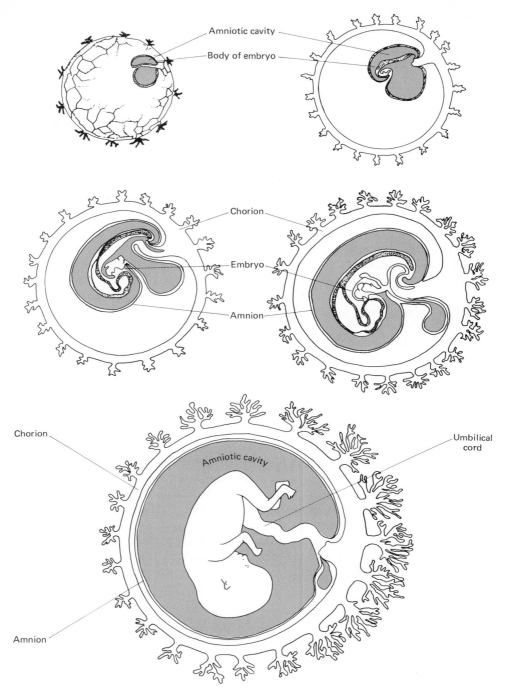

Figure 16-1 The development of the embryo showing the derivation of the birth membranes. The amnion originates as the lining of the small amniotic cavity and eventually comes to surround the fetus. The chorion surrounds the amnion. (B. M. Patten. *Foundations of Embryology*, McGraw-Hill, New York, 1958.)

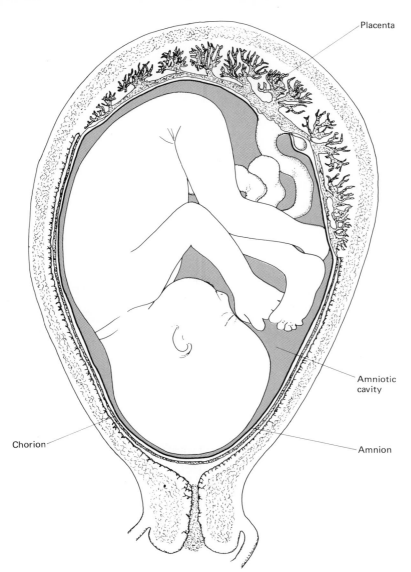

Placenta

Figure 16-2 A more advanced fetus, showing the relationship of the placenta, chorion, and amnion. (B. M. Patten. *Foundations of Embryology*, McGraw-Hill, New York, 1958.)

Amniotic cavity

Chorion

Amnion

how these three types of tissue may individually surround a fetus at the time of birth.

Since both monozygotic and dizygotic twins may develop as separate embryos, they may develop with separate amnions, chorions, and placentas (Figure 16-3*A*). MZ twins may have a single chorion (DZ twins never do), and within that chorion may have either double (Figure 16-3*B*) or single (Figure 16-3*C*) amnions. From this it is clear that the identification of twins as monozygotic or dizygotic is not unambiguous unless there is a single chorion. The frequencies of some of these combinations are given in Table 16-2. The determination at the time of birth is subject to considerable error (unless the twins are unlike-sexed) and

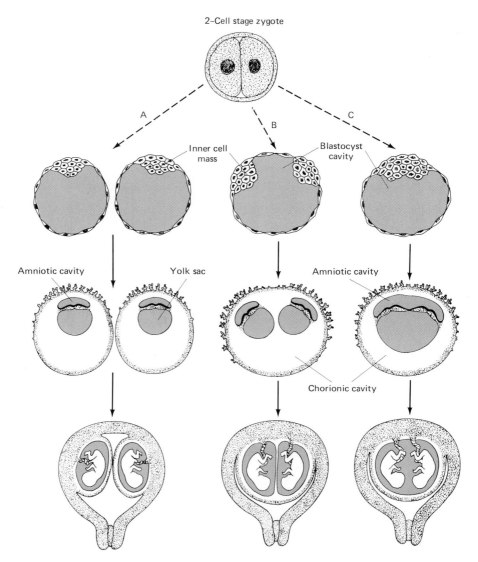

Figure 16-3 Schematic diagrams showing the distribution of the birth membranes in monozygotic twinning. *A.* The separation of the two cell masses is complete, so that each embryo has its own amnion, chorion, and placenta. Sometimes the two placentas will fuse to form a single one. This sequence is also characteristic of dizygotic twinning. *B.* The two inner cell masses separate, but within a single blastocyst cavity. Each embryo then has its own amnion, but both are within a single chorion. *C.* The inner cell mass does not separate until later so that both embryos are included within a common amnion and chorion. (J. Langman. *Medical Embryology*, Williams & Wilkins, Baltimore, 1969.)

very often twins misclassify themselves, primarily on the basis of the obstetrician's diagnosis at the time of their birth. One study shows that about a quarter of all MZ twins believe they are dizygotic, whereas a fifth of the DZ twins consider themselves monozygotic.

Table 16-2 The frequencies with which MZ and DZ twins are found within a single or a double chorion. This table shows that the type of twinning can be definitely established only if there is a single chorion, that twinning type being MZ. (From J. H. Edwards, in *Birth Defects: Orig. Art. Ser.*, ed. D. Bergsma. *New Directions in Human Genetics.* Published by Williams & Wilkins Co., Baltimore, for The National Foundation—March of Dimes, White Plains, N.Y., Vol. I, (2):79, 1965.)

Type of Twinning	Number of Chorions		Totals
	Single	Double	
MZ	20%	10%	30%
DZ	0	70%	70%
			100%

It was noted above that the frequency of MZ twinning is fairly constant from one group to another and the frequency of DZ twinning quite variable. It would appear that the chance of the separation of the zygote, or its early products, into two cell masses to produce MZ twins does not depend much on the genetic constitution of the individuals involved, whereas the probability of multiple ovulation, which would present the opportunity for two different eggs to be fertilized by two different sperm, does.

THE ALGEBRAIC ANALYSIS OF TWINNING FREQUENCIES.   It would be useful to be able to analyze twin data as to zygosity on a mathematical rather than biological basis. This would make it possible to extract information from crude birth records and come to some conclusions about the frequencies of the different kinds of twin births in the overall population.

This approach was suggested by the German physician Weinberg many years ago. Basically it is very simple. Since the ratio of male to female births is about 50:50 (see Chapter 18), DZ twins stand a 50 percent chance of being of unlike sexes. On the other hand, MZ twins must be like-sexed. Therefore, if we look at any birth data on twins, we should see an excess of like-sexed twins (coming from the MZ class), and we should be able to make use of this excess to estimate the numbers of the two types of twins.

Suppose, for instance, that in a population of 1,000 twin births, there were 200 unlike-sexed twins and 800 like-sexed twins. What proportion of MZ and DZ births would this distribution represent? The answer is very simple. Corresponding to the 200 unlike-sexed (boy/girl) twins which must be of DZ origin, there should be another 200 twins of like sexes (girl/girl or boy/boy) also of DZ origin. If this total of 400 DZ twins is then subtracted from the total of 1,000, we are left with 600 like-sexed twins who must represent the MZ twins. We would therefore conclude that in this particular population of twins 60 percent are monozygotic and 40 percent dizygotic.

It is possible to look at census birth data and calculate the probable frequencies of MZ and DZ twins, knowing no more than the number of like- and unlike-sexed twins. This has been done by a number of workers to show the change in the frequency of the different kinds of births in the Black and White populations in the United States. One such analysis is shown in Figure 16-4. From this it appears that the frequency of MZ twinning in both races does not change appreciably as the age of the mothers changes but that the frequency of DZ twinning increases markedly as the mother reaches middle age, more than quadrupling in frequency from age 17 to 38, with a small drop thereafter. At the highest level, the percentages of twin births are almost 1.3 percent for the White population and 1.2 percent for the Black population.

All of the figures thus far quoted underestimate the number of twin pregnancies. The frequency of miscarriage or stillbirth of one member of a twin pair is much higher than that of single fetuses, and such pregnancies will not be classified as multiple. Furthermore, in many cases both twins may be lost very early in pregnancy. Finally, it should be kept in mind that in about one-eighth of all twin births at least one of the members fails to survive, even when born in a hospital. It would appear that the frequency of the initial event may be much higher than now calculated.

HIGHER MULTIPLE BIRTHS.   The frequencies of triplets and quadruplets are very much less than that of twins. According to a simple rule originated by Hellin, and sometimes called Hellin's rule, the incidence of twins is 1 in 89, the observed incidence of triplets about 1 in $(89)^2$, and quadruplets about 1 in $(89)^3$. These figures might make sense if all fertilizations were independent, with a 1 in 89 probability of ovulation and fertilization of each additional egg after the first. However, because of the variable proportion of twinning and multiple births of monozygotic origin confounding the statistics, as well as the very greatly increased hazard of mortality of embryos and fetuses in multiple pregnancies,

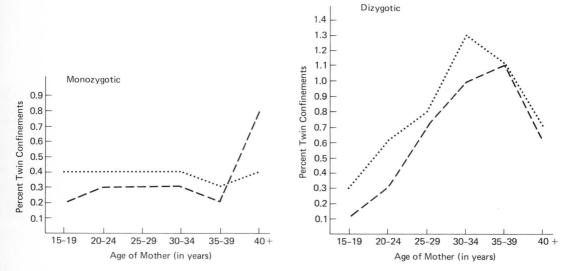

Figure 16-4   The change in frequency of twin births as a function of the age of the mother. The monozygotic frequencies for both Black and White races appear to be fairly constant at all ages; the dizygotic frequencies increase sharply to about age 40. (N. Myrianthopoulos. *Am. J. Hum. Genet.*, November 1970.)

this rule can be considered an interesting rule of thumb with a fortuitous agreement with the observed data.

INHERITANCE OF THE TWINNING TENDENCY.   Studies of the Yorubas in western Nigeria, where polygamy is practiced, have made it possible to get additional information about the contributions not only of the woman but also of the man to the twinning tendency, since it is possible to compare the frequency of twinning among the various wives of one man. The very high twinning frequency in this group is a distinct advantage in this analysis. These studies show (1) that mothers who have previously had twins do not have any greater tendency toward twinning than those who have not, (2) that fathers do not contribute to the twinning tendency (since the other wives of their polygamous union do not have any more twins than average), and (3) that mothers who are themselves twins do not have a higher tendency toward twinning than those who are not twins.

The total information on the inheritance of twinning is diffuse and vague. There do exist pedigrees with a concentration of twin births that cannot be attributed to chance alone. On the other hand, the sibs of parents of twins show only a very slight increase above average in the probability of their having twins also. Since DZ and MZ twinning have completely different biological causes, one might expect that a genetic predisposition to twinning would result in a clustering of one or the other but not both types of twinning within the same family pedigree. In those few cases in which there is an indication of a slight clustering of twins, they appear to be randomly distributed among the two types.

One interesting exception is that of a familial fraternal twinning tendency in which the twins are unequal in size. It appears to depend on the genotype of the embryo, a dominant gene having been transmitted by either parent. This is explained by the assumed failure of the embryo carrying the dominant to influence hormone production that normally prevents a second egg fertilized a month or more later from implanting successfully after the first has been implanted. The occurrence of a second implant months after a first is known as *superfetation*.

Generally speaking, however, the evidence for a twinning tendency is weak, and if any conclusions at all can be drawn it is that the dizygotic twinning tendency may have a small polygenic component, expressing itself by way of the mother of the twins, and not the father.

FETAL LOSS IN TWINNING.   It would be expected that the frequency of twinning would go down in a population in which there is a high rate of fetal loss, since, for a twin pair to be born, both members must survive. All populations carry recessive lethal genes that will be brought together more often under conditions of inbreeding, resulting in embryonic of fetal death. Because the loss usually occurs very early, the pregnancy may not even be noticed. If two populations are compared, one with an embryonic loss of 10 percent and the other with an overall embryonic loss of 20 percent, the difference could go unnoticed because there is now no good method to measure early embryonic loss, and, in any case, these losses are only a fraction of the total zygotic losses found in all populations. However, the chance of both twins surviving in the first population would be $.9 \times .9$ or $.81$ and $.8 \times .8$ or $.64$ in the second. Thus the frequency of DZ twinning, with its dependence on the genetic constitution, will appear to be about 30 percent higher for the first population than for the second.

American Indians have a relatively high rate of inbreeding compared to other groups, so it is possible that their lower twin frequency may be explained in this way. It should be pointed out, however, that such loss, of genetic origin, should affect the frequency of DZ twinning only. Because MZ twins have identical genotypes, they would both be affected or not in the same proportions as single children, with no net change in the MZ twinning frequency. Perhaps the quite variable DZ but relatively constant MZ birth frequencies from one population to another (Table 16-1) are a reflection of some fetal loss due to genetic factors.

Similarly, it has been observed that there is a low frequency of trisomy-21 in twin births. An embryo with trisomy 21-has only a one-third chance of surviving to birth, so the chance of two such indentical twins surviving simultaneously is very much reduced. The data show that among twin pairs with at least one affected by trisomy-21 there are far fewer identical pairs than would be expected.

Diagnosis of Zygosity

PHYSICAL SIMILARITIES.   If one were challenged to make a list of specific physical characteristics—facial features, for instance—with a genetic basis, it would be difficult to name half a dozen. For this, it might appear that the total number of such features is quite limited. However, a comparison of the features of two identical twins, often so alike that their parents and sibs may confuse them, leads to the opposite conclusion—that the genetic composition of a person is effective in determining even the most trivial aspects of physical appearance.

It should come as no surprise, then, to discover that measurements of physical attributes show MZ twins to be much more alike than either DZ twins or non-twin sibs. (It should be noted that for all comparisons of MZ twins with either DZ twins or sibs the latter two must be limited to like-sexed pairs to avoid the obvious difficulties in comparing boys versus girls. In all subsequent discussions of comparison, it should be taken for granted that both DZ twin and sib data come from like-sexed individuals, unless otherwise stated.) Thus, the average difference in height between sibs as they reach the same age is 4.5 cm, or about 2 inches; for DZ twins about the same, 4.4 cm; but for MZ twins only 1.7 cm (Figure 16-5). Fingerprint patterns of MZ twins are quite similar, and on the basis of this criterion alone zygosity can be diagnosed correctly in about 86 percent of all cases. Other physical characteristics show varying degrees of likeness. Weight, which might be expected to be more dependent on environmental influences, is not much more alike in MZ than in DZ twins.

PHYSIOLOGY.   Virtually all aspects of physiology are more similar in MZ twins. Blood pressure, breathing rate, changes in heartbeat and breathing after exercise, and electroencephalographic patterns are strikingly similar. DZ twin girls show, on the average, a difference of about a year in the time they begin menstruating, whereas MZ twins differ by less than 3 months. Curiously, left-handedness, although generally more frequent in twins, appears more often in fraternal than in identical twins.

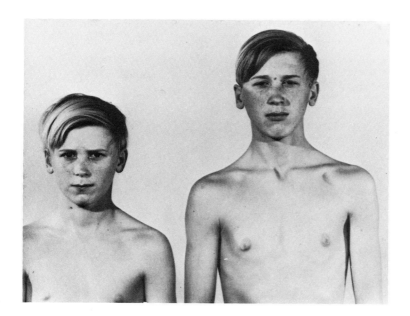

Figure 16-5 A pair of twins, superficially somewhat dissimilar, particularly in height, but alike in all other criteria to the extent that the probability of monozygosity is practically 1. This is a nice illustration of the danger of dogmatic generalizations based on appearances only. (S. Pruzansky, M. Markovic, and D. Buzdygan. *Acta Genet. Med. Gemellol.* **19**:225, 1970.)

INDEX OF SIMILARITY. When two twins are alike with respect to having a given characteristic, they are said to be *concordant;* if different, they are *discordant*. A convenient index, the frequency of concordance, is calculated by dividing the number of twin pairs in which both have the characteristic by the number in which at least one has the characteristic. Thus, if 83 MZ twin pairs are examined for the presence of club feet and 50 pairs show it not at all, 20 pairs have one member affected, and 13 both, the frequency of concordance in this hypothetical example would be 13/(20 + 13) or .39.

Frequencies of concordance for some common developmental defects are given in Table 16-3. From these figures, we would be justified in concluding that, although developmental variables play the more important role in their determination, the genetic determination (probably, in most cases, polygenic) cannot be disregarded.

BLOOD GROUPS. In addition to the information on the birth membrane and physical similarity (including sex), other criteria may be weighed in the judgment of zygosity. The blood groups are particularly helpful when the parents are so constituted as to produce a variety of phenotypes in their progeny. Clearly if two twins are of group O and both their parents are O, the similarity is of no use in determining zygosity, since *all* of the children in that sibship will be of group O. On the other hand, if one of the parents is A and the other B, then the similarity of O children would be helpful because the parents, who must be heterozygous *AO* and *BO*, could produce A, B, and AB children as well as O. Dissimilarity with respect to any one of the blood groups—ABO, MN, Rh, etc.— would lead to the conclusion of dizygosity immediately; identity with respect to all the blood groups would suggest monozygosity with a very high probability. Since half of all DZ twins are of opposite sex and the distribution of the alleles of the blood groups is such that 80 percent of the remaining half are likely to

Table 16-3 Estimated rates of concordance for some congenital malformations in MZ and DZ twins. (S. Hay and D. Wehrung, *Am. J. Hum. Genet.*, **22**:6, 1970.)

Congenital Malformations	MZ Pairs			DZ Pairs		
	Total Pairs	Concordant	Percent Concordant	Total Pairs	Concordant	Percent Concordant
Cleft lip, with or without cleft palate	51	9	17.6	84	2	2.4
Cleft palate	15	6	40.0	42	2	4.8
CNS malformations	112	6	5.4	110	2	1.8
Congenital heart disease	69	5	7.2	60	2	3.3
Positional foot defects	99	19	19.2	192	4	2.1
Polydactyly	47	20	42.6	106	6	5.7
Reduction deformities	27	2	7.4	32	0	0.0
Down's syndrome	5	4	80.0	46	0	0.0

differ in one or more major blood group antigens, 90 percent of all DZ twins will be clearly DZ on the basis of these two criteria.

CHROMOSOME POLYMORPHISMS. The new staining techniques that reveal chromosome bands and other longitudinal differentiation have revealed widespread permanent structural variations in otherwise normal chromosomes. When these variants are found segregating in sibships, they may be used as genetic markers like any other inherited trait, and identical twins should be concordant in this regard also. In one study involving 16 like-sex twin pairs, the determination of zygosity on the basis of chromosome variants agreed in every case with the determination based on the blood groups.

BIOCHEMICAL IDENTITY. One interesting test for identity based on subtle biochemical differences has been reported in a reputable medical journal. A trained bloodhound can readily distinguish between sibs and DZ twins, but will confuse the scents of MZ twins. Unfortunately, this sort of test, apparently about 100 percent effective, might be difficult to implement in the average run-of-the-mill genetics clinic. In any case, such extreme measures may not be necessary; an experienced observer can make a correct zygosity diagnosis with an accuracy of 90 to 95 percent on the basis of physical features alone.

ANTIGENIC SIMILARITY. Other tests for zygosity depend on differences in antigen specificities likely to be found in DZ twins but not in MZ twins. One of these is the mixed lymphocyte test, described in Chapter 9. If lymphocytes from

MZ twins are mixed in a test tube, there should be no effect on mitosis since they should carry the same antigens, whereas the lymphocytes from DZ twins should induce mitosis, just as a mixture of those from sibs does. A more critical test, quite informative but more difficult to perform, involves making reciprocal skin transplants between two sibs. Skin, one of the most difficult of tissues to get to take, will be rejected by the nonidentical host in almost every case, but will usually be accepted when the graft is from an identical twin.

Unusual Twin Events

CHIMERAS.   Very often biologically important observations are made by chance during routine clinical tests. An illustration of this comes from a blood-typing laboratory in England.

When a certain Mrs. K. in Sheffield was blood-typed, it was found that she had two types of blood cells, about three-fifths of type O and the rest A. These results prompted the investigators to ask a simple question: was she a twin? Her answer was yes, but given with some surprise since her twin brother had died 25 years earlier.

Why did the investigators ask her this question? Because the most likely explanation for her condition was that she was one of a twin pair and that during embryonic development cells of the other twin (group A) became incorporated with her blood-forming system, which happened to be of group O. In fact, by further testing the fraction of her cells that were A, it was possible to describe the antigenic properties of eight additional blood groups of the long-deceased twin brother. Curiously, he happened to be more similar in these particular groups to an older sister than he was to his twin. The invasion of A cells could not have been very extensive, except for erythroblastic tissue, because those cells must have had a Y-chromosome (coming from a male twin), but there were no evidences of the masculinization of Mrs. K., expected if the migrating Y-bearing cells had had any developmental impact. In fact, Mrs. K. had married and had had a child.

Several such cases have since been found in which both twins were still alive. Each one had some cells of the blood groups of the other twin and appeared to have normal sex development in cases of unlike-sexed twins. It is likely that the interchange of blood cells occurred rather late in development by way of a fusion of blood vessels. This event is well known in cattle, in which DZ twins interchange cells in 90 percent of the cases; when the sexes are different, the hormones from the male cause the female to develop into a sterile, somewhat malelike type called a *freemartin*. Such an effect of one sex on the other in developing twins is unknown in humans.

When cells of two demonstrably different origins combine to produce one individual, as in the case of Mrs. K., the result is called a *chimera*. This is a distinctly different phenomenon than mosaicism. A mosaic consists of two different types of cells, originally the same at the stage of the zygote, but becoming genetically different subsequently by some event such as mutation, by nondisjunction in one cell line, or by X-chromosome inactivation.

How common chimerism is in humans is unknown. Studies of twin pairs for evidences of cell interchange between them suggest a very low frequency; however, it is possible that some cases classified as mosaicism, and explained by mutation or nondisjunction, may in fact have arisen by cell migration. For instance, there are persons some of whose cells appear to be XX and others XY. Possibly there existed two twin zygotes initially of either sex; after one twin acquired cells from the other, the second twin embryo might have failed to become a viable fetus, or, for that matter, might not have been formed in the first place if the two cell masses fused to give a single embryo. Such a single birth would not be recognized as a case of chimerism, and if subsequent inspection showed two different kinds of cells to be present the person would simply be referred to as a mosaic.

ABNORMAL DEVELOPMENT.   In some cases, the division of the dividing cells into two different groups occurs rather late in development. The embryos may then have some structural feature in common, to produce what have been called "Siamese twins." Depending on the nature of the common elements, the twins may (or may not) be separated safely by surgery. Such twins are monozygotic.

SUPERFECUNDATION.   Among the unusual circumstances surrounding the birth of twins are those few cases in which it has been shown that the two twins must have had different fathers. One such case involved a mother with twin son and daughter. It was shown that the girl twin was A and MN, inconsistent with the groups of the mother (O and M) and her husband (B and M). However, the boy twin was B and M, consistent with those groups. A boarder at the residence proved to be A and MN, consistent with the groups of the girl twin but not the boy. The twins, then, seem to have had different fathers. This phenomenon, the production of members of the same multiple birth by different fathers, well known in domestic animals, is known as *superfecundation.*

MZ TWINS OF DIFFERENT PHENOTYPES.   It is to be expected that if one member of an identical twin pair carries some unusual genetic trait the other will as well. This is implicit in the term "identical." Identical twins who both had Down's syndrome, or both Klinefelter's syndrome, or both Turner's syndrome have been reported, as well as three instances of twins who had both Down's and Klinefelter's syndromes simultaneously.

In rarer instances, twins who satisfy all the criteria for monozygosity are strikingly different. Sometimes one appears to be male, the other female. In principle, the origin of these cases is simple. Mitotic nondisjunction may occur in an XY zygote during early development (Figure 16-6), giving rise to two cell lines, one XY and the other X0 (having lost the Y). Later separation of the cell lines into two different masses may occur in such a way as to produce two embryos, one a normal male and the other a female with Turner's syndrome, of MZ origin. In other cases, MZ twins are like-sexed but one is normal and the other chromosomally abnormal. For instance, if the zygote were originally XX, then the loss of one of the X's could give rise to a twin pair, one a normal female and the other having Turner's syndrome. This is actually the most common type of chromosomally unlike MZ pair. Since the mitotic event gives rise to a clone of cells that will tend to be located together, the cell masses separating later will preferentially include either normal or abnormal cells, although it would not be

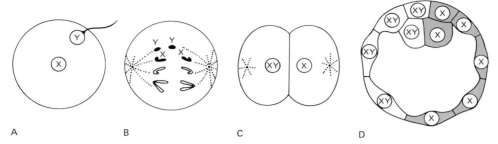

A   B   C   D

Figure 16-6 A hypothetical scheme to account for the origin of a mosaic embryonic mass of cells by chromosome loss during the first cleavage. A Y-chromosome is lost during mitosis so that some of the resulting cells of the XY embryo are X0. If this clump of cells now separates into two groups, one of constitution XY and the other X0, two monozygotic twins may develop, of opposite sexes. (From *Principles of Human Genetics,* Third Edition, by Curt Stern. W. H. Freeman and Company. Copyright © 1973.)

surprising to find (as is, in fact, the case) that such different MZ twins are sometimes mosaics (Figure 16-7). Curiously, one of the first unlike MZ pairs to be described consisted of a male with short stature whose lymphocytes were X0 and a Turner's syndrome female who was a lymphocyte mosaic for X0 and XY cells. It can be surmised that the male was also a mosaic, with some cells also of XY composition. In this case, the cells observed almost certainly did not include all the types actually present in the mosaic.

In half a dozen cases, MZ twin pairs have been described in which one was normal and the other had Down's syndrome. The course of events in this case is probably the following (Figure 16-8): After one or more normal mitoses of the zygote, a nondisjunctional event in a dividing cell gives rise to two different cell lines, one with three chromosomes 21 and the other with one. There are then three different cell lines present. The monosomy-21 cell type reproduces slowly or not at all and is eventually lost. The splitting of the cell masses made up of normal cells and trisomy-21 cells then occurs, creating twins along lines that results in one of them having primarily, if not entirely, disomy-21, and the other trisomy-21.

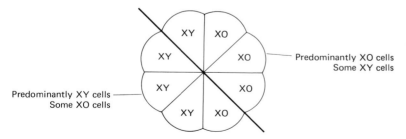

Predominantly XY cells
Some XO cells

Predominantly XO cells
Some XY cells

Figure 16-7 A schematic diagram showing how the cleavage of the cell mass into two might be along a plane that would make both the products mosaics. Clearly, since the cleavage might be in any direction, producing masses with unequal numbers of cells, the extent of mosaicism, if any, in each twin can be extremely variable.

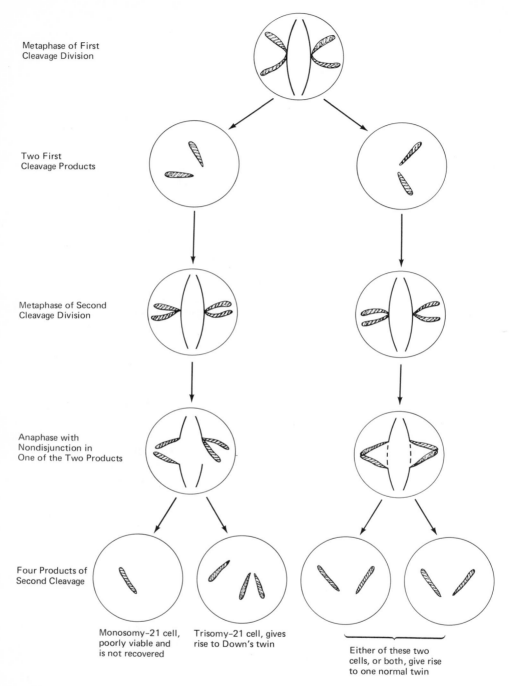

Metaphase of First
Cleavage Division

Two First
Cleavage Products

Metaphase of Second
Cleavage Division

Anaphase with
Nondisjunction in
One of the Two Products

Four Products of
Second Cleavage

Monosomy–21 cell,
poorly viable and
is not recovered

Trisomy–21 cell, gives
rise to Down's twin

Either of these two
cells, or both, give rise
to one normal twin

Figure 16-8   A series of early cleavage divisions, with nondisjunction of chromosome 21, giving rise to two MZ twins, one with trisomy-21 and the other normal.

Disease Conditions in Twins

INFECTIOUS DISEASE. Comparisons of degrees of concordance (similarity) in MZ versus DZ twins for various disease conditions give some indication of the degree of genetic cause of those diseases (Table 16-4). In these comparisons DZ twins are divided into two groups: like-sexed (DZ_1) and unlike-sexed (DZ_2). In this table, the first three pathological conditions show only a slight increase, if any, between MZ and DZ twin types, from which it can be concluded that the genetic contribution to the disease is far outweighed by other (environmental) factors. For the remaining conditions, even when known to be caused by infection (such as tuberculosis), there is a definite increase in concordance rates in MZ over DZ twins, suggesting a genetic predisposition toward that disease.

ALLERGY. For allergic disease, a clear genetic component is indicated (Table 16-5). Out of some 500 twin pairs, 76 were found to have at least one affected member. When these were then classified according to concordance and zygosity, most of the discordant pairs were dizygotic and most of the concordant pairs were monozygotic. Furthermore, when the pairs with an allergic member were then examined for the incidence of migraine headache, 23 cases were found compared to only one case in a control set of 76 pairs of nonallergic twins. This result strongly implies a basic connection between migraine headache and allergy.

CANCER. Although there are many relatively rare genetic diseases that are responsible for cancerous growths (neoplasms) affecting specific organs or areas of the body, the possible inheritance of a predisposition to cancer as a more general and widespread disease is a particularly vexing question. There is no doubt that a few families, as rare exceptions in the medical literature, show high incidences of cancer with an inheritance pattern that appears to be that of a simple single-locus effect. Furthermore, the probability of a second occurrence of a

Table 16-4 Concordance rates for various disease conditions in monozygotic twins (MZ), like-sexed dizygotic twins (DZ_1), and unlike-sexed dizygotics (DZ_2). (B. Harvald and M. Hauge, *Genetics and Epidemiology of Chronic Diseases*, H.E.W., PHS Publ. No. 1163.)

Type of Disease	MZ		DZ_1		DZ_2	
Verified cancer, same site	8/160	.050	8/335	.024	4/321	.012
Verified cancer, all sites	17/160	.106	43/335	.128	36/321	.112
Coronary occlusion	20/102	.196	24/155	.155	10/133	.057
Peptic ulcer	30/144	.208	24/208	.115	18/243	.074
Arterial hypertension	20/80	.250	10/106	.094	4/106	.038
Cerebral apoplexy	22/90	.224	16/148	.108	22/179	.123
Tuberculosis	50/135	.370	42/267	.157	36/246	.146
Rheumatic fever	30/148	.203	12/226	.036	14/202	.069
Rheumatoid arthritis	16/47	.340	2/71	.028	8/70	.114
Bronchial asthma	30/64	.469	24/101	.238	22/91	.242
Diabetes	36/51	.706	12/55	.218	10/55	.182

Table 16-5 Breakdown of twin pairs, in which one member suffers from allergy, by concordance and zygosity. (M. Milani-Comparetti, *Acta Genet. Med. Gemellol.*, **19**:240, 1970.)

	MZ	DZ	Total
Discordant	8	30	38
Concordant	28	10	38
Total	36	40	76

cancer in a family in which one member has already been affected is often doubled or trebled over the chance of a new occurrence in a family previously unaffected. Indeed, in the case of breast cancer, the chance may increase by as much as ninefold when other members of the family have had the disease. If this seems like a striking increase, it should be remembered that the overall incidence in the population is quite low and that even a ninefold increase does not actually amount to a substantial risk for any one individual.

In any case, in line with the evidence from concordance in identical twins (Table 16-4), it can be stated with some assurance that the genetic predisposition to cancer is relatively weak and that the major cause must be elsewhere. It is generally agreed that the primary cause is a change in the cell from an advanced to an embryonic state, and that the signal for this change may come from a variety of causes, including perhaps interactions between the normal components of a cell and any one of a large number of viruses. Furthermore, it is certain that different types of cancer may have different origins and quite likely that even the same type, as breast cancer, may be familial in some cases and not in others, the distinction being that the genetic component would be of greater importance in the familial than in the sporadic cases. Gross chromosome aneuploidies are found in some malignant growths; generally these are interpreted as the consequence of wild uncontrolled mitosis characteristic of the cells, rather than the cause of it. One exception to this statement is the specific alteration of chromosomes 22 and 9 found in *myelogenous leukemia* which affects the cells that give rise to the bone marrow; this will be discussed in Chapter 22.

References

BULMER, M. G. 1970. *The Biology of Twinning in Man.* Oxford: Clarendon Press.

EDWARDS, J. H. 1965. The application of knowledge. *Birth Defects: Orig. Art. Ser.*, **I(2)**:79.

HARVALD, B., and M. HAUGE. 1965. Heredity factors elucidated by twin studies. In *Genetics and the Epidemiology of Chronic Diseases.* Washington, D.C.: Department of Health, Education, and Welfare.

HAY, S., and D. WEHRUNG. 1970. Congenital malformations in twins. *Am. J. Hum. Genet.*, **22**:662–78.

LANGMAN, J. 1969. *Medical Embryology.* Baltimore: Williams & Wilkins.

MILANI-COMPARETTI, M. 1970. Allergy and disease in twins. *Acta Genet. Med. Gemellol.*, No. 19. Rome: Istituto di Genetica Medica e Gemellologia.

MITTLER, P. 1971. *The Study of Twins.* London: Penguin.

MORTON, N. E., C. S. CHUNG, and M. P. MI. 1967. *Genetics of Interracial Crosses in Hawaii.* Basel: Karger.

MYRIANTHOPOULOS, N. 1970. An epidemiologic survey of twins in a large, prospectively studied population. *Am. J. Hum. Genet.*, **22:**611–29.

NEWMAN, H. H., F. N. FREEMAN, and K. J. HOLZINGER. 1937. *Twins: A Study of Heredity and Environment.* Chicago: University of Chicago Press.

TURPIN, R., and J. LEJEUNE. 1969. Monozygotic twinning and chromosome aberrations (heterokaryotic monozygotism). In R. Turpin and J. Lejeune, eds., *Human Affliction and Chromosomal Aberrations.* Oxford: Pergamon.

Questions

Useful terms: polygenic, isogenic strain, dizygotic (DZ), monozygotic (MZ), amnion, chorion, placenta, superfetation, concordance, discordance, chimera, superfecundation.

1. Why may the science of genetics be justifiably criticized for its approach to the inherited nature of common human traits?
2. How do inbred lines in lower animals serve a useful function in genetic analysis? What common human phenomenon serves as a substitute?
3. What is the relative frequency of monozygotic and dizygotic twins? With what variables does this relation change?
4. Are twins almost always aware of the particular class they belong to? Why not? How often are they wrong?
5. How does the number of the different kinds of membranes surrounding the fetus differ in the two kinds of twins?
6. Suppose that 1,000 pairs of twins attend a twin convention and that there prove to be 900 of like sexes and only 100 pairs of unlike sex. What would your estimate of the frequency of the two types of twins be in the sample, assuming a sex ratio of 1:1?
7. If there is a set of twins in your sibship, and you also have a pair of twin cousins, does this make it likely that you yourself will probably have twins among your children? Would your answer be different if it is specified that twins were monozygotic rather than dizygotic?
8. Why are there fewer twin pairs with trisomy-21 than might be expected?
9. If you wanted to make a comparison of some characteristic in monozygotic and dizygotic twins, would you need a larger sample of one of the types over the other to select from? Why?
10. What kinds of characteristics of twins can be used to make rather positive determinations about the type of twinning?
11. How can a chimera for the blood groups be detected? What is the difference between a mosaic and a chimera?
12. Can twins who are identical by all of the usual criteria still be grossly different physically? How can this happen?
13. What does the study of concordance in twins tell us about the genetic basis for cancer?

17

Behavior and Intelligence

Genotype and Behavior

Many persons regard the human at birth to be virtually devoid of behavioral characteristics, except, possibly, for a few innate drives like maternal concern for offspring and interest in the opposite sex. Variations from the norm, according to this idea, are entirely environmental in origin. This point of view is consistent with the democratic ideal that "all people are created equal" and that the maximization of opportunity for all and improvement of social conditions will inevitably produce a society free of ignorance, crime, and mental illness. Such a view denies the existence of any substantial genetic contribution in accounting for the differences between individuals. For this reason, the burden of proof might rest with those proposing any substantial effect of the genotype. In this section we shall bring together evidence concerning different behavioral characteristics, some more convincing than others, in favor of a genetic contribution.

There is no difficulty finding specific cases in which the genotype of an individual affects his behavioral characteristics. A wide variety of genetic conditions lead to obvious differences in the behavior of the affected person. At one extreme are those cases in which the behavior of an individual is profoundly affected by genetically simple conditions, such as phenylketonuria, caused by homozygosity for a single recessive gene, or trisomy-21, the addition of a single chromosome. In such cases, development may be so defective that the deficiency of mental ability affects the individual's capability to perform satisfactorily throughout his entire lifetime.

There are other clearly genetic conditions with effects not quite so drastic but still with a profound influence. Hereditary deafness is one such that, as a secondary effect, may reduce language development, academic capabilities, and normal interaction with other people. Other inherited characteristics operate at a more subtle level to lead to a diminished participation in normal activities. Any genetic condition that leads to a physical disease, a lowered vitality, a reduced capability of performing physical endeavors, or an inability to perform adequately in academic life (for instance, poor eyesight) will affect behavior in the long run. It might even be argued that the sex of an individual is a genetic determinant of behavior. These may not seem to be good examples of a genetic component to behavior, however, because they are either too abnormal or too much a part of the normal aspects or ordinary human life.

Instructive behavioral examples can readily be found in domesticated animals. For instance, in the relatively short period of time that dogs have been intensively inbred, man has been able to produce different strains with quite distinct behavioral characteristics, with certain strains known for their specific psychological traits. The terrier is known as an aggressive type, the sheep dog as one with a natural instinct for herding, and the German shepherd as a good watchdog, and behavioral differences among strains could be extended to include most, if not all, strains of dogs (Figure 17-1). We would not, however, expect to find such widely differing innate characteristics in the human population, if for no other reason than that intensive selection for specific characteristics has not taken place and intermixing of different groups of people over a long period of time

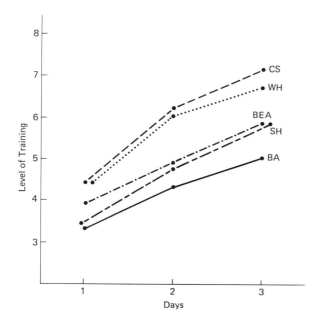

Figure 17-1 Differences in response of five different strains of dogs to obedience training. From the top, the strains include the cocker spaniel (*CS*), white-haired fox terrier (*WH*), beagle (*BEA*), Shetland sheepdog (*SH*), and basenji (*BA*). (J. P. Scott and J. L. Fuller. *Genetics and the Social Behavior of the Dog,* University of Chicago Press, Chicago, 1965.)

has diluted any tendency toward concentration of specific characteristics in any one group that might theoretically have been possible.

Differences attributed to nationalities, such as greater industriousness of one or stronger emotion of another, are undoubtedly more of a fictional than a factual origin (and are likely to be due primarily to nongenetic factors to the extent that they exist). For evidence of a genetic basis for behavioral characteristics, we must look elsewhere.

Except for the pathological conditions of the sort noted above, giving rise to severe deficiencies in behavior, there are no known genes that are specifically responsible for honesty or dishonesty, activity or lethargy, violent or calm temperament, or other such traits, even though the early literature on "human genetics" consisted almost entirely of such pedigrees (Figure 17-2). It has been mentioned previously that achondroplastic dwarfs appear generally to be somewhat happier than other individuals of similarly small stature; if this is true, it represents a rare exception to the preceding generalization with respect to the effects of single genes.

Abnormal Chromosome Constitution

When an individual is characterized by a deviation from normal diploidy, the first effect to be noticed is a depression of intelligence. The well-known exception to this involves additions and subtractions of the X-chromosomes, for which there is a special explanation in terms of X-chromosome inactivation. Deviations from the balanced autosomal state are responsible for other characteristics that affect the individual's behavior. Since these are in the range of pathological types, we shall not consider them further in this connection.

THE XYY MALE. Of more interest is the controversial effect of additional Y-chromosomes on the behavior of males. For each study showing a correlation of this chromosomal condition with abnormal aggressive tendencies, another can be found denying such a correlation. At this stage, it is possible only to present evidence on both sides of this unresolved situation.

Studies on XYY individuals suggest that these males are quite likely to be involved in behavioral problems related to the impulsive, exaggerated performance of a deed that is carried out without regard for the long-range consequences. This behavior very often leads to intrusions on other people or their property followed by arrest and imprisonment. Comparisons of XYY males with ordinary males, however, show that among those institutionalized the XYY prisoners had convictions more frequently for crimes against property and less frequently for crimes of violence against individuals. It should be noted also that although some noninstitutionalized XYY males are indistinguishable from normal, other exhibit an antisocial behavior pattern which goes well beyond the normal range, this happening more frequently than it does with XY males.

Although the incidence of XYY newborn infants has been variously estimated from about 1 in 1,000 to 1 in 3,000, the incidence of XYY individuals in some institutions for the criminally insane and mentally defective delinquents, based on surveys from Europe, the United States, Canada, and Australia, is about 1 in 50. On the other hand, a recent study of 2,538 inmates of all types of institutions in England has not revealed a significantly greater incidence of XYY males than expected on a random basis.

THE XXY MALE.   Males with Klinefelter's syndrome (XXY), of birth frequency one in 500, are found in psychiatric institutions at a frequency of one in 100. They show an increased incidence of mental retardation and are found as mentally retarded institutionalized males also with a frequency of one in 50. They are therefore much more likely to be characterized by some major behavioral disability than the average person, and when they are classified according to the kind of misdeed responsible for their incarceration their defect is often one of failure in accomplishing some task, or an inability to do so, a pattern different from that of an XYY.

EFFECT OF ABNORMAL Y-CHROMOSOMES.   In several instances, unusually large Y-chromosomes have been reported, and at the same time it has been suggested that an aggressive or violent behavioral pattern of the males carrying them is associated with this Y (Figure 17-3). The obvious implication is that perhaps this large Y has been formed by the joining of two smaller normal Y's, either by translocation or by an insertion of one Y segment into another, thus making these individuals effectively XYY. Various morphological Y-chromosome types

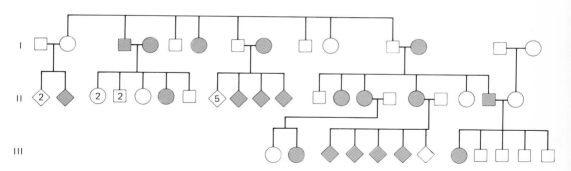

Figure 17-2   A pedigree published in the early days of genetics purporting to show the inheritance of violent temper, with the following explanation:

"Family 27. The patient, III,8, aged 21 years, early showed wilfulness and 'quick temper'. Placed in an institution her 'outbursts of temper' gave trouble. She gets depressed, and longs for the fascination of the life of the demimonde. None of her four brothers shows her temper.

"The father began early to go on sprees; when drunk he is so violent that he is like a madman and has often had to be locked up. One of his sisters is of quiet disposition and so is probably one brother, but three sisters have a violent temper. Of one sister four of the five children have quick tempers and of another sister one of the two daughters has a bad temper.

"Of the father's father not much is known. Two of his sibs had violent tempers; and two of them married women of violent temper and had children with bad tempers. A sister (I,2) of unascertained morals and temper has one child (one out of three) who has a quick temper.

"The patient's father's mother was quick-tempered.

"The mother is gentle, overlenient, easily led and her parents were easy-going."

(C. Davenport. The feebly inhibited. I, Violent temper and its inheritance, *J. Nerv. Ment. Dis.*, **42**:593, 1915.)

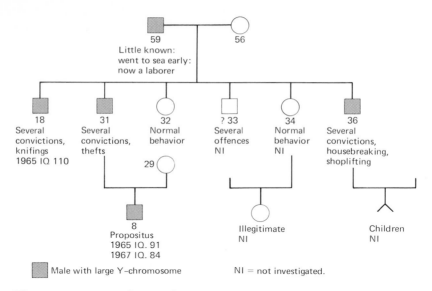

Figure 17-3 A pedigree of recent vintage showing a concentration of antisocial behavior in one kindred. It has been possible to examine the chromosomes of the males indicated by solid symbols, and all have unusually large Y-chromosomes. (P. W. Harvey, S. Muldal, and D. Wauchob. Antisocial behavior and a large Y-chromosome, *Lancet,* **7652:**888, 1970.)

have been described. With quinacrine staining techniques, some are large and stain intensely, others are small and stain hardly at all. In general, it can be said that there is no definite association of either the size or the staining properties of the large single Y-chromosome in a male with the characteristics of those individuals having extra Y-chromosomes. Of course, it is always possible that each of these altered Y's has a different composition, and, irrespective of size or staining, it is in those in which certain loci on the Y-chromosome are effectively doubled that we would get the equivalent of two Y's in the same cell.

It should be emphasized that in none of these cases of Y-chromosome variations are the tendencies described necessarily expressed. In the vast majority of instances, the characteristics of individuals with these abnormal chromosomes appear to be quite within the normal range and they do not run afoul of the law or in any way appear grossly different from their peers. In the first place, it must always be kept in mind that a specific genetic condition depends for its final biological expression on the other genes in the genotype, many apparently unrelated, and these are quite different from one zygote to the next. Then, there are innumerable uncontrollable variables that can modify the course of development and switch it from one course to another. Finally—and this is particularly true for behavioral traits—the impact of other persons, social conditions, and so forth during early life must have a profound influence on the way that a genetic condition potentially affecting behavior will be expressed, or if it is to be expressed at all. It is invariably a gross oversimplification to maintain that a human with a certain genotype will inevitably show some characteristic behavior pattern. Therefore, rather than simply concluding that an extra chromosome may be responsible for abnormal behavior of specified types, we say that the genetic trait predisposes the affected individual to behave in a certain way, provided that appropriate stimuli are present at crucial points during a person's life. This is a point implicit in all subsequent discussions of behavioral traits.

Psychoses

Many students of gross behavioral abnormalities in humans are of the opinion that these illnesses are functional, caused by a traumatic disturbance in the individual's life which triggers a mental breakdown, and that the susceptibility is based entirely on the person's environment and not his genetic constitution. Others have suggested that there is an underlying genetic predisposition to psychosis, making certain persons more liable than others to mental illness as the result of stressful circumstances. We shall consider the kinds of evidence that have been interpreted in favor of a genetic predisposition toward psychoses.

MANIC-DEPRESSIVE PSYCHOSIS.   Manic-depressive psychosis is one in which there exist alternating moods of depression and elation, with periods of normal behavior interspersed, and may consist of cases in which either the manic or the depressive periods predominate. The "bipolar" type that involves both is interesting because it appears to have a more clearly inherited basis than most mental diseases. It has been known for a long time that this disease tends to occur in families. Because females are more frequently manic-depressive than males, sex-linked dominance has been hypothesized, since females, with two X's, would stand twice the chance of having a relatively rare X-chromosome dominant allele than males, who have only one X. Although manic-depressive psychosis occurs with a frequency of about 1 in every 20 in the population, the concordance in identical twins can be as high as 90 percent. When this disease was found in a number of families who happened to carry also an allele for red/green color-blindness, or Xg, a sex-linked blood-group factor, the disease appeared to be transmitted to the progeny associated with the same allele as in the affected parent, indicating an X-chromosome location of that allele.

In addition to the cases of manic-depressive psychosis that appear to have a strong familial tendency and possibly a sex-linked inheritance, there are other cases, in particular of the nonbipolar type, that are sporadic and therefore not amenable to such a simple explanation.

SCHIZOPHRENIA.   Schizophrenia is a personality disorder in which the affected individual has difficulty distinguishing between reality and the products of his imagination. There is a much higher concordance in monozygotic twins than in dizygotics (25 to 40 percent versus 4 to 10 percent). Those who advocate a simple genetic basis that might lead, under stressful circumstances, to a manifestation of the disease have suggested a large variety of mechanisms including a simple dominant, a simple recessive, and multiple genes, all with variable penetrance.

The enzyme monoamine oxidase, an important enzyme in nervous system chemistry and also present in blood cells, proves to be present in less than normal levels of activity in schizophrenics. That this low enzymatic activity has a genetic rather than an environmental basis is shown by the fact that identical twins are concordant with respect to the reduced amount. In one study of 13 pairs of genetically identical twins, in which only one of each twin pair was known to have schizophrenia, all had low levels of this particular enzyme. Other biochemical compounds as well have been reported to be deficient in schizophrenics.

In all of these cases, a real criticism can be raised with respect to the higher concordance of monozygotic twins because of the greater psychological and cultural similarities of monozygotes over like-sexed dizygotes. Of some help in this connection are those cases of monozygotic twins separated shortly after birth and raised by different families. In 15 such cases there was at least one affected twin, ten were concordant, and five discordant, once again suggesting a genetic basis for the predisposition toward schizophrenia.

In another study of about 50 children who were removed from their schizophrenic mothers before 1 week of life and put into foster homes, about half developed schizophrenia or suffered from other mental disorders. In a control group of about the same number, only one in five showed any mental abnormality.

A large number of studies have been made of schizophrenia and other neuroses; these agree in showing a much higher concordance rate in monozygotic than in dizygotic twins (Figure 17-4).

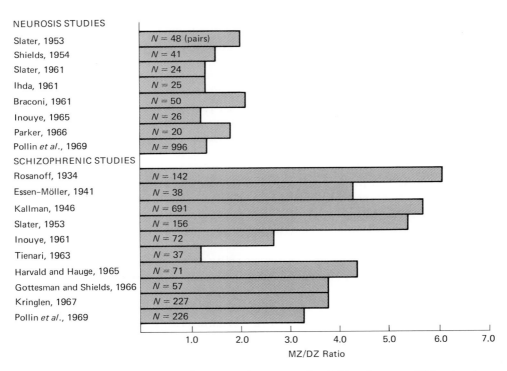

Figure 17-4 The ratio of the concordance rate for MZ twins over DZ twins in a number of studies throughout the world for schizophrenia and other neuroses. (W. Pollin. The unique contribution of twin studies to the elucidation of nongenetic factors in personality development and psychopathogenesis, *Acta Genet. Med. Gemellol.*, **19:**301, 1970.)

CRIMINALITY.   Although it may be true that in cases of aneuploidy a person's behavioral characteristics, particularly decreased intelligence, may make him more likely to become involved in an antisocial act, it seems nonsensical to suggest that such a vaguely defined characteristic as criminality might be determined by a simple dominant or recessive gene. Yet this is precisely the suggestion that was made by proponents of the eugenics movement in this country as well as in certain Western European countries shortly after the rediscovery of Mendel's work. Mendelian principles were applied directly to humans, with socially undesirable traits unequivocally described as being determined by simple dominant or recessive factors and attributed to the lower classes (the immigrants and the poor), to other nationalities, and to other religious and ethnic groups. It is probably not a harsh judgment to conclude that this point of view may have been motivated by a desire to justify the supremacy of the wealthier White Anglo-Saxon Protestants of the period.

Two families, the Jukes and the Kallikaks, are now considered to be the classic examples of this distortion of genetic principles. From their pedigrees it could be seen that not only criminality but also alcoholism, licentiousness, and other antisocial activities were considered simple dominants or recessives and therefore appropriate for pedigree analysis. It hardly needs to be pointed out that missing from this kind of analysis is the question of the environment in which these less fortunate individuals found themselves and the effects of dire poverty on the human condition. Fortunately, pedigrees of this sort are now largely forgotten, except as illustrations of the extremes that characterized the early days of eugenics.

It might be concluded that it would be difficult to find evidence for genetic predisposition to "criminality." This is not entirely true. Table 17-1 lists eight different studies comparing the concordance of criminality in monozygotic and dizygotic twins, where, in each case, one member of a twin pair was known to have been convicted of a crime. From each of the studies there is a greater frequency of identical twins similarly involved than of fraternal ones.

In a study involving about 6,000 pairs of twins born in Denmark in the period from 1891 to 1910, in which both twins survived at least to 15 years of age, it was shown that there is a correlation with respect to conviction for a crime or for a minor offense. If one of the male twins is convicted of a crime, the chance of a monozygotic twin also being convicted is .53, but if the male twins are dizygotic only .22. If one of the male MZ twins is convicted of a minor offense, the probability that the second twin will be similarly convicted is .235, but only .08 in DZ twins.

Analyses of this sort raise the perplexing problem of the extent to which identical twins share not only the same genetic composition but also a common set of childhood experiences, possibly developing a greater affinity for each other during the period of infantile and childhood development that is then accentuated by overt and subtle tendencies of their parents, peers, and others to regard each pair as a single individual rather than two separate entities. It has even been suggested that perhaps they are joined by some mystical bond that only they experience and that makes them inclined to a high degree of concordance (Figure 17-5).

It is generally true that when evidence bearing on human psychological traits is based on twin studies it is always possible to interpret this on the basis of a

Table 17-1 Comparison of monozygotic twins and like-sexed dizygotic twins with regard to behavioral difficulties, delinquency, and criminality; data taken from eight different studies. (E. Slater and V. Cowie, *The Genetics of Mental Disorders,* University Press, London, 1971.)

		MZ Pairs Conc.	Disc.	DDZ Pairs Conc.	Disc.	Concordance Rate as a Percentage MZ	DZ
Rosanoff, Handy, and	Males	80	14	32	36	88	47
Plessett (1941)	Females	39	4	29	35	91	45
Adult criminality, sexes undifferentiated							
Lange (1929)		10	3	2	15	77	12
Le Gras (1933)		4	—	—	5	100	0
Kranz (1936)		20	11	23	20	65	53
Stempfl (1936)		11	7	7	12	61	37
Borgström (1939)	Adult	3	1	2	3	75	40
Yoshimasu (1965)		14	14	—	26	50	0
Hayashi (1967)	Juvenile	11	4	3	2	73	60

person's own preconceptions and personal philosophy. In fact, it is common to find that otherwise serious and objective scientists become highly emotional and colorful in their choices of words in controversies of just this sort, when unambiguous evidence is minimal.

Intelligence

There are many who believe that all humans are, mentally, equally endowed at birth and that differences apparent later in life result from environmental variables. Others maintain that mental capabilities are entirely determined by the individual's genetic constitution, environmental influences playing a minor role at best. This is not a trivial question. It forms the basis for divergent educational philosophies, which in turn determine the nature of the education considered appropriate for young children. This problem of maximizing educational opportunities in the face of wide variations in academic aptitude has occupied the attention of educational psychologists for many years, and, considering the widespread debate on this issue, it can hardly be considered resolved at the present time.

The controversy over the extent of the genetic component in determining behavior discussed previously (such as in the case of the XYY) is mild compared to that surrounding discussions of the genetic basis for intelligence. The difficulty begins with the definition of "intelligence" itself, since it is regarded differently by people in different fields—and even within a given field. In any case, it must be a complex, multidimensional aspect of human capability and behavior, with a large number of interrelated factors. Surely certain of these factors must be more

Figure 17-5 Concordance in identical twins. The physical similarity of the twins, being genetic, is expected, and so is the similarity of dress, which is environmental. The striking feature to this photograph, however, is something else—without having been given any instructions about how to hold their hands, each twin pair has unconsciously put them in a characteristic position. (Courtesy of K. Fredga, University of Lund.)

important than others—more important in some cultures than in others, more important to some people than to others, and even more important at one time in a person's life than at some other time.

At the outset, it should be conceded that there do exist factors that, in some families, are responsible for a generalized depression of intelligence. This is nicely shown in Figure 17-6, which gives the relative intelligence of the sibs of individuals who are somewhat retarded compared to the sibs of those who are severely retarded. The sibs of the severely retarded tend to be quite normal; apparently the latter may carry some specific chromosomal or recessive defect usually not found in another sib. The mildly retarded, however, tend to have sibs similar to themselves, suggesting that this degree of retardation is characteristic (in some cases) of the kinship and is genetically determined. Discussion of the extent of a genetic contribution to intelligence might better avoid both of these conditions, and focus on the variations in the population generally considered to be "normal."

THE INTELLIGENCE QUOTIENT.   The idea of testing large numbers of people for various characteristics originated with Frances Galton, who was the first to do this on a large scale, in a systematic manner, and with a definite purpose in mind —statistical analysis. It is Galton whom the schoolchild must thank for the innumerable batteries of tests that are given (often without warning) throughout the school year. Galton collected data on a large number of physical and psychological characteristics in humans. He examined the biographies of those persons

293

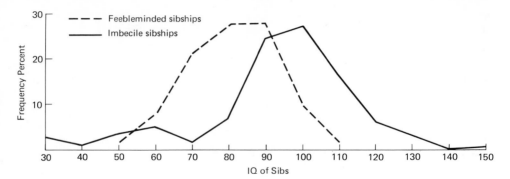

Figure 17-6 A comparison of the relative intelligence of the sibs of the mildly retarded and the severely retarded, with the curious result that the sibs of the more severely retarded are more normal. (J. A. F. Roberts. The Genetics of Mental Deficiency, *Eugen. Rev.*, **44**:71, 1952.)

in England considered to be eminent, and discovered that many of them were close relatives. (He himself was a cousin of Charles Darwin.) From evidence of this sort, he concluded that the characteristics leading to eminence had a biological basis and were genetically inherited. He did not seem to be as concerned as we might be today with the possibility of inheritance by way of class or wealth (rather than genes) in nineteenth-century England.

During the early part of this century, a French educator named Binet constructed a test for the purpose of predicting the success of French schoolchildren as they proceeded through academic life. Since the tests he designed were based on the French elementary school curriculum, they proved, as might have been expected, to be excellent predictors. From these tests came the concept of an intelligence quotient, or IQ: the age level at which the student performed on the test, divided by his or her actual age, then multiplied by 100. Thus, if a student of age 10 performed at the level expected of a 12-year-old, the IQ would be 12/10 × 100, or 120. Although the test has been greatly modified since its inception, because of the changing fashions in education, the basic principle remains the same today.

The IQ is a simple numerical score that is supposed to reflect a person's "intelligence," and, it goes without saying, high intelligence is a characteristic greatly desired by many people and richly rewarded by society. Concomitantly, low intelligence is considered to be an inferior attribute. For these reasons, in any discussion of the nature of intelligence a person is likely to accept or reject data or arguments, not on the basis of their validity, but rather on his own personal prejudices and preconceptions.

Characteristics of living things that show a continuous distribution around an average—human height is a good example—tend to be found in a "normal" distribution (Figure 17-7). The greater the deviation from the average, the less frequent the class. This bell-shaped curve of the relative frequencies of the different classes in a typical population with variation applies when there are a large number of chance factors combining to produce the end result. Consider, for instance, the results of tossing a coin a hundred times. While the most likely result is 50 heads and 50 tails, there is a considerable but less likely chance of getting 49 heads and 51 tails, or vice versa, and still less of getting 48 of one and 52 of the other. Should someone perform this simple 100-toss test a million times and plot the frequency of the different results, he would come close to obtaining a *normal curve*. The normal distribution is a mathematically sound concept; the science of statistics as it developed was based primarily on the properties of the normal distribution.

Galton argued that this distribution should hold for psychological characteristics, since they are based on many complex physiological factors acting in different directions. It should be realized, however, that the fitting of this curve to IQ scores is artificial. The data obtained from the tests fit this distribution only because the test scores for any age group are adjusted to conform to the normal distribution. In other words, one unit of score at the low end of the scale might represent something completely different from a similar difference at the upper end of the normal distribution. For this reason, arithmetical manipulations of IQ test scores, even such a procedure as taking simple averages, may be open to question, and this qualification should be kept in mind later in this discussion when comparisons are made using averages of test scores.

ENVIRONMENTAL FACTORS AFFECTING IQ TEST PERFORMANCE. The scores from IQ tests, the scores obtained prior to any manipulation, may be affected by many factors. Most people would agree that the environment in which a child grows up must have a great influence. Families in which there is a strong emphasis on reading, booklearning, arithmetic, and general conversation will produce children much more competent to handle IQ tests than will families in which these educational advantages are absent. It is not surprising that the children of the professional classes in general score higher on IQ tests than children of other socioeconomic groups. It has been found that 7-year-olds in Scotland are an average of 11 months ahead of other British children with respect to reading ability; Scottish parents more often read to their children than do parents elsewhere in the British Isles, and Scottish teachers introduce the elements of reading

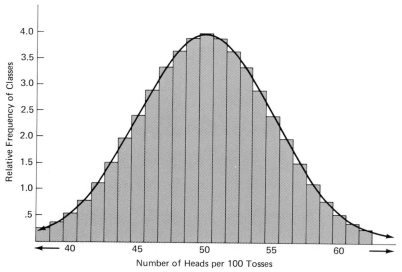

Figure 17-7 The normal curve. If every person in the world tossed an unbiased coin 100 times, and the results were recorded on graph paper, the distribution would appear much like this. The average number of heads (and tails) would be 50, and the expected frequencies of the other possibilities are given by the relative heights of the bars on the graph.

earlier than do English or Welsh teachers. Therefore, in every discussion of IQ performance, the question of environmental background becomes a crucial one.

Educational psychologists have always been painfully aware of the validity of the criticism that a child's performance on these tests will inevitably be modified by his environment, and may depend on many extraneous factors. Attempts have been made to construct tests that would be independent of the culture in which the child is found, so-called culture-fair tests (Figure 17-8). Whether such tests completely eliminate the effect of the child's cultural background may be open to question, but they undoubtedly have many advantages over the usual written tests.

MULTIPLE BIRTH AND BIRTH ORDER INFLUENCES. It has been found that the mean verbal reasoning scores are 100.1 for 48,913 persons of single births, 95.7 in 2,164 cases of twins, and 91.6 in 33 sets of triplets. In addition, when two identical twins have different birth weights, the smaller of the two will usually have the lower IQ in childhood. These observations suggest that retarded pre-natal development or immature condition at the time of birth is responsible for these low performances. On the other hand, when one of a twin pair is stillborn or dies shortly after birth, the remaining twin scores just about as high as individuals from single births. Perhaps the lower scores of twins may be explained by the observation that they are limited in vocabulary and use primitive sentence construction, a characteristic that becomes more acute as the children age from

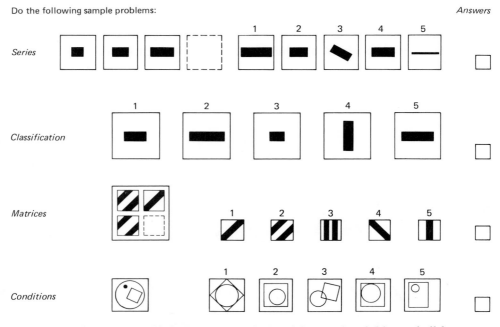

Figure 17-8 A "culture-fair" test, one designed for use by children of all languages and cultures. (From the Culture Fair Intelligence Test, Scale II, © 1949, 1957 by the Institute for Personality and Ability Testing, Champaign, Illinois. Reproduced by permission.)

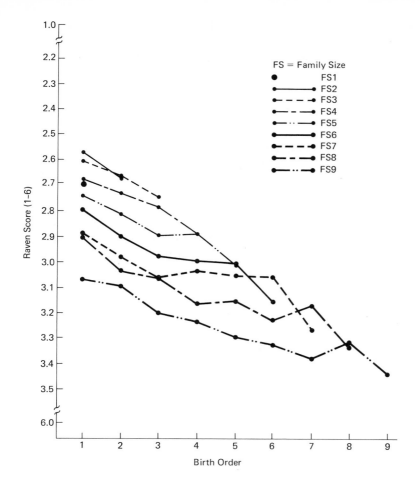

Figure 17-9 The drop in IQ performance in successive births; data from men in the Netherlands born during World War II and immediately thereafter. (L. Belmont and F. A. Marolla. Birth order, family size and intelligence. *Science,* **182:**1097, 1973.)

2 to 5 years, although at later ages they tend more toward the norm for their ages, possibly a result of schooling.

The performance on IQ tests appears to be highest for the firstborn, and drops with succeeding births (Figure 17-9). This was found in a study of 400,000 men 19 years old born in the Netherlands from 1944 to 1947, who were tested primarily to study the effects of the Dutch famine of 1944 to 1945 on mental and physical development. Interestingly, the prenatal life under famine conditions for many of these men produced no measurable adverse effects on their later intellectual performance.

One explanation for both of these results is that a child's mental development during the formative years may depend on the degree of close association with adults, as a result of which he may more closely conform to adult standards of vocabulary, logic, behavior, and performance. However, when the birth is multiple, or when there are several children in the family, the children learn to play and communicate with each other and to satisfy their needs at a less sophisticated level, so that their intellectual development, based on adult standards, may be somewhat retarded. This effect might be accentuated for twins, whose identical age would increase the empathic bond, lessening their dependence on verbal communication.

EVIDENCE OF GENETIC FACTORS AFFECTING IQ TEST PERFORMANCE.   When children are adopted into foster homes, their performance on IQ tests is, as expected, largely determined by the economic standing and professional orientation of the adopting family, so that children adopted into the higher-income categories do better on these tests than those adopted into the lower-income categories. Children who are adopted into the higher-income categories, however, do not reach the level of aptitude of those who are the natural children in those families. One interpretation of this is that the natural children are genetically constituted to do better than the adopted ones, since adopted children will represent a more randomly selected (and therefore average) group than the natural children of the professional class. On the other hand, the adopted children may have undergone psychological trauma with the change of parents early in life. Another interpretation, of course, is that the adopted children may be subject to a certain amount of discrimination by their foster parents, who unconsciously may favor their natural children in their allocation of time and attention. Curiously, it appears that children who are adopted by the lower economic groups tend to perform somewhat better than the natural children in those same groups. A simplistic interpretation of this is that the capabilities of the average adopted child are somewhat superior to those of the natural children, and that even in the less academically inclined environment they are able to do better than the natural children.

Perhaps a more convincing demonstration than the preceding is found when the correlation of the IQ's of adopted children is made in a different way, by comparing the adopted children with their natural parents and foster parents. If the IQ is primarily environmentally determined, the highest correlation should be between the child and the adoptive parents with whom the child grows up. This, however, is not the case (Figure 17-10). There appears to be little or no correlation with adoptive parents, but a clear correlation with the natural parents.

EVIDENCE FOR GENETIC BASIS FOR IQ PERFORMANCE FROM TWIN STUDIES.   The comparison of monozygotic twins with like-sexed dizygotic twins can be applied in this case. This comparison, however, leaves much to be desired, since it can be argued that as monozygotic twins grow up they may behave alike for reasons related to their obvious similarity. A particularly valuable source of information might be cases of identical twins who were separated shortly after birth and brought up under different family conditions. The comparison of such twins with like-sexed dizygotic twins who were brought up within their natural families would minimize the influence that identical twins might have on each other. The first extensive study was made by a team including a geneticist, a psychologist, and a statistician from the University of Chicago, who, in 1936, arranged for such separated identical twins to come to Chicago for a free visit to the World's Fair in return for allowing themselves to be subjected to intensive physical and psychological tests. Another large-scale study was made in England after cases of separated identical twins were uncovered by way of a plea through a television program on the BBC. In all, almost 120 sets of identical twins separated shortly after birth have been investigated intensively.

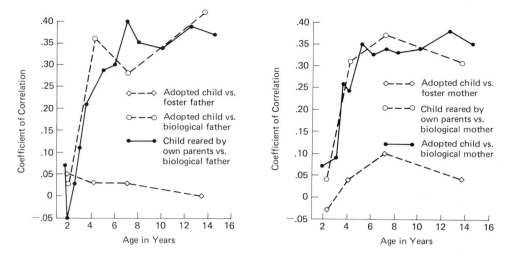

Figure 17-10 Correlation of the IQ of children with the educational level of actual and foster parents, showing the higher correlation with real than with the foster parents. (M. P. Honzik. Developmental studies of parent-child resemblance in intelligence, *Child Dev.*, **28**:215–28, 1957.)

All of these studies agree in their conclusions that separated MZ twins are more similar in their performances on IQ tests than are DZ twins who grew up in the same family (Figure 17-11).

This conclusion is sometimes expressed in terms of "heritability," i.e., the extent to which the genetic background makes a contribution to the variability of a given characteristic. Thus, if identical twins are always completely concordant with respect to some ordinarily variable characteristic (as they might for eye color), then we would say that the heritability is 1, or 100 percent. On

Category		0.00 0.10 0.20 0.30 0.40 0.50 0.60 0.70 0.80 0.90	Groups Included
Unrelated persons	Reared apart		4
	Reared together		5
Foster parent-child			3
Parent-child			12
Siblings	Reared apart		2
	Reared together		35
Twins — Two-egg	Opposite sex		9
	Like sex		11
Twins — One-egg	Reared apart		4
	Reared together		14

Figure 17-11 The grouped results of 52 studies indicating the correlation in IQ performance when the persons tested have various degrees of relationship. (L. Erlenmeyer-Kimling, and L. F. Jarvik. Genetics and intelligence: A review, *Science*, **142**:1477–79, 1963.)

the other hand, if they are discordant as often as two people taken at random from the overall population might be (as is the case for identical twins being affected by certain infectious diseases), the heritability would be considered to be 0. The information from MZ twins leads to the conclusion that the heritability of IQ performance in the samples tested is about 60 to 80 percent.

However, in view of the variability in testing procedures, the uncertainties in the mathematical treatment of the results, and the cultural and environmental differences separating socioeconomic and ethnic grouping, a serious question can be raised about the general applicability of this figure except, perhaps, to indicate that there is some genetic component to IQ performance.

References

BODMER, W. F., and L. L. CAVALLI-SFORZA. 1970. Intelligence and race. *Sci. Am.,* **223:**19–29.

CAVALLI-SFORZA, L. L., and W. F. BODMER. 1971. *The Genetics of Human Populations.* San Francisco: Freeman.

ERLENMEYER-KIMLING, L., and L. JARVIK. 1963. Genetics and intelligence: a review. *Science,* **142:**1477–79.

GOTTESMAN, I. I., and J. SHIELDS. 1972. *Schizophrenia and Genetics: A Twin Study Vantage Point.* New York: Academic Press.

KARLSSON, J. L. 1970. Genetic association of giftedness and creativity with schizophrenia. *Hereditas,* **66:**177–82.

LUDMERER, K. M. 1972. *Genetics and American Society: A Historical Appraisal.* Baltimore: Johns Hopkins Press.

MEDAWAR, P. B. 1959. *The Future of Man.* New York: Basic Books.

MONEY, J., V. LEWIS, A. A. EHRHARDT, and P. W. DRASH. 1967. IQ impairment and elevation in endocrine and related cytogenetic disorders. In Z. Zubin and G. Jervis, eds., *Psychopathology of Mental Development.* New York: Grune & Stratton.

MORTON, N. E. 1972. Human behavioral genetics. In *Genetics, Environment and Behavior.* New York: Academic Press.

NEWMAN, H. H., F. N. FREEMAN, and K. J. HOLZINGER. 1937. *Twins: A Study of Heredity and Environment.* Chicago: University of Chicago Press.

SHIELDS, J. 1962. *Monozygotic Twins Brought up Apart and Brought up Together.* London: Oxford University Press.

SLATER, E., and V. COWIE. 1971. *The Genetics of Mental Disorders.* London: Oxford University Press.

Questions

Useful terms: manic-depressive psychosis, schizophrenia, intelligence quotient (IQ), normal distribution.

1. Do genetic factors ever affect behavioral differences? If your answer to this question is positive, does it follow that the behavior of a person is determined by his genotype?

2. What is the evidence that an extra Y-chromosome (either in XYY or in XXY) may modify behavior?

3. Is there any evidence for an inherited basis for manic-depressive psychosis or schizophrenia?

4. Some of the strongest arguments for a genetic basis of behavioral characteristics come from a study of monozygotic twins. What is the strongest argument against the use of such data?

5. What is the history of the development of intelligence testing?

6. What is meant by IQ?

7. Most biological characteristics that show continuous variation from one extreme to another, like height, are distributed in a population roughly according to a normal curve. Can the same be said for intelligence?

8. Is it possible to design tests that cannot be criticized on the basis that a child's family and cultural background may have an influence on his performance?

9. Is there a simple explanation for the relatively poor performance of children in larger sibships on IQ tests?

10. How can data from adopted children provide us with an argument for a genetic component to intelligence?

11. You have your own view of the genetic basis of intelligence. What kind of evidence would you require to show that your point of view, whatever it is, is incorrect?

18

The Sex Ratio

No biological event is so faithfully recorded, on such a vast scale, as the birth of a human. Not only the time and place of the birth, but also the sex and condition of the newborn and information about the family are recorded in detail tens of thousands of times each day throughout the world. Transferred to punched cards, this information can be compiled into tables relating one characteristic of a birth to one or more others. The sex of the newborn child has been examined repeatedly as a function of virtually every conceivable variable describing human births, and in many cases surprising relationships have been uncovered.

The Numerical Value of the Sex Ratio

From what we know of meiosis, it would seem that exactly half of the products of male meiosis should contain an X-chromosome, and half should contain a Y, leading to an expectation of precisely equal numbers of male and female births. Anyone who follows the birth notices (which invariably designate the sex of the newborn) in his daily newspaper realizes that even with a limited count of new births, say a hundred or so, males generally predominate. In fact, larger numbers, derived from health agencies of countries all over the world, show a marked departure from equality, always in the direction of an excess of male births. For the United States, figures published by the Bureau of Vital Statistics show that about 51.4 percent of all births are male. This ratio of males to the total is sometimes expressed as the decimal .514, or as the number of male births per hundred female births, in this case about 105 males to 100 females (or 105:100). This ratio at birth is called the *secondary* sex ratio, as distinguished from the *primary* sex ratio, the hypothetical ratio at the time of fertilization.

Variations in Secondary Sex Ratio

ETHNIC ORIGINS. Table 18-1 lists a selected number of countries for which reliable statistics in quantity are available. They clearly demonstrate that the ratio varies from one group to another. Although the differences are clearly not large, the numbers on which they are based are, so that we can be sure that many of the differences are real. A few trends suggest themselves. Countries in the vicinity of the Mediterranean Sea (Algeria, United Arab Republic, Iran) have high ratios. The Oriental countries, Taiwan and Japan, come next. The United States and the U.S.S.R. have very similar ratios, lower than the preceding, and the lowest ratio is found in Jamaica. Although the numbers in the last case are small, they agree in direction with the figures for the non-White population of the United States, with a sex ratio of less than 103 males : 100 females. It is obvious that none of these countries represents an unmixed racial group; the non-White category in the United States is predominantly Black, but includes a good fraction of Orientals as well as other ethnic groups. Similarly, the population of the U.S.S.R. includes a large number of distinct ethnic groups. Korea is unusual; it has a very high sex ratio, 115.3 males : 100 females. An earlier high

Table 18-1 Sex ratios in countries for which extensive data are available. (From World Health Organization and U.S. National Bureau of Vital Statistics Publications.)

Country	Years (inclusive)	Males	Females	Sex Ratio
Korea*		32,230	27,942	.5767
Philippines	1963–67	2,118,622	1,929,783	.5233
Iran	1965–68	2,196,831	2,024,163	.5205
Cuba	1965–66	265,751	253,637	.5194
Algeria	1964–65	530,014	494,411	.5174
U.A.R.	1963–67	3,138,796	2,929,225	.5172
Taiwan	1963–67	1,049,721	987,449	.5153
Indonesia	1964	1,537,444	1,446,173	.5153
Japan	1963–67	4,369,092	4,127,508	.5142
U.S.A. (White)	1963–68	9,575,338	9,071,982	.5135
Madagascar	1964–67	444,612	422,062	.5130
U.S.S.R.	1963–68	13,278,678	12,611,562	.5129
Pakistan	1965–67	5,118,000	4,913,000	.5102
Jamaica	1963–64	68,406	66,142	.5084
U.S.A. (non-White)	1963–68	1,890,834	1,843,763	.5063

*For explanation of high figures for Korea, see text.

figure reported for that country had been criticized on the grounds that perhaps a higher social value placed on male births led to a higher reporting of male than female births. The figures in the table can be considered unusually reliable because the workers reporting this specific result were aware of this criticism and made a special effort to get accurate figures.

SEASONAL VARIATION.   A graph of available United States sex ratio data from 1915 to 1948, based on the month of birth, follows the pattern shown in Figure 18-1. These changes are quite convincing because the main trends, the lowest values in February, October, and March, and the highest from May to July, with another high point in January, are similar in independent sets of data. Although the difference between the months with the highest and lowest proportions of male births is only a quarter of 1 percent, the total numbers involved in the analysis are so large that the differences are quite significant. If differences in temperature at the time of conception are in some way responsible for this seasonal variation, shifting the month of birth as shown in that figure by 9 months would more directly relate the sex ratio to the temperature.

Since the seasons in the southern hemisphere differ from those in the northern by 6 months, one might anticipate that if changing temperature is in some way responsible for the changing sex ratio a similar study in that part of the world would show a similar curve, except that the axis would be shifted by 6 months.

Figure 18-1 Monthly variation in the secondary sex ratio. Three independent sets of data are plotted. In each case the 0-line is the mean value for each set, and the deviations are in units of .001 percent. (H. M. Slatis. Seasonal variation in the American live birth sex ratio, *Am. J. Hum. Genet.*, **5:**28, 1953.)

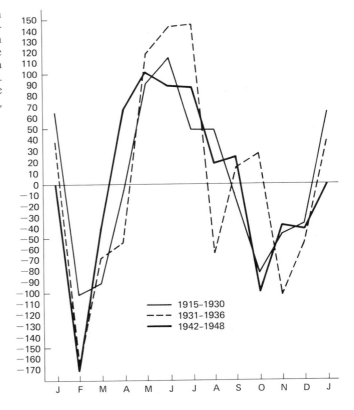

For Australia, data giving sex of child by month of year are not yet sufficiently numerous to determine decisively whether such a shift exists, but they do suggest such a reversal.

PARENTAL AGES.   It has been known for a long time that children born to young mothers are more often males than are those born to older mothers (Figure 18-2). Since the ages of spouses tend to be correlated, it is not surprising that a simple graph of the sex ratio of children as a function of the father's age should show a similar result. It should be noted that these changes, although quite definite, are numerically small (a decrease of only about one male birth per thousand total births for each increase of 5 years in age of the parents) so that the probability at any particular birth is only minutely affected. It would not be wise for parents wishing a female child to postpone having offspring on the grounds that they would have a greater chance at an older age!

Since increasing male and female ages are both correlated with a decrease in sex ratio, it can reasonably be questioned whether one of these two correlations occurs only because of the similar ages of spouses. To decide this point, it is necessary to analyze data that give the sex of the child and the two parental ages for each birth. The statistical method involved essentially compares the sex ratio when the mother is old and the father young, and vice versa. This comparison shows that it is the older father, not mother, who is more directly correlated with the decreasing sex ratio in older parents. This is a surprising result, since one would naturally imagine that if the proportions of males and females at

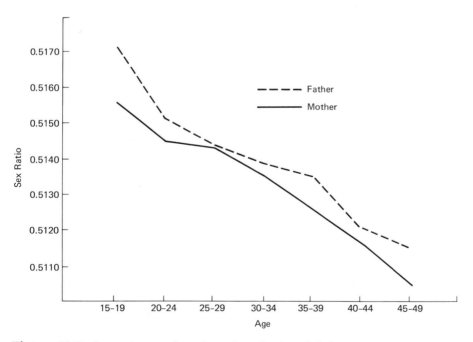

Figure 18-2 Sex ratio as a function of mother's and father's age. (E. Novitski and L. Sandler. The relationship between parental age, birth order and the secondary sex ratio in humans, *Ann. Hum. Genet.*, **21**:123–31, 1956. Used by permission of Cambridge University Press.)

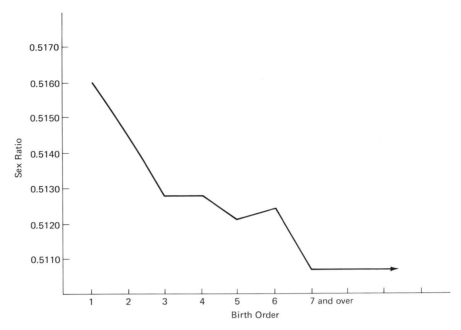

Figure 18-3 Sex ratio as a function of birth order. (E. Novitski and L. Sandler. The relationship between parental age, birth order and the secondary sex ratio in humans, *Ann. Hum. Genet.*, **21**:123–31, 1956. Used by permission of Cambridge University Press.)

birth is to be correlated with any external variable, that variable would be the age of the mother, for this variable itself should be more closely correlated with the direct biological changes affecting the differential survival of males versus females.

BIRTH ORDER.   Children who are first-born tend to be males more often than do those born later (Figure 18-3). This relation shows a steady drop in sex ratio from the first to the third child, with less change occurring thereafter. A question arises again of whether the apparent paternal age effect is really a birth order effect. Perhaps both are dependent on some one other more fundamental cause. Data on this question are limited in extent because all the features surrounding a birth are coded and condensed by the statistic-collecting agency to fit on a computer-punched card. Since there are many other more important data such as time and place of birth, in hospital or at home, attendance by a physician or midwife, single or multiple birth (and what type), and so on, it is only rarely that a specific tabulation will include simultaneously the sex of the newborn, the age of each of the parents, and the order of the child in the family. Such figures have been made available by the National Office of Vital Statistics of the United States for the year 1955. One analysis of these limited data agrees with earlier ones in indicating no relationship of sex ratio changes with the mother's age, but significant relationships with the father's age and birth order; another analysis using a different statistical method suggests only a correlation with birth order. If the paternal age effect is spurious, and represents only a secondary correlation with the birth order of the child, it becomes difficult to rationalize the finding that sex ratio changes are more closely related to the father's than to the mother's age. Birth order not only should be more closely correlated with mother's than father's age but also must have its basic biological mechanism operating through the female rather than the male parent. For the present, it appears reasonable to assume that there are at least two factors, paternal age and birth order, involved in sex ratio changes with time in the production of a sibship.

THE CHANGE IN SEX RATIO DURING WARTIME.   A number of workers have noted, and have been puzzled by, the increase in sex ratio during years of war. One explanation advanced to account for this is that "nature attempts to make up for the deficiency of males, caused by losses during the war." It is not clear how "nature" could sense this deficiency, and immediately proceed to put into effect some mechanism to compensate for it. This explanation is so weak that many people have been led to question the validity of the basic data. Nevertheless, during World War I the ratio of male to female births went up in Great Britain, France, Germany, and even a nonbelligerent country, the Netherlands. Of the countries involved in that war, Italy and Finland were the only two that showed no increase. During World War II the secondary sex ratio for England reached the highest level ever recorded for that country since 1841 when birth registration was initiated. In Figure 18-4 are reproduced the sex ratios of the White population of the United States during the period from 1935 to 1968. It is clear that there is a slight rise during World War II. Although the sex ratio increase is not very great, its reality becomes more obvious by comparing the ratios for the years 1941 to 1947, individually, with the ratios for ten years before or after World War II. The ratio goes up again in 1968, and this timing coincides with the American involvement in Vietnam.

OTHER INTERESTING CORRELATIONS.   Studies have been made of the sex ratio of births with respect to a large number of other variables. These include the socio-economic status of the parents, the ABO blood group of the parents, legitimacy

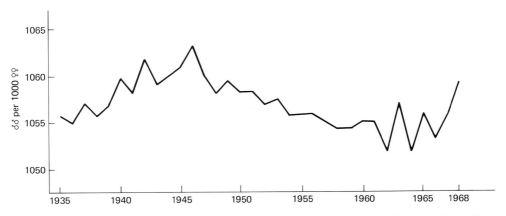

Figure 18-4   Variation of the sex ratio in the United States showing the unexplained increase in male births during World War II. (Data from U.S. Bureau of Vital Statistics.)

or illegitimacy of the birth, urban or rural residence, and the occupations of the parents. One recent paper by an Air Force fighter pilot claims a significant increase in the proportion of females born to jet fighter pilots compared to jet transport pilots or pilots who have been grounded! Another report from Japan suggests that people with a diet largely of fish have a different sex ratio of progeny than those not on such a diet. Although the data on which these various studies are based appear to be significant statistically, other similar studies have failed to yield the same results and so these relationships cannot be considered to be completely verified.

One reservation must always be kept in mind in interpreting these relationships. Because the shifts in sex ratio are relatively slight, large numbers must be used in order to reach results that are statistically valid. These large numbers can be obtained only by combining subsets of data that may have different characteristics and that may be represented unequally in the overall total population. Thus, if there is a paternal age correlation with sex ratio, and if more young males have progeny during wartime than ordinarily, the sex ratio would be expected to rise, giving the otherwise inexplicable wartime effect. Similarly, if birth order is highly correlated with sex ratio changes, and if more first children are born during wartime, then the sex ratio during wartime would be higher. This confounding of relationships by heterogeneity in the data is a danger inherent in all such analyses and is well recognized by all who try to analyze birth data.

Biological Basis for Sex Ratio Changes

Of the variables already discussed that seem to be associated with sex ratio changes, it is clear that none of them can represent primary causes; at best, they can be only secondary correlations with some more direct biological phenomenon. There is no conceivable way that a state of war, the season of the year, parental age, or birth order, as such, can directly modify the sex ratio. We must look for some more basic mechanism that might itself be subject to change as these external conditions change. As a hypothetical example, if one of the primary events is the immunization of the mother by a male embryo or fetus against subsequent male embryos, the loss of males in later births would cause the sex ratio to drop, creating the birth order effect, and then a shift in the relative proportions of first versus higher-order births during war time could shift the sex ratio upward. Unfortunately, this proposal explains the wartime change only in direction, i.e., an increase, but not in magnitude; the magnitude of the increase is greater than can be accounted for by the known changes in paternal ages or greater number of first children born. As one set of investigators succinctly put it: "A belief in the wisdom of Divine Providence in increasing the proportion of males born at such times is not essential to the rationalization of this relationship, although it must be admitted that no more satisfactory explanation is yet available."

DISTURBANCE OF MEIOSIS.   Although our simplest conception of meiosis in the male indicates that precisely half of all sperm will be X-bearing and the other half Y-bearing, we know from work with experimental animals, particularly

Drosophila and the house mouse, that unequal numbers of the two kinds of progeny, expected equally frequently from a heterozygote, may be regularly produced by individuals of certain genetic constitutions. In about ten different species of *Drosophila*, males with a certain kind of X-chromosome will produce only progeny with that X-chromosome (and therefore females) with no, or very few, progeny with the Y-chromosome. In another species, *Drosophila affinis*, males of another genetic constitution will produce only male offspring and no females. In ordinary laboratory strains of *Drosophila*, the sex ratio is almost always significantly lower than 50–50; i.e., there is an excess of females. In experiments designed to pinpoint the source of these deviations in *Drosophila*, it appears that the ratio characteristic of any given line depends, generally, on the genetic composition of the male and, more specifically, on the strain from which the Y-chromosome is derived.

Such cases, in which unequal proportions of progeny of the two types are produced from a heterozygote as a result of a meiotic peculiarity, are now put under the general classification of *meiotic drive*. As far as the relative frequencies of the two types of zygotes are concerned, the same proportions would appear if the two kinds of sperm are produced in unequal numbers in the first place, or if the two kinds are produced in equal numbers but are unequally effective in fertilizing eggs, as long as the determination of that characteristic of inequality takes place at meiosis. With the advent of quinacrine staining, it has been possible to distinguish X- and Y-bearing sperm, and it would appear, from present limited observations, that the frequencies of the two types are about the same, although small deviations (or larger deviations in a smaller number of people) sufficient to account for observed sex ratio deviations are not excluded. At the present time, we know of no good example in humans in which a genetic factor disrupts the presumed equal proportion of the two sexes at the time of zygote formation and instead causes the production of predominantly one sex or the other.

POSTMEIOTIC EFFECTS.   It has been suggested that conditions within the female reproductive tract may favor one type of sperm over another, and, in fact, a few striking cases have been reported in which the sex ratio in humans has been apparently altered by the application of external agents (usually acid vs. alkaline solutions) to the female tract either before or at the time of insemination. For a number of reasons, such studies need further verification under conditions more controlled and with larger numbers than those now reported; for the time being they must remain in doubt.

The main reason for maintaining a healthy skepticism about such effects is that animals, unlike plants, produce male gametes with biological characteristics independent of the haploid genetic content of the sperm itself. No case has yet been found in which a gene in the sperm functions as it does when in the nucleus of a cell, although there are some cases where events happening in the latter phases of spermatogenesis modify the subsequent behavior of the sperm (as in meiotic drive). The properties of the sperm, their motility, longevity, and effectiveness in fertilization, seem to be properties that are inherited from the diploid cells from which they are derived and not from the genes that they themselves carry in the haploid condition. In fact, the sperm head is quite minute, with a volume not much larger than that required to hold the chromosomes in a packed and condensed state, and it is not clear how chromosomes in such condition, without the benefit of the complex extrachromosomal machinery that normal gene function depends on, could function in the usual sense. Indeed, in *Drosophila* it has been possible experimentally to contrive genetic constitutions such that a male will produce sperm lacking virtually all of their normal haploid complements, and these sperm are competitive with normal ones in fertilizing an egg.

Another explanation based on a prefertilization phenomenon is that of relative sizes of the two sperm. Because the Y-chromosome is smaller than the X-chromosome, a Y-bearing sperm is probably somewhat smaller than an X-bearing sperm. In fact, this has been observed in insects, in which the differences between the male- and female-determining haploid complements are much greater and the size differences therefore more observable. Based on this presumed difference, it can be imagined that the Y-bearing sperm with its smaller size wins the "race" to get to the egg. However, there is no reason to believe that this difference in size would alter its motility. Another imaginative suggestion is that since adult males (with a Y-chromosome) are stronger than females (without a Y) a Y-bearing sperm must be stronger than a non-Y-bearing sperm. Whether motility itself is of great importance in the transport of the sperm to the egg is debatable. There is no evidence that the sperm has a motion directed toward the egg, although it is possible that greater random motion on the part of some sperm might give them a slightly greater chance of entering the oviducts. Once in the oviducts, their spasmodic contraction, and the beating action of the cilia in them, may play an important role in sperm transport. Furthermore, it is unlikely that the first sperm to reach the egg is the effective one in fertilization. It seems more likely, as with other animals, that large numbers of sperm are in the immediate vicinity of the egg at the time that one effects fertilization.

Another possibility is that the X- and Y-bearing sperm differ in the ease with which they accomplish fertilization after making initial contact with one egg. This possibility seemed almost impossible to test experimentally a short time ago, but with the means of distinguishing X- and Y-bearing sperm using quinacrine staining it is now feasible.

DIFFERENTIAL IMPLANTATION EFFECTS. It has been suggested that X- and Y-bearing sperm are functionally present in equal frequencies, producing equal numbers of XX and XY zygotes, but that at the time of implantation of the developing egg in the uterine wall, about 4 days after fertilization, the two types have different probabilities of successful implantation. If a larger number of male than female zygotes were successfully implanted, this could give a high secondary sex ratio. Why more of one sex than the other might be implanted is not clear. One suggestion is that an antigen-antibody reaction might hinder implantation of embryos of one sex compared to the other; however, such a mechanism as ordinarily envisaged would depend on prior immunization, i.e., a prior birth, and could not account for the fact that the birth most deviating from the simple 50-50 expectation is the first in the family. Another suggestion is that male zygotes, with their Y-chromosome, are more unlike the mother than female zygotes, and that this difference is responsible for antigenic similarity which, to account for the excess of male births, would have to favor the implantation of male zygotes.

Embryonic and Fetal Loss

The simplest and most attractive explanation for the excess of males at birth is that the sex ratio at the time of fertilization, the primary sex ratio, is precisely one to one, but that disproportionate loss of females during development leads to their underrepresentation at birth. Innumerable studies have been undertaken to determine the prenatal sex ratio.

DIAGNOSING EMBRYONIC SEX.   Some early studies gave sex ratios of stillbirths that were much higher than those of live births (Table 18-2). Unfortunately, these ratios were in precisely the direction opposite to that required by the embryonic loss hypothesis. A note of caution must be registered here, however, for the morphological differentiation of the two sexes may be ambiguous during the first half of intrauterine life, the distinction being all the more difficult in younger embryos. In any case, these data do not support the supposition that a regular excess of female loss occurs during pregnancy, and it would be necessary to postulate a massive loss of female embryos of age 2 months or less, at a time when the visible morphological distinction between the sexes is almost impossible to make.

NUCLEAR SEXING.   A more accurate determination of sex may be provided by nuclear sexing, i.e., checking for the presence of the sex chromatin ordinarily present in the female and not in the male, as well as for the presence of the fluorescing body indicating the presence of a Y-chromosome. Except for anomalies such as X0 or XXY chromosome constitutions, the diagnosis should be more reliable than morphological sexing. The totals of studies by six different groups of investigators for the first 4 months of gestation, by month, are given in Table 18-3.

The study of spontaneous abortions in an attempt to infer the primary sex ratio has its limitations. Obviously they are a suspect class, since they may not represent a random sample of all embryos and fetuses, but rather are those that have not survived to term. If males are less viable than females in the embryonic stage, as we know they are at all other stages of life, one would expect more males among spontaneous abortions. Therapeutic abortions provide a better sample for determining the embryonic sex ratio. Nevertheless, among the studies that have been made, one reports a sex ratio of 0.62 in 125 embryos, another 0.60 in 1,083, a third 0.70 in 223, and a fourth 0.55 in 209 cases. However, it is possible that these high observed ratios do not reflect the true sex ratio but, since the cells

Week of Gestation	Males	Females	Sex Ratio
20–23	4,331	3,157	.5784
24–27	4,073	3,494	.5383
28–31	4,331	3,801	.5326
32–35	5,389	4,802	.5288
36	2,940	2,633	.5275

Table 18-2 Sex ratios of stillbirths by period of gestation and sex determined by morphology (U.S. White population, 1966 and 1967). (U.S. Bureau of Vital Statistics.)

Table 18-3 Sex ratio of embryos and fetuses determined by nuclear sexing of spontaneous abortions, by month of gestation. (Condensed from A. C. Stevenson, Sex Chromatin and the Sex Ratio in Man, in *The Sex Chromatin*, K. L. Moore, ed., Saunders, Philadelphia, 1966.)

Month of Gestation	Males	Females	Sex Ratio
1	85	51	.62
2	457	230	.66
3	633	236	.72
4	199	94	.67

from spontaneous abortions are often unsuitable for cytological analysis, instead result from misclassification when the failure to observe an X-chromatin body in an XX cell leads to the false identification of that embryo as XY.

CHROMOSOMAL SEXING. One way to get more precise data would involve karyotyping. However, because of the greater effort and expense involved, data have been gathered only very slowly. Spontaneous abortions have received more attention than therapeutic abortions, since it is more important, medically speaking, to discover why an embryo or fetus did not survive to term than it is to examine the chromosome complement of an embryo which probably would have developed into a normal infant. It can be predicted that karotype studies made after amniocentesis will provide important data in this matter.

In any case, several statements can be made from investigations of the sex chromosome composition of embryos spontaneously aborted. X0 embryos, which are sex chromatin negative like males, but female in sex, constitute about 10 percent of such embryos. (These would constitute a major source of error in nuclear sexing.) Otherwise, the overall frequency derived by adding the small number of observations from several studies suggests an equality of the two sexes, or even a small excess of XX's over XY's. One study, for instance, gave 296 XX's to 281 XY's. The discrepancy in the results obtained when the sex of the embryo is determined by the presence or absence of the X-chromatin body versus that obtained by karyotyping will undoubtedly be resolved in the near future. Staining for the Y-chromatin body as well as the X-chromatin body should remove much of the doubt about the exact classification of cells.

It may be hypothesized that the primary sex ratio is 100:100, but that excessive loss of female zygotes and embryos very early in development leads to the observed excess of males later on. However, it would require about twice as many female as male deaths in those abortions occurring very early in pregnancy to change a primary sex ratio of .500 to a 28-week ratio of .515. This great an excess of early female embryonic deaths over male embryonic deaths seems quite unlikely.

RECESSIVE SEX-LINKED LETHALS. If new recessive sex-linked lethal mutations arise spontaneously, they might alter the sex ratio since they would eliminate any males (but not females) carrying them. It is curious that this class of mutational change, well known in *Drosophila*, has never been identified in man. A heterozygous female, half of whose sons would die, would produce sibships with twice as many females as males, and half of those daughters would repeat the high female ratio in their progeny.

The main answer probably rests on the nature of dosage compensation in the human, quite different from that in *Drosophila*. In *Drosophila*, the ratio between the X-chromosomes and autosomes in the male, 1:2, and in the female, 2:2, is apparently equalized by the doubling of the action of the X-chromosome in the male, so that both sexes operate at a 2X:2A level. According to the hypothesis of X-chromosome inactivation in the human, however, just the reverse happens: one of the two X-chromosomes of the female becomes inactive so that both sexes are functionally 1X:2A. Because the decision as to which X-chromosome in the female is to become inactive is made during early embryonic life, and all subsequent cells derived by mitosis from the cells present at that time remember how the decision was made, the female consists of a mosaic, patches of the body with one X-chromosome functional and other patches with the other functional. If one of the X-chromosomes should carry a lethal gene, all those cells in which that X becomes the active X would suffer from the lack of a normal gene and, in most cases, die, and the death of these cells would probably kill the embryo.

In any case, the concept of the simple sex-linked recessive lethal is not directly applicable to the human genotype, for the inactive-X condition makes the female effectively hemizygous for sex-linked genes just as is the male. Attempts to measure the frequencies of "recessive sex-linked lethals" in man must be concerned with a set of lethals capable of surviving in a mosaic female, this set comprising some unknown but small fraction of the total loci on the X capable of mutation to lethals.

ANTIGEN-ANTIBODY REACTIONS.   A more acceptable explanation of long standing for sex ratio changes postulates a maternal-fetal incompatibility, according to which an antigen present in the developing embryo causes the formation of an antibody in the mother, which then diffuses into the embryo with fatal results. If it is postulated that embryos differ in susceptibility depending on their sex, this would account for a deviation in the secondary sex ratio from 1:1, and the steady shift toward more females as more children are born would simply reflect the increased immunization of the mother with successive pregnancies.

Why there should exist differential mortality with respect to sex as a consequence of mother-fetus incompatibility is not clear. To circumvent this difficulty, several investigators have proposed that the Y-chromosome may possess antigenic properties so that with the birth of male children the mother becomes immunized (to some small extent) against later male children. This would account for the drop in sex ratio as family size decreases. Recent evidence confirms the existence of an antigenic locus on the Y-chromosome; if it can be shown to induce antibodies in the female with a subsequent loss of male embryos, this will constitute a major advance in our understanding of secondary sex ratio changes.

An antigenic locus on the X-chromosome might also provide the basis for an antigen-antibody reaction affecting fetuses of the two sexes differently. The Xg^a locus has been implicated in differential mortality of the sexes. Such a reaction could work in the following way: the mother-embryo incompatible combination comes from Xg(a−) mothers and Xg(a+) fathers, since the mother lacks the Xg antigen whereas all her female progeny carry it, inheriting it from the father, and none of the male progeny carries it (Figure 18-5).

Figure 18-5 Mating of an Xg(a⁻) woman and an Xg(a⁺) man, showing that all daughters will be positive and all sons negative, suggesting the possibility that an antigen-antibody reaction could develop that would preferentially affect females.

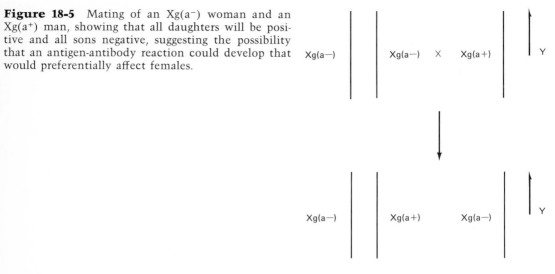

From such incompatible matings in a population of Amish, an isolated religious group in the eastern United States, it has been found that, among 394 first-born, 234 were males, giving a sex ratio of 0.59. Furthermore, in 230 second births after a daughter was born, 148 were male, with a sex ratio of 0.64. These results may be explained on the assumption that female embryos are incompatible with the mother and are rejected in some cases. If this rejection occurred very early in development, it would not be detected in the sex ratios determined by examination of stillborn infants or embryos from therapeutic or spontaneous abortions, which would show an excess of males. When the incompatible female embryo is brought to term (according to this hypothesis), the mother is sensitized so that the probability of the second embryo being rejected, if it is female, is increased.

Although a maternal-fetal antigen-antibody reaction dependent on X-chromosome antigens makes an attractive hypothesis, it cannot account for all of the observations associated with secondary sex ratio changes. For instance, why does the sex ratio go down with succeeding children when loss of female embryos should cause it to go up?

Sex Ratio in Sibships

If a couple has five boys and their sixth child is also a boy, a question immediately arises as to the probability that they carry a genetic condition predisposing them to male children. When the number is greater than half a dozen or so, the event of the birth of still another male child might be considered newsworthy. To evaluate this properly, we should ask ourselves how often we might expect such an event to happen.

For simplicity, we will take the probability of the birth of a male to be $\frac{1}{2}$. The probability of six children, all of male sex, is $(\frac{1}{2})^6$ or $\frac{1}{64}$. We should also keep in mind that in all likelihood our attention was called to the event not because all six children were males but because all six were of the same sex. In all fairness, then, we should multiply the probability by 2, to get $\frac{1}{32}$, the probability of like-sexed sibships of six. Since the number of families with six children is very great, this apparently unlikely event is really quite common, since like-sexed sibships would be found in one case out of every 32.

Although it is true that in very rare cases pedigrees have been unearthed in which there appears to be a great excess of one sex over the other, these are not representative of the situation in any family for which we have authenticated records over several generations. There is, unfortunately, no record of the sex of the progeny when a male from one of those very uncommon lines producing only males marries a female from a rare line producing only females! Whereas there can be no doubt that somewhere in the human population there exist genetic factors that shift the probability of producing a given sex considerably away from $\frac{1}{2}$, the difficulties surrounding all human genetics investigations preclude their isolation and investigation experimentally. Small family size, which aggravates the problem of ascertaining the sex ratio in a sibship, is likely to become progressively more common in the near future.

PERSONAL DECISIONS. In many countries, a premium is placed on the birth of a son and it can be surmised that in some cases family size will continue to grow until at least one male is born, and that after the birth of the male(s) the production of offspring will be discontinued. Since we have reliable data showing that the sex ratio changes with successive children, limiting the number of children will undoubtedly cause the sex ratio to increase, by an amount that will be almost imperceptible, to be sure, but readily predictable from Figure 18-3.

Our concern here is of a somewhat different sort. Can parental preferences for one sex or the other markedly change the sex ratio in the population? Although it would seem intuitively clear that a strong preference for a boy, for instance, would cause all families to have at least one boy and therefore cause the overall sex ratio to shift upward sharply, this is not the case. Consider the following analogy. Take a coin and toss it several thousand times, recording the results as a series of H(ead)s and T(ail)s, the H representing a male birth and T a female birth. Now make a slash after every H, or, if preferred, after every few H's. For our argument, we can postulate that the groups of letters between slashes represent sibships, all with at least one male birth which also terminates the sibship. Yet it is also clear that the overall sex ratio (H:T ratio) is 1:1.

What does change, in our extreme example, is the distribution of the sexes in sibships of different sizes. All sibships of size one that have been deliberately terminated have a sex ratio of all males, no females. Many studies of the sequences of sexes in families have been made, and, although some of the reports are contradictory, it does appear to be the case that families with two children of different sex are more likely to stop having additional children than are families with two children of the same sex. It can be reasonably presumed that if a couple have two children of different sex they are more likely to be satisfied by the outcome; if they have two of the same sex, they may try again, hoping the third will be different.

The Control of Sex of Progeny

It was not many decades ago that the possibility of controlling the sex of progeny was considered in the realm of science fiction, along with such impractical dreams as television and space travel. The impact of such control will be two-fold: Parents wishing to have a child of a specific sex may have it as the first child, and not depend on chance alone, whch can (and does) postpone the desired event until the second, third, or some much later event, as our previous example with a random series of H's and T's demonstrates. The obvious immediate result will be a decrease in average family size. Also, if the two sexes are not preferred by equal numbers of families, the sex ratio will shift, with a predictable change in the structure of society. Certainly just about every aspect of society would change if there were, say, three times as many males of reproductive age as females, or, conversely, if there were three times as many females as males. The consequences of each type of imbalance can be left to the imagination, but it can be argued that they would be so far-reaching and inevitable that any move at the present time on the part of a sizable segment of the population to shift the proportions of the sexes from the present approximate equality to some other value would be met with profound social disapproval and probably legal prohibition.

This is not to suggest, however, that research in these directions should not continue or that under special circumstances such control would not benefit society. At the level of the individual, we can consider the case of the woman who, for acute medical reasons, may have one and only one child, and who has a strong interest in a specific sex. At the population level, such sex control could be used by overpopulated countries which, after due consideration of its undesirable aspects, decide to adopt it as a measure of size control.

PREZYGOTIC DETERMINATION. Probably the only type of control that might be generally acceptable would depend on the separation of the two types of sperm, X- and Y-bearing, by some means, followed by artificial insemination by the desired type, or the provision of conditions for preferential fertilization of the egg by one of the types.

Separation of the sperm into two types has been attempted on a number of occasions in the past. This effort has been based on the assumption that since the X- and Y-chromosomes have different masses (as well as differing in other respects) there might be a corresponding difference between the sperm carrying them. Centrifugation has been tried, as has electrophoresis. In the latter case, a mixture of sperm is placed in solution between two electrically charged plates with the hope that one kind of sperm may migrate to one plate, the other type to the other. Results of these efforts have been ambiguous, partly because of the difficulty of the test of effectiveness, which is to artificially inseminate animals with such a treated sample and count the proportions of the sexes produced. Now that it is possible to identify individual sperm with or without a Y-chromosome easily, we may look forward to a resurgence of interest in experiments in this area, with new and imaginative methods proposed for the separation. Differential staining of the two sperm types by quinacrine itself gives one approach to this problem, except that present staining methods kill the sperm.

One method repeatedly described in the literature involves exposure of the sperm to different conditions within the female genital tract, with the expectation that some conditions will be conducive to fertilization of an egg by an X-bearing sperm. In the past, these conditions were reportedly brought about by certain diets (high lettuce, low meat, etc.); more recently it has been suggested that the conditions may be controlled by acid or alkaline douches. Based on a relatively small number of cases, the treatment appears effective (even though sperm, in vitro, do not show any motility differences or other special peculiarities in media with various degrees of acidity). In any case, any doctor who prescribes any procedures for assuring the birth of children of a specific sex may be certain that half of his clients will be satisfied.

Whether any of the current reports of the control of sex of offspring will prove valid remains to be seen. In the meantime, with the introduction of so many new ways of differentiating between X- and Y-bearing sperm, reports of success will continue to appear regularly in the daily press. We should continue to regard these with cautious skepticism.

References

BARLOW, P., and C. G. VOSA. 1970. The Y chromosome in human spermatozoa. *Nature*, **226:**961.

EDWARDS, A. W. F. 1966. Sex-ratio data analysed independently of family limitation. *Ann. Hum. Genet.*, **29:**337–47.

ETZIONI, A. 1968. Sex control, science, and society. *Science*, **161:**1107–12.

JACKSON, C. E., J. D. MANN, and W. J. SCHULL. 1969. Xgᵃ blood group system and the sex ratio in man. *Nature*, **222:**445–46.

KANG, Y. S., and W. K. CHO. 1962. The sex ratio at birth and other attributes of the newborn from maternity hospitals in Korea. *Hum. Biol.*, **34:**38–48.

KIRBY, D. R. S., K. G. McWHIRTER, M. S. TEITELBAUM, and C. D. DARLINGTON. 1967. A possible immunological influence on sex ratio. *Lancet*, 139–140.

NOVITSKI, E., and L. SANDLER. 1956. The relationship between parental age, birth order and the secondary sex ratio in humans. *Ann. Hum. Genet.*, **21:**123.

SLATIS, H. M. 1953. Seasonal variation in the American live birth sex ratio. *Am. J. Hum. Genet.*, **5:**21–33.

WEIR, J. A. 1962. Hereditary and environmental influences on the sex ratio of PHH and PHL mice. *Genetics*, **47:**881–98.

Questions

Useful terms: primary sex ratio, secondary sex ratio, birth order, meiotic drive.

1. With what variables does the sex ratio at birth appear to change?
2. Are there any cases known in other species where deviations from a 1:1 sex ratio can be found in modifications of the process of meiosis?
3. Make a list of the different kinds of postmeiotic effects that might be responsible for a deviation in the sex ratio from 1:1.

4. Why cannot a simple explanation based on an immune reaction between the mother and the fetuses of one of the sexes satisfactorily explain all sex ratio deviations from equality?

5. How may the sex of an embryo miscarried very early be reliably ascertained?

6. Explain how sex-linked lethal genes could theoretically cause a disturbance in the sex ratio, and explain why this is probably not important in man.

7. Articles in the daily newspaper frequently feature one-sexed sibships. How frequently should a family with six boys and no girls occur?

8. Assume that there are 20 million families in a country and that 1 percent of these have eight children. What is the expected total number of sibships of size eight that will be single-sexed?

9. If all parents curtailed family size after the birth of a male child, would this have any influence on the overall sex ratio in the population? Is the same conclusion true if parents could decide the sex of their offspring prior to conception?

19

Genes in Populations

Use of Population Analyses

The beginning student ordinarily thinks of the study of human genetics as primarily an analysis of human pedigrees. It may come as something of a surprise, then, to learn that one of the most powerful tools for genetic analysis is an algebraic test of the frequencies of genotypes and phenotypes in the overall population, quite independent of any family relationships.

One would have thought, for instance, that the multiple allelic basis for the ABO blood groups depended on the observation that certain combinations of parents produce some kinds of progeny but not others: that an AB parent and an O parent could produce offspring either A or B but never AB or O. Actually, the one-locus three-allele hypothesis was arrived at in a completely different way. Originally it was believed that there were two loci involved, an A locus and a B locus with a dominant allele at each responsible for the production of the corresponding antigen on the cell surface. Our present belief that these blood groups are determined by multiple alleles was suggested in 1925, after an algebraic analysis of the frequencies of the four different types in the overall population. We shall make a detailed examination of this analysis later in the chapter. First we must understand the theoretical basis for population analyses.

The production of progeny representing a new generation is subject to an almost infinite number of complications: the dictates of social customs and taboos, limitations of choice by the stratification of society by class, wealth, ethnic, religious, or other distinctions, the unpredictability of personal decisions, to say nothing of the biological unknowns in fertilization and development. At first sight, it might appear that any attempt to predict frequencies of homozygotes or heterozygotes from one generation to the next would be certain to fail. This is not the case: The mathematical expressions for the distributions of most alleles in populations are extraordinarily simple and equally useful.

The Concept of Random Associations of Alleles

As an illustration, assume that a blood group exists with two alleles, A and A′, with no discernible effect and recognizable only after testing with rare antisera. Only a few persons would ever be aware of their genetic composition—whether AA, AA′, or A′A′—and they would have no reason to let this information enter into their decision to have (or not to have) offspring; their spouses would be equally unconcerned about their genetic composition at this locus. For this reason, we can bypass all the complications mentioned above and look at the production of a new individual (as far as this locus is concerned) as the combination of one randomly selected allele present in the sperm with a second random allele present in the egg. If the frequency of the allele A′ is 10 percent, the other 90 percent being A, then the chance that the sperm will carry A′ is 10 percent or one tenth. The same would be true for the egg.

Another way of visualizing the random distribution of alleles in the gametes, and the consequences after zygote formation, is to consider the following analogy. Imagine that there are two containers, one with a large number of sperm, the other eggs. The proportions of sperm with the allele A and sperm with A′ in the first container are the same as those in the general population; the same is true for the two alleles in the eggs in the second. Zygotes are then formed by the combination of randomly selected sperm and eggs. If we know the relative proportions of gametes with A and A′, we can predict the frequencies expected for the genotypes of the zygotes.

The Hardy-Weinberg Rule

Mendel showed that the ratio of the three genotypes in the F_2 after a cross of $AA \times A'A'$ is $1AA:2AA':1A'A'$. The question of the relative proportions of the three genotypes AA, AA', and $A'A'$ from the population standpoint came up only a year or two after the rediscovery of Mendel's paper, when a number of researchers pointed out that the Mendelian ratio of $1:2:1$, derived from experimental matings, did not describe the genotype frequencies of alleles in the general population, where the proportions of parental genotypes were unknown. In 1908 an English mathematician named Hardy and a German physician named Weinberg independently suggested the algebraic basis for this distribution of genotypes. As we have shown in Chapter 7, the argument is quite simple and straightforward, requiring nothing more than the knowledge of high school algebra. Unfortunately the generally accepted name for this formula is the Hardy-Weinberg law, which sounds more forbidding than it really is. We shall refer to it as the Hardy-Weinberg rule.

This rule is based on the simple principle of probability that the chance of two independent events happening simultaneously is the product of their individual probabilities, times the number of different arrangements. Take a pair of alleles, A and A'. First, an egg is randomly selected from a large pool containing the proportion p of allele A, and q (or $1 - p$) of allele A'. In other words, the chance of reaching into the pool and getting an egg with allele A is p, and the chance of getting one with A' is q. The same applies with a second pool of sperm. The probability that the zygote formed by the two chosen gametes will have genotype AA is $p \times p$, or p^2. The chance of getting an egg and a sperm both with A' is $q \times q$, or q^2. For the heterozygote AA', the chance the egg has A and the sperm A' is pq; or the sperm might be A and the egg A', which also has the probability of pq. So the total expected frequency of heterozygotes is $pq + pq$, or $2pq$.

Therefore, the algebraic binomial distribution $p^2 + 2pq + q^2$ corresponds to the expected distribution of the genotypes AA, AA', and $A'A'$. From this it is clear that if we know the frequencies of the alleles at any locus we can easily calculate the expected frequencies of the different genotypes and their accompanying phenotypes. As an example, if the allele S' for sickle-cell anemia has a frequency of .1, and the normal allele S has a frequency of .9, then the genotypes $S'S'$, SS', and SS should be found in the relative proportions $(.1)^2 : 2(.1)(.9) : (.9)^2$, or $.01 : .18 : .81$.

RESTRICTIONS ON RANDOMNESS. In some cases, our assumption of complete randomness in selecting eggs and sperm is obviously not valid. If, for instance, brunettes have a predisposition to marry brunettes, and blondes to marry blondes, then zygote formation with respect to hair color will not be random in the population. This kind of preference in mating based on phenotype is referred to as *assortative mating*. Similarly, if a group of people form an inbreeding unit and do not freely exchange genes with the population at large, as is the case for many religious and ethnic groups throughout the world, they represent an *isolate* and cannot be included without qualification with the rest of the population. In the same way, it would be unreasonable to consider two sibs as two randomly selected people from the population—we know that they have a specific nonrandom genetic relationship to each other.

It must be true for the vast majority of inherited characteristics, however, that the alleles at the loci responsible are combined essentially at random. For example, most people do not know their own blood group, with respect either to ABO or to any of the other dozen or more cell-surface antigen types. Even if this were general knowledge, it seems unlikely that an appreciable number of impending marriages would be cancelled because the bride's and groom's cell-surface antigens did not seem suitable. A population involving random mating with respect to a set of alleles or loci is said to be *panmictic* (noun: *panmixis*) for that set.

Algebra of the MN Blood Groups

The best way to show that this algebraic rule actually holds for human populations is to examine a simple allelic case in which available data give the frequencies of all three genotypes. The MN blood groups serve as an excellent example. At a single locus there are two major alleles, M and N. The alleles are *codominant*—that is, the heterozygote MN has both the M and N antigens. Since fewer than one couple in a million are aware of their MN types, assortative mating cannot possibly affect the randomness in this case. With available test sera, it is simple to determine whether individuals are phenotypically of group M (MM), group N (NN), or group MN (MN). This has been done for several populations (Table 19-1). For each population studied, a very good estimate of the allelic frequencies within that population can easily be made. Since all of the loci of homozygous M individuals and half of the MN heterozygotes carry M, we need simply to take the frequency of M homozygotes and add to it half the frequency of the heterozygotes to get a direct measure of the frequency p of the M allele. Similarly, the frequency of the N allele, q, can be computed as the frequency of N homozygotes plus half that of the heterozygotes. (Alternatively, since there are only two alleles that must add up to 100 percent, $q = 1 - p$.)

Given those two frequencies, it is now possible to use the binomial distribution to arrive at the expected frequencies of both types of homozygotes as well as the heterozygotes. This has been done in Table 19-1 on the line labeled "Calculated." It can readily be seen that there is astonishingly close agreement between the observed and the expected frequencies. The skeptic might wonder if there is not an element of circular reasoning in performing this calculation, since the figures on which the agreement is based come from the data themselves. It need only be pointed out that a wide variety of populations might have allele frequencies of .6 and .4—including the extreme case in which 60 percent are MM and 40 percent are NN—but only the single distribution of 36 percent MM, 48 percent MN, and 16 percent NN precisely fits the binomial expectation for those allele frequencies.

The preceding illustration in Table 19-1 shows that for the MN blood-group alleles panmixis can be assumed to exist in the selected populations. Such a demonstration gives us confidence in applying this simple algebraic formulation to other genetic characteristics in the human population, for otherwise we might be led astray by considerations such as social taboos or stigmas, laws governing marriage, or family relationships that might be imagined to have a significant effect in modifying genotypic frequencies in the population.

Table 19-1 The percentages of the MN groups in three populations selected because they represent populations in which the two alleles are found with quite different frequencies.

Population	Number		*M*	*MN*	*N*	p_M	q_N
German	16,000	Observed	30.6	49.1	20.3	.552	.448
		Calculated	30.4	49.5	20.1		
Melanesian (New Guinea)	1,148	Observed	1.0	9.6	89.4	.058	.942
		Calculated	.3	11.0	88.6		
Eskimo (Greenland)	739	Observed	87.4	12.2	0.4	.935	.065
		Calculated	87.4	12.2	0.4		

Calculation of Gene Frequencies from the Homozygote Frequency

Let us apply the algebra to the case of a simple recessive condition. Cystic fibrosis is a simple autosomal recessive disease that cripples about one in every 1,600 Whites (much less commonly Blacks and Orientals), causing death usually before age 18. In total numbers of persons affected, it is the most prevalent genetic disease in the United States. It is therefore important to know how frequently individuals in the population carry this disease in the heterozygous state.

If the frequency of the homozygous recessive, q^2, equals $1/1,600$, q is equal to its square root, or $\frac{1}{40}$, and p then equals $1 - \frac{1}{40}$ or $\frac{39}{40}$. $2pq$, the frequency of heterozygotes, equals $2 \times \frac{39}{40} \times \frac{1}{40}$, or about $\frac{1}{20}$. Thus, from knowing the frequency of the homozygous affected individuals, we can very easily compute the probable proportion of heterozygotes in the population.

The chance that two individuals (such as a married couple) taken at random from the population are both heterozygous will be $\frac{1}{20} \times \frac{1}{20}$ or $\frac{1}{400}$. The chance that such individuals will produce a homozygous recessive child is $\frac{1}{4}$, so the overall chance of the production of an affected child is $1/400 \times \frac{1}{4} = 1/1,600$, in agreement with the original statement of the frequency of occurrence. (That is, in fact, a circular calculation, since it amounts to saying that $2pq \times 2pq \times \frac{1}{4} = q^2$, if p is approximately equal to 1.)

FREQUENCY OF Rh INCOMPATIBILITY.   In some cases it is important to estimate the frequency with which certain kinds of children are produced. For instance, it is interesting to compare the actual frequency of erythroblastosis fetalis (less than 1 percent) with a theoretically expected frequency based on the calculated number of Rh-positive children from Rh-negative mothers.

If Rh$^-$ women represent about 16 percent of the population, then it follows that all eggs of these women—16 percent of the eggs in the population—are potentially capable of producing Rh-incompatible children. Rh$^+$ children of Rh$^-$ mothers, however, must have an R allele from the father. The chance that an allele is R is quickly calculated. If rr homozygotes are found with a frequency of 16 percent, then $q^2 = .16$ and $q = .4$. Since the frequency of the r allele is .4, the frequency of R must be $1 - .4 = .6$. Therefore, the chance that an r egg from an rr mother (.16) will be fertilized by an R sperm (.6) is about 10 percent (.16 $\times$ .6 $=$ .096). Since erythroblastosis actually occurs with a frequency of less than 1 percent, we can say that erythroblastosis is found less than one tenth as often as theoretically expected. It was, in fact, the pursuit of the factors that prevented Rh incompatibility from occurring with the theoretically expected frequency of 10 percent that led to the therapy of Rh immunization.

The ABO Blood Groups

The classic example of the usefulness of the algebraic approach in solving genetic problems is given by the analysis of the ABO blood groups. In 1910 two Polish immunologists suggested that these groups were determined by two independent

Table 19-2 A listing of the genotypes that would be responsible for the four ABO blood-group phenotypes on the assumption that there exist two loci, one responsible for the A antigen and the other for B, with one dominant allele at each locus being sufficient for the expression of its antigen.

Genotype	Phenotype
AABB	AB
AaBB	
AABb	
AaBb	
AAbb	A
Aabb	
aaBB	B
aaBb	
aabb	O

loci, an *A* locus and a *B* locus (Table 19-2). Although this was a reasonable hypothesis at the time, it has since been replaced by the generally accepted hypothesis of one locus with three alleles. There appears to be one unambiguous way of distinguishing the two hypotheses, as shown in Table 19-3. An individual of group AB, according to the early theory, having married someone of group O, could produce progeny of groups A, B, AB, or O. According to the three-allele hypothesis, however, their progeny can only be A or B, never AB or O. The distinction seems clearly demonstrable. Curiously, if the progeny of AB × O matings are tabulated (Table 19-4), we see that prior to 1925 the data favored the first hypothesis, since both AB and O individuals are found among the progeny of such matings.

DISPROOF OF THE TWO-LOCUS HYPOTHESIS. We shall now show that the two-locus hypothesis is untenable, using only population data. The algebraic argument is simple. If A and B are independent, it should be possible to treat them as independent algebraic quantities. To do this we apply the elementary principle of probability that the chance of two independent events occurring simultaneously is the product of their individual probabilities. That means that

Table 19-3 The most obvious genetic test of two different hypotheses of the genetic determination of the ABO blood groups. According to the two-locus hypothesis, the progeny of a marriage of AB × O could give all four phenotypes (if the AB parent happened to be doubly heterozygous), whereas according to the one-locus three-allele hypothesis the progeny could be only A and B.

	Two-locus Hypothesis	One-locus Three-allele Hypothesis
Phenotypes of parents	AB × O	AB × O
Possible genotypes of parents	AaBb × aabb	AB × OO
Possible genotypes of progeny	AaBb, Aabb, aaBb, aabb	AO, BO
Possible phenotypes of progeny	AB, A, B, O	A, B

Table 19-4 The phenotypes of offspring recorded from matings between O and AB over a period of time, showing how the observations changed as the theory of the genetic determination changed. (A. H. Sturtevant, *A History of Genetics*, Harper & Row, New York, 1965.)

Published Records	O	A	B	AB	Number of Papers
Up to 1910	2	2	2	3	1
Up to 1925	27	80	59	24	18
During 1927–1929	2	228	234	1	6

the frequency of persons in the population of group AB (with both antigens simultaneously) should equal the product of the frequencies of all those with the A antigen times the frequencies of those with the B antigen. This is not, however, simply the frequency of those in group A times the frequency of those in group B, because AB individuals also have the A antigen and the B antigen, so this class must be added to both.

$$\text{Frequency of AB} = (\text{frequency of all A}) (\text{frequency of all B})$$
$$= (A + AB) (B + AB)$$

This means that for any population, irrespective of the frequencies of the blood groups in it, we should get a good agreement by making this simple multiplication. This has been carried out in Table 19-5. There is clearly a wide discrepancy: For the French population, the expected frequency of AB, .059 is almost twice as large as the observed, .034, and the discrepancy is also very great for the calculations with the Japanese data.

Clearly, something is wrong. We might suspect the data to be unreliable, possibly because of statistical variability in the number used, or because of errors in blood-group typing. Perhaps the populations tested were not panmictic within themselves with respect to the ABO blood groups, or there may have been some unknown factor, such as low viability or lethality, causing the observed

Table 19-5 Blood-group frequencies for several populations, showing that the test for independence of the A and B groups fails.

Group	Number	Frequencies			
		O	A	B	AB
French	103,242	.441	.435	.090	.034
Japanese	240,204	.323	.364	.228	.085

Calculations To See Whether AB = (A + AB) (B + AB)				
Group	A + AB	B + AB	Calculated AB: (A + AB) (B + AB)	Observed AB
French	.469	.124	.058	.034
Japanese	.449	.313	.141	.085

Figure 19-1 A checkerboard showing the expected population frequencies of the ABO blood groups based on the assumption that two independent factors, A and B, determine those groups. This formulation leads to a logical absurdity and forces us to assume instead that there are three alleles at a single locus. The squares are numbered for convenience in referring to the expressions given in the text; the expected frequencies are given in small letters and the genetic constitutions in capital letters.

Gametes from Female Parent

Gametes from Male Parent	AB p	A q	B r	O s
p AB	p^2 AB 1	pq AB 2	pr AB 3	ps AB 4
q A	pq AB 5	q^2 A 6	qr AB 7	qs A 8
r B	pr AB 9	qr AB 10	r^2 B 11	rs B 12
s O	ps AB 13	qs A 14	rs B 15	s^2 O 16

frequencies to deviate from the expected. Actually, the real reason is that the basic genetic assumption on which our algebra is based, i.e., that there are two independent loci involved, is faulty.

Let us look at the data in still another way. Suppose that there are four kinds of gametes in the population pool, those with the AB factors, those with A, those with B, and those with neither. We may represent all combinations of these gametes in a checkerboard (Figure 19-1) to produce all possible zygotic types. Since we know that the A and B factors are inherited as dominants, we can assume that any zygote receiving one A factor (or more) will have the A antigen, receiving one B or more will have the B antigen, and receiving at least one of each will have both A and B antigens. Furthermore, we can specify the expected frequency for each zygotic combination by multiplying the marginal frequencies, which we represent algebraically by p, q, r, and s. Since we have four observed zygotic types in the human population, we should find it possible to work back from the observed population zygotic frequencies to the theoretical frequencies of gametes that best fit the data. Examination of the phenotypes in the 16 cells in Figure 19-1 shows the following relationships to hold:

1. Cell 16 = group O = s^2
2. Cells 11 + 12 + 15 + 16 = B + O
$$= r^2 + 2rs + s^2$$
$$= (r + s)^2$$

3. Cells 6 + 8 + 14 + 16 = A + O
$$= q^2 + 2qs + s^2$$
$$= (q + s)^2$$

From the extensive data for the French population in Table 19-5 we see that the frequencies are

$$O = .441 \qquad A = .435 \qquad B = .090 \qquad AB = .034$$

Using the relations shown in Figure 19-1 giving the relative frequencies expected for the O, A, and B groups, and substituting the numerical values for these groups, we can rewrite the equations as follows:

1. Frequency of group O, $.441 = s^2$
2. Frequencies of groups B + O, $.090 + .441 = .531 = (r + s)^2$
3. Frequencies of groups A + O, $.435 + .441 = .876 = (q + s)^2$

From 1, $s = \sqrt{.441} = .664$

From 2, $r + s = \sqrt{.531} = .728$ $\qquad r = .728 - s = .064$

From 3, $q + s = \sqrt{.876} = .936$ $\qquad q = .936 - s = .272$

We have now calculated the frequencies of the gametes carrying the A, B, and O factors. There is one class left, that carrying both A and B, occurring with frequency p. However

$$p + q + r + s \text{ must equal one}$$

$$\text{therefore } p = 1 - q - r - s$$

$$p = 1 - .272 - .064 - .664$$

$$p = 0!$$

In other words, the AB gamete type has a zero frequency. Could this be a mistake? We can make a simple check, since we have not yet used any information provided by the AB group itself. If $p = 0$, then the first row and first column of cells also equal zero. The only source of AB zygotes is from cells 7 and 10, and each of these has an expected frequency of qr, for a total of $2qr$. We have just calculated q to be .272 and r to be .064.

$$2qr = 2(.272)(.064) = .035$$

This agrees very well with the observed population value of .034; we may feel reassured that our calculations, and reasoning, are not in error.

(The student is asked to verify for himself that this startling result is true for ABO frequencies from other large populations as well.)

Why does the gamete type carrying both A and B together have a zero frequency, according to our calculations? It is possible to think of reasons. Perhaps gametes with A and B together are lethal or do not produce viable zygotes. This, however, is not an acceptable explanation because such a loss of alleles would have led to the gradual reduction of the A and B alleles in the population over many centuries, so that they should now be eliminated. Other complicated genetic explanations can be similarly devised, and similarly refuted.

The correct explanation is also the simplest: Gametes with both A and B simultaneously are nonexistent because in AB individuals who theoretically might produce such gametes, the A and B factors are always found on homologous chromosomes and so they separate from each other during meiosis. Furthermore, since they are never put together on the same chromosome strand by crossing over to give a single chromosome of constitution AB, we can infer that they are at the same locus, that is, they are allelic. Furthermore, since there is a third property, O, which is an alternative to A and B, we further deduce that there is a third allele, O, necessary to complete the genetic explanation.

In this way, from population data alone and without recourse to pedigree studies, we can conclude that the early two-locus hypothesis is incorrect and that three alleles at one locus fit the data better. It was a simple calculation like this that caused Bernstein in 1925 to challenge the two-locus hypothesis and propose instead the one-locus, three-allele hypothesis.

VALIDITY OF THE THREE-ALLELE HYPOTHESIS. Subsequent work has proved Bernstein to be correct—published data after 1927 show a precipitous drop in number of AB and O offspring from AB × O parents (Table 19-4). Just why the early results included so many impossible progeny is not at all clear; perhaps the antibodies available for testing in those days were not as discriminating as those available now, or perhaps the technicians responsible for typing regularly misclassified individuals of group AB as being either A or B, or groups A and B as being O. In any case, this shift in the data after 1927 to agree with the one-locus, three-allele hypothesis is a very nice example of the closer agreement of experimental results with theory when the observer "knows" that certain results are impossible!

At the present time, the algebraic test for consistency is considered a necessary step in the formulation of any new genetic hypothesis, when such a test is feasible. Any genetic interpretation, however well it might agree with pedigree data, cannot be considered acceptable if it involves algebraic inconsistencies in a population analysis, as the two-locus hypothesis of the ABO blood groups does.

Time Required for Equilibrium to Be Established

If one population of humans with one set of allele frequencies meets another set, and complete panmixis occurs, the genotypes of the next generation will be found in the proportions $p^2\,AA : 2pq\,AA' : q^2\,A'A'$, where p and q are the allele frequencies in the combined population. Therefore, in this case, it has taken only one generation to reach Hardy-Weinberg equilibrium.

In actual practice the stratifications of populations prevent complete panmixis. Data from European countries with superficially homogeneous populations will show different blood group frequencies from one section of a country to another, with each of the sets of data internally consistent with the Hardy-Weinberg equilibrium, a relic no doubt of the historical immobility of populations. Adding together data simply because they are taken from the same national or ethnic group can lead to disagreement with the theoretical expectations of the frequencies of the genotypes. For this reason, the populations used for calculations in this chapter have been chosen from a specific province (Nord) in France and one city (Miyasi) in Japan. On the other hand, it is surprising that studies of gene frequencies in highly heterogeneous populations, such as the Whites of New York City, show reasonable approximations to the Hardy-Weinberg expectations.

There are a few special circumstances in which the equilibrium is delayed for one or more generations, even with panmixis. These cases are found when our simple analogy of gametes randomly selected from pools of each sex having the same allele frequencies breaks down. For instance, if a group of males all of type M should mate with a group of females all of type N, then the F_1 will necessarily all be of group MN; the alleles would not be in equilibrium in the F_1, but would in the following generation.

Equilibrium for Linked Loci

It is a logical extension of the Hardy-Weinberg rule that, with panmixis, equilibrium will readily be reached for more than two alleles at the same locus, as in the ABO blood groups, or even for sets of alleles at independent loci. It may not be obvious that similar considerations apply to two different loci located on the same chromosome. As an example, it is a common fallacy to assume that the locus for brunette-blonde hair color and the locus for brown-blue eyes are found on the same chromosome. According to this idea, some population groups would have the brunette and brown alleles on the same chromosome, others would have blonde and blue alleles on the same chromosome. People who are blue-eyed brunettes or brown-eyed blondes would therefore carry the less common recombinant chromosomes. Were this in fact the case, after a number of generations (the exact number depending on the closeness of the two loci on the chromosomes and the randomness of mating), crossing over in heterozygous individuals would give rise to recombinant chromosomes, some carrying the brunette and blue alleles together, others the blonde and brown, and eventually the alleles at the two loci would be randomly associated in any population study. Any apparent association in the population at that point would have to be attributed to some other cause—in this case, to a single locus responsible for a generalized pigment reduction affecting both hair and eyes. The test provided by the Hardy-Weinberg rule is a very powerful tool for many genetic hypotheses, but it is useless by itself in determining linkage between two loci. For that type of analysis, specific pedigree data are necessary.

Variations in Equilibrium Frequencies

A corollary to the principle that equilibrium is arrived at usually within one or at most a few generations is the idea that such an equilibrium, once attained, will be repeated in each subsequent generation with the same frequencies of genotypes. The English mathematician Hardy developed the simple point of algebraic expectations after a statistician asked why, if brown eyes are dominant to blue, did they not take over the entire allelic population, resulting in the inevitable elimination of blue eyes? That query clearly resulted from confusing the meaning of "dominant," which simply describes the phenotype of the heterozygote, with some erroneously imagined genetic advantage to that particular allele in the course of evolution.

FACTORS CHANGING ALLELE FREQUENCIES. In a panmictic population, the genotypic frequencies will shift when the relative frequencies of the alleles change. By far the most important cause of changes in allele frequency is *selection*, operating differently on different genotypes. If one of the genotypes is "superior" (i.e., produces more offspring), then the alleles in that genotype will increase at the expense of those in other genotypes. The converse argument holds for an "inferior" genotype. In this way, *natural selection* will increase the frequency of advantageous alleles and decrease that of deleterious ones.

Mutation from one allele to another will, theoretically, change gene frequencies, but mutation rates are so low (approximately one per million) that the effect of mutation every generation will not be as great as that of selection. The great significance of mutations in a population is that they must provide the genetic variability on which selection can act.

When a population is very small, the gene frequencies in one generation may not coincide with those in the next because random selection of a small number of gametes from the population of gametes may just happen to result in a higher or lower percentage of an allele than in the previous generation. This is basically a statistical, or chance, phenomenon, and is known as *genetic drift*. Because genetic drift is relatively insignificant in populations as large as, say, a thousand people, this chance variation can have played an important role only when the human population consisted more of smaller isolates than it does now, although, to be sure, genetic drift may still be effective in certain small groups—religious or ethnic isolates, for instance.

These factors will affect gene frequencies in a population over long periods of time. It should be noted, however, that except for a few cases such as sickle-cell anemia, in which one of the genotypes has a much reduced viability, they do not produce detectable deviations in the Hardy-Weinberg equilibrium at any one time. Our calculations of expected frequencies of genotypes—of the blood groups, for instance—are based on data too limited to show any effects of these additional disturbances.

References

BERNSTEIN, F. 1925. Zusammenfassende Betrachtungen über die erblichen Blutstrukturen des Menschen. *Z. Abst. Vererbungsl.*, **37**:237–70.

HARDY, G. H. 1908. Mendelian proportions in a mixed population. *Science*, **28**:49–50.

STURTEVANT, A. H. 1965. *A History of Genetics.* New York: Harper & Row.

VON DUNGERN, E., and L. HIRSCHFELD. 1911. Ueber gruppenspezifische Strukturen des Blutes. III. *Z. Immunitaetsforsch. Exp. Ther.*, **8**:526–62.

WEINBERG, W. 1908. Ueber den Nachweis der Vererbung beim Menschen. *Jahresh. Ver. Vaterl. Naturkd. Wuerttemb.*, **64**:368–82. (The essential section of this paper is translated in Stern, C. 1943. The Hardy-Weinberg law. *Science*, **97**:137–38.)

Questions

Useful terms: assortative mating, isolate, panmixis, Hardy-Weinberg rule, codominant, binomial distribution, equilibrium frequency, natural selection, mutation, genetic drift.

1. Make a list of characters for which there is probably some degree of assortative mating. Make another list for which assortative mating probably does not occur.
2. Can you suggest several reasons why large populations might divide themselves into isolates?
3. Is the population of the United States panmictic?
4. How is it possible to show that populations of humans found in different parts of the world are panmictic?
5. Would populations that consisted of one third of each of the two homozygotes and one third of the heterozygote classes be panmictic?

6. Using the analogies of gametes found in a common pool from which specific alleles are selected at random, it can readily be shown that irrespective of the frequencies of genotypes in any generation the Hardy-Weinberg equilibrium will be set up after one generation of panmixis. Is it necessary to show that this equilibrium will continue for the next generation?

7. If the frequency of the recessive allele responsible for a specific type of albinism is one in 20,000, what is the frequency of heterozygotes in a population? Would you use the same calculation if the question referred to the frequency of color-blindness in a population?

8. Can you show very simply, without the use of complicated mathematics, that the population data for the distribution of the ABO phenotypes do not agree with the hypothesis of two independent loci, an A and a B locus?

9. Why was it necessary to await the mathematical analysis of population data to refute the two-locus hypothesis of genetic determination of the ABO blood groups, although this should be immediately apparent from the kinds of children produced in marriages of AB × O?

10. What are the factors responsible for altering frequencies of alleles from one generation to the next? Why are such changes of any significance?

20

Multiple Allelic Systems and the Blood Groups

More than 250 different erythrocyte cell-surface antigens have been identified so far, determined by more than a dozen well-defined loci, some of which have large numbers of allelic differences. Two of these loci, ABO and Rh, have been discussed earlier. In this chapter we shall consider their properties in more detail, as well as those of other loci.

The ABO or ABH System

BIOCHEMISTRY.   The basic antigen in the ABO complex of antigenic specificity is a protein to which are attached modified sugar molecules, derivatives of galactose and glucose (Figure 20-1). This protein may be visualized as having dozens of the sugar chains extending out from its surface. Thus the chemical composition of the sugars can determine the antigenic specificity of the molecule. These combinations of sugars and proteins are called glycoproteins.

The gene "*H*" is responsible for an enzyme that adds another sugar, fucose, to the basic molecule. This locus is independent of the ABO locus. The *H* substance, produced by the *H* gene, is then acted upon by the *A* or *B* allele, or both, to produce further derivatives. The *B* allele adds a galactose to the chain; the *A* allele adds instead a modified galactose, abbreviated to gal-NAC (Figure 20-2). The *O* allele is inactive and adds nothing.

THE H SUBSTANCE.   The H substance is antigenic; that is, it can stimulate production of the anti-H antibody. However, the H substance is found in most

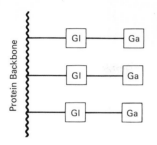

Figure 20-1 The basic structure of the molecule responsible for the antigenic specificity in the ABO blood group system. A galactose molecule (Ga) and a glucose molecule (Gl) together are attached at intervals along the protein backbone to produce a combination known as glycoprotein.

people, since almost everyone is either *HH* or *Hh*, and it is recognized by the antigen-antibody system as a normal component of the body. Therefore, no antibody is ordinarily produced.

THE BOMBAY PHENOTYPE. Persons homozygous for the recessive allele *h* do not manufacture the H substance; therefore, they do not produce the AB antigens, which depend on the H substance as a precursor. Persons who are homozygous *hh* will superficially appear to be of group O, irrespective of their allelic composition at the AB locus. Such homozygotes were first found in Bombay, India, in 1952, as two individuals who needed transfusions after blood loss; the *h* allele is often referred to as the *Bombay allele*. Figure 20-3 shows a pedigree of a family with this allele: the propositus appears to be O, although her AB origin and offspring both indicate that she must have carried the B allele and must have been of the genotype BO, *hh*.

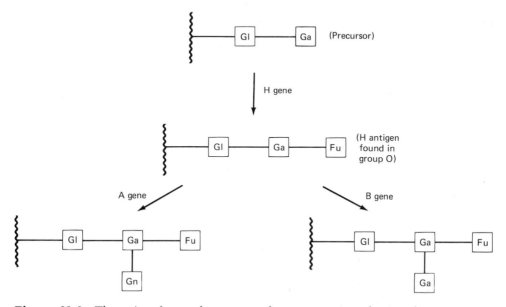

Figure 20-2 The series of steps that convert the precursor into the A and B antigens. The *H* gene, present in almost all people, adds a fucose molecule (Fu) to the galactose, whereupon the *A* gene, when present, adds a modified galactose (Gn) and the *B* gene another galactose (Ga) to the galactose already present.

Figure 20-3 Pedigree of a family in which the mother appeared to be group O (although her mother was AB), and who, with her A husband, produced three AB children. It seems clear that she must have carried the *B* allele, which was not expressed; she was homozygous *hh*.

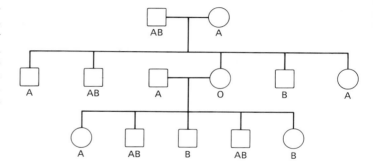

Persons with the Bombay phenotype have cells that are not agglutinated by either anti-A or anti-B antisera, suggesting that they are of group O, but neither are their cells agglutinated by anti-H, which would be expected of group O cells. Their serum contains anti-H as well as anti-A and anti-B antibodies. The wisdom of performing crossmatches of the donor and recipient blood prior to a transfusion is well illustrated by the problem that arises when the recipient has the Bombay phenotype. A simple agglutination test of Bombay erythrocytes with A and B antisera would prove negative, indicating that the recipient is of group O. If, however, O blood were transfused, agglutination would occur because the anti-H antibody present in the *hh* individual would react with the H antigen normally found on the erythrocytes of group O donors. Crossmatching prior to transfusion would reveal this potential danger.

THE LEWIS SYSTEM.   Another locus, Lewis (*Le*), manufactures an enzyme that can add a fucose molecule to the glucose (Figure 20-4). When this addition is made to the precursor, a molecule is produced with an antigenic property called Le^a, but when the addition is made to the H substance, which already has one fucose added to the galactose, a molecule with two fucoses is produced. This molecule then loses both its H and Le^a antigenic properties and acquires a new property, Le^b. The recessive allele *le* is inactive. The *Le* gene does not alter the A and B specificity so that a person's ABO blood group remains the same, independent of his or her *Le* constitution.

THE SECRETOR LOCUS.   Still another locus, *Se*, is involved not in determining a specific antigen but rather in allowing the *H* gene to express itself in certain tissues and not in others. Eighty percent of the population have an *Se* allele and produce the A, B, and H substances in some watery secretions, such as saliva; such persons are known as secretors. The remaining 20 percent are homozygous *se se* and do not secrete these substances. The A, B, and H antigens on the erythrocytes are unaffected.

The interaction between the *Le* and the *Se* loci is more complicated, and still not completely understood. It appears that the Le^b antigen occurs only in individuals who are secretors, i.e., who have the *Se* allele.

OTHER ALLELES OF THE ABO SYSTEM.   The two major A-type alleles are A_1 and A_2. These can be distinguished because normal anti-A antiserum can be shown to have two major types of antibodies, one of which agglutinates the cells of

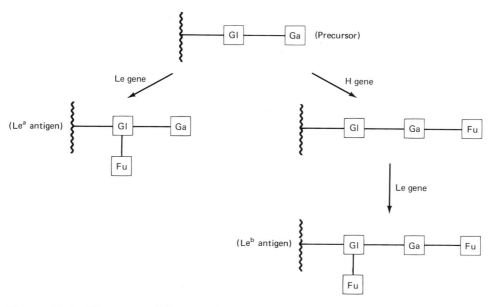

Figure 20-4 The action of the Lewis locus. A molecule of fucose (Fu) is added to the glucose (Gl). When the addition is made to the precursor, directly, Le^a antigen is produced, but when the addition is made to the H substance the Le^b antigen is produced.

some individuals with an A allele but not others, whereas the second agglutinates the erythrocytes of all persons with an A allele. The allele on the cell type which is agglutinated by both antibodies is called A_1; cells with A_2 are agglutinated by only one of these two antibodies. About 75 percent of all A alleles are A_1; most of the rest are A_2. A_4 is an antigen affected by serum from O individuals but not from B. A_3, A_5, A_6, and A_0 have been reported independently by various workers as very weak alleles; in some cases they may represent the same weak allele. A rare phenotype known as A_m has little or no A antigen on the erythrocytes, although it is present in normal amounts in the watery secretions. A_m^h is similar to A_m in that the erythrocytes show a weak A reaction; in addition, the H antigen is diminished or absent in those cells, although the secretions are normal with respect to both the A and H antigens. Minor variations of the B allele have also been reported.

The MN System

The ABO groups were discovered before any of the other groups because of the adverse and frequently fatal reaction after the transfusion of the blood of one person into another. This agglutination is immediate because each person naturally carries the antibodies for those antigens of the AB system not present. This is the case, however, only for the AB antigens; for other blood groups, antibodies are not naturally found to the antigens not present, and in many cases they can be induced in humans only with difficulty.

More than a quarter of a century (1900 to 1927) elapsed between the discovery of the ABO groups and the description of a new independent system, the MN groups. Since humans do not usually carry antibodies to the MN antigens, the corresponding antibodies had to be induced in experimental animals.

TECHNIQUE TO TEST FOR NEW BLOOD GROUPS.   Testing for new blood groups is accomplished in the following way: the erythrocytes of a person who has been completely characterized with respect to all known blood groups are injected into an experimental animal—in this instance, a rabbit (Figure 20-5). Several injections at intervals of a week or more will increase antibody production in the rabbit. Serum is then removed from the animal; it contains large numbers—undoubtedly many dozens—of antibodies newly induced by the human tissue. To the serum containing all of these antibodies are added erythrocytes from another person identical to the first in all known groups. Practically all of the newly formed antibodies will be absorbed at the reaction sites on the erythrocytes; these will include antibodies to many antigens peculiar to humans and not rabbits. There will remain in the serum, however, antibodies corresponding to any antigens possessed by the person who provided the erythrocytes for the initial injection, but not by the person providing the erythrocytes for the absorption. These will be evident because the apparently completely absorbed serum will still be capable of agglutinating the original erythrocytes. Tests can then be made to determine the possible genetic basis of the new antigen responsible for the unexpected antibody.

In this way, it was found that all persons could be categorized in three antigenic classes, quite independent of the ABO groups both serologically and genetically. These classes were called *M, N,* and *MN:* the first class carried the new antigen *M,* the second *N,* and the third carried both.

GENETICS OF THE MN GROUPS.   Individuals of group M are homozygous for the allele called *M* (using *L* as the basic symbol, $L^M L^M$—or, by our abbreviated symbolism, *MM*); group N individuals are homozygous *NN,* and MN persons are heterozygous *MN.* Subsequently it was found from some studies in Sydney, Australia, and in England, that associated with the MN groups there exists a second antigenic property with two alternative forms, *S* and *s.* These two properties also have a codominant relationship, although the symbolism unfortunately implies a dominant-recessive relationship. Table 20-1 gives the frequencies

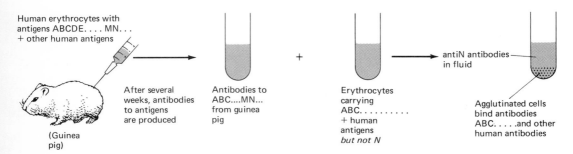

Human erythrocytes with antigens ABCDE. . . . MN. . . + other human antigens

After several weeks, antibodies to antigens are produced

(Guinea pig)

Antibodies to ABC....MN... from guinea pig

+

Erythrocytes carrying ABC. . . . . + human antigens *but not N*

antiN antibodies in fluid

Agglutinated cells bind antibodies ABC. . . . .and other human antibodies

Figure 20-5 The technique for testing for additional erythrocyte antigens, leading to the discovery of the MN groups.

	MN Group		
	M	**MN**	**N**
With S antigen	74	60	35
Without S antigen	26	40	65

Table 20-1 The percentages of individuals carrying or not carrying the S antigen in the three MN groups.

of the presence or absence of the S antigen in individuals of the three MN types. From this it can be seen that they are not randomly associated: three quarters of type M, but only one third of type N, are S-positive. This statistical association implies a biological one as well; the simplest conclusion is that both antigens are produced by a single locus and that there are four common alleles, *MS, Ms, NS,* and *Ns,* occurring with the relative frequencies of 0.25, 0.28, 0.08, and 0.39, respectively.

Note that if a person is heterozygous *MN* and at the same time *Ss* it cannot be stated without further information whether that composition is *MS/Ns* or *Ms/NS.* However, this information can readily be obtained in certain cases by inspecting the composition of that person's progeny. Thus, if a man with the M, N, S, and s antigens marries a woman with M and s antigens (or any other homozygous combination), the combinations found in their progeny will reveal the constitution of the heterozygous parent (Figure 20-6).

The fact that the MN property is inherited with the Ss property without the formation of any new combinations is strong evidence either that they are located on the same chromosome, so close together that they are not ordinarily separated from each other by crossing over in meiosis, or that they are different properties of the same gene. Evidence for this latter possibility will be presented below.

NULL PROPERTIES OF THE MN GROUP. Variants exist for all of the properties described above. The more interesting are the null alleles. One of these, *Mg,* produces an antigen that reacts with neither anti-M nor anti-N antibody. Another, *SU,* affects the S property by producing an antigen that reacts with neither anti-S nor anti-s antibody. Finally, as might be predicted if these two antigenic properties were different manifestations of the activity of a single gene, there exists an allele *Mk* that produces neither the M nor the N antigen, nor the S nor s antigen. Thus a person of composition *Ns/Mk* would be classified

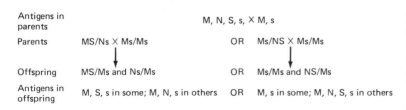

Figure 20-6 The method for resolving the problem of allelic constitution of a person with all four antigens *MNSs.* If that person produces offspring by a homozygous mate, the composition of the progeny will reveal whether the allelic combination was *MS/Ns* or *Ms/NS.*

as having only the N and s antigens—that is, would appear to be homozygous *Ns*, with unusual results in pedigrees.

For instance, if a person classified as *Ns* married one of type *MS* and produced a child of group *MS*, a question of mistaken paternity might be raised since *Ns/Ns* × *MS/MS* should produce offspring all showing the four antigens (*Ns/MS*) and not just two. On the other hand, if the first parent were *Ns/Mk* and the other *MS/MS*, half the progeny would be *Ns/MS* as ordinarily expected, and the other half the unusual type *MS/Mk*, which would show only two of the four expected antigens.

OTHER ANTIGENIC PROPERTIES.   Several other antigenic properties transmitted along with the *L* locus have been reported. Two, Hunter and Henshaw, are named after the persons in whom they were first detected. Three others have the symbols *Mi^a*, *Vw*, and *Vr*. These are relatively rare, found in only a few families, and their relationship to the M and S properties remains to be clarified. They are mentioned only to illustrate the great diversity of allelic forms at these loci.

The Rh System

Hemolytic disease of the newborn caused by the Rh factors may be regarded as the action of a gene, *R*, which produces in the fetus an antigen quite effective in inducing antibodies in *rr* mothers. However, as research on this system has progressed, more and more antigenic properties determined by it have been uncovered. It is now undoubtedly the most extensive erythrocyte antigenic system known in humans. The complications of the Rh locus have been exaggerated by the parallel development of two different systems of nomenclature, each with its strong advocates among the professional immunologists.

THE ALLELIC INTERPRETATION.   If the Rh system were viewed as one with multiple alleles, it could be considered to be composed of eight common types, half of them having the characteristics of the *R* allele and the other half *r*. Five distinctly different types of antisera can be produced. These are not completely independent, but have curious, overlapping specificities depending on the alleles in the blood producing antiserum. Table 20-2 shows a list of the alleles, along with their frequencies in U.S. Whites and their reactions with those antisera. If this table is deeply revealing to the professional immunologist, it is not to the novice, who would have to be moved by a spirit of generosity to describe it as anything other than incomprehensible.

THE SUBLOCUS INTERPRETATION.   The difficulty of relating the reactions of the various antisera to the antibodies of an apparently random selection of the Rh groups was largely circumvented in an ingenious scheme proposed by the English statistician R. A. Fisher, who in 1944 pointed out a simple relationship of the reactions between the antigens and the antibodies as they were known at that time. If it is assumed that there is not a single gene involved but that rather there are three closely linked ones, usually inherited as a unit (of three subloci), then we can relabel the loci *C*, *D*, and *E* (the symbols *A* and *B* having already been taken up by the ABO groups), such that at each of the three subloci there exists

Table 20-2 The eight major alleles of the Rh blood groups, their frequencies, and their reaction with certain antisera carrying some of the antibodies of the Rh group.

Alleles	Frequencies in U.S.A.	Anti-Rh0	Anti-rh'	Anti-rh''	Anti-hr'	Anti-hr''
R^0	2.7	+	−	−	+	+
R^1	41	+	+	−	−	+
R^2	15	+	+	+	+	−
R^z	0.2	+	−	+	−	−
r	38	−	−	−	+	+
r'	.6	−	+	−	−	+
r''	.5	−	−	+	+	−
r^y	.01	−	+	+	−	−

the possibility of at least two antigenic properties: C or c, D or d, E or e. On this basis, the alleles may be represented as in Table 20-3. The D antigen is the one whose antibody, anti-D, is responsible for erythroblastosis. The antisera have also been relabeled in Table 20-3 as specific for one of the antigenic properties. When the antigens and antisera have been thus relabeled, the table of interactions becomes more comprehensible: a positive reaction occurs when the antigen and antisera are tagged with the same symbol. It is unfortunate that this system, fairly widespread and commonly adopted in England, and more understandable to the layman, should use capital and lowercase letters, implying dominance and recessiveness, when all of the antigens except *d* commonly appear and the alleles are therefore codominant. *D* has been selected as the initial letter in this triplet nomenclature because an allele has been found that appears to lack all four antigenic properties C, c, E, and e. This has been interpreted as a small deficiency that removed the *C* and *E* subloci. It is logical to assume that two elements removed simultaneously would be adjacent to each other, since this would

Table 20-3 The alleles and the antisera of the Rh groups relabeled according to the *CDE* sublocus interpretation. Note that a positive reaction occurs when the allele and the antiserum carry the same symbol. The table duplicates the information on reactivity in Table 20-2 but is more comprehensible because the symbols are changed.

Alleles	Anti-D	Anti-C	Anti-E	Anti-c	Anti-e
Dce (R^o)	+	−	−	+	+
DCe (R^1)	+	+	−	−	+
DcE (R^2)	+	−	+	+	−
DCE (R^z)	+	+	+	−	−
dce (r)	−	−	−	+	+
dCe (r')	−	+	−	−	+
dcE (r'')	−	−	+	+	−
dCE (r^y)	−	+	+	−	−

require just one simple event in the linear arrangement *DCE*. This, then, implies that *D* is to one side of *C* and *E*, rather than between them, and the accepted order is arbitrarily taken as *DCE*. Placing *D* first has another advantage—it calls attention to the most important antigenic property of this locus, that responsible for hemolytic disease of the newborn, HDN.

Serious criticism has been raised against the *DCE* nomenclature. Certainly single alleles may have several antigenic properties; it is not necessary to postulate closely linked genes or subgenes to account for them. There are not simply two possible variants at each of the three hypothetical subloci to give 2^3 or eight alleles—on the contrary, more than 40 antigenic properties are now known. On the other hand, the single allelic representation, generally favored by American immunologists for theoretical reasons, is awkward because it is not obvious from the nomenclature which combinations of antigens and antibodies will and will not react.

Since we are not concerned in this elementary text with the complexities of antigen-antibody reactions found at this locus, we shall compromise by using a single letter, *R* or *r*, to substitute for *D* and *d*, respectively, and to indicate the probable single-gene nature of this locus, with subsequent symbols involving *C, c, E, e*, or such additional letters as may be needed for a complete description of other antigenic properties. Thus *DCe* becomes *RCe*, and *dCE* is *rCe*. This nomenclature has the practical advantage of providing for unlimited flexibility of terminology, since letters now used exclusively for other blood groups could, if desired, be used here unambiguously.

Even with the simplest system of designating alleles, the problem remains of specifying genotypes for single individuals, but we must depend on relationships in a sibship, or in a pedigree. Suppose, for instance, that an Rh-positive person also carried the antigens for *C, c, E*, and *e*. Since *r* acts as a null allele with no antigenic activity, that person might be either homozygous for *R* or heterozygous *Rr*. This person, then, might be any one of the following six possibilities:

1. *RCE/Rce*   4. *RCe/rcE*
2. *RCe/RcE*   5. *RcE/rCe*
3. *RCE/rce*   6. *Rce/rCE*

Additional information is necessary to make a decision. For instance, if it were known that the mother of the person above was Rh-negative, with the C and E antigens but not c or e, that mother would necessarily be *rCE/rCE*. Since this allelic type would be passed on to all of her children, including our problem case, the genotype of the latter would then have to be the sixth type.

OTHER ALLELES.   More than 40 variants at the *R* locus are known—some of these have unusual and interesting characteristics. One, mentioned earlier, is deficient for the antigenic properties involving *C, c, E*, and *e*, and can be represented as *R--*. An even more unusual "null" type allele is one with no antigenic property, *r--*. Since the effect of this allele is to make heterozygotes appear homozygous, the phenotypes of some of the progeny will appear to be inconsistent with those of the parents, just as in the case of homozygotes for the *h* allele in the ABO group.

The Xg Blood Group

An erythrocyte antigen was discovered during investigations of a man in Grand Rapids, Michigan, who suffered from severe nosebleeds. It is of particular interest because the locus responsible is on the X-chromosome. The allele determining the presence of the antigen is designated as Xg^a; the alternative allele, for which there is no known antigenic property yet, is Xg. A simple count of males in England and the United States with (65 percent) and without (35 percent) this antigenic property gives us an immediate estimate of the allele frequencies in the population. The estimate of the proportion of homozygous Xg females is 0.35×0.35 or about 0.12. This is very close to the observed frequency of females who lack the Xg^a antigen in these populations.

TURNER'S SYNDROME WITH Xg^a. As might be expected, Turner's syndrome females with one X-chromosome would show an incidence of sex-linked recessive traits more like that of the male than that of the normal female. In fact, it was first suspected that Turner's syndrome females might have only one X-chromosome precisely because of their typically high "male" frequencies of sex-linked disorders, such as color-blindness. Cytological verification came later.

When the X-chromosomes of the father and mother carry different alleles, it may be possible to determine the origin of the single X-chromosome in X0 females, and in more than a hundred such females it has been shown that about one-fourth inherit their X from the father and three-fourths inherit it from the mother. Since the former cases arise when the egg lacks an X-chromosome, i.e., when nondisjunction occurs in oogenesis, and the latter from nondisjunction during spermatogenesis, we can say that most cases of Turner's syndrome arise from failure of the sperm, not the egg, to carry the X-chromosome. This is unlike the case of trisomy-21, in which the nondisjunctional event appears to occur during oogenesis. It should not surprise us, then, to learn that the greatly increased frequency of trisomy-21 as the mother's age increases does not hold for Turner's syndrome, which is not primarily maternal in origin. Turner's syndrome does not increase with the father's age either, perhaps reflecting a basic difference in the course of germ cell formation in the two sexes, namely, the prolonged dictyotene stage, probably responsible for nondisjunction in the female, and the absence of such a dormancy stage in the male, with the continuous production of mature sperm from spermatogonial cells throughout the reproductive life of the male.

ORIGIN OF KLINEFELTER'S SYNDROME MALES. The parent in whom the nondisjunctional event takes place to give rise to Klinefelter's syndrome is not so easily determined. The frequency of phenotypes for sex-linked characters such as color-blindness and Xg^a in Klinefelter's syndrome is not the same as in females, although Klinefelter's syndrome males have two X-chromosomes like females, but intermediate between the female frequency and the male frequency.

Klinefelter's syndrome could arise by fertilization of an XX egg by a Y sperm, by fertilization of an X egg by an XY sperm, or by mitotic nondisjunction of the X-chromosome in a normal XY zygote yielding an XXY embryo. If an XX egg results from second-division nondisjunction in oogenesis, then the two X's will usually be identical and resultant XXY progeny will have malelike frequencies of sex-linked recessives. Mitotic nondisjunction during early cleavage divisions will also give rise to XXY's with identical X's. On the other hand, first-division nondisjunction in oogenesis and first-division nondisjunction in spermatogenesis are characterized by femalelike frequencies of sex-linked recessive phenotypes. Because the observed frequency of common sex-linked recessive phenotypes is intermediate between the high frequency found in males and the much lower frequency found in females, it is reasonable to assume that Klinefelter's syndrome males originate from at least two different types of nondisjunction.

The ABO Incompatibility

If immunization of the mother to the Rh antigen can give rise to HDN, then why does the ABO group not react similarly? In fact, one might expect that in some mother-child combinations the results would be disastrous since the maternal antibody would already be present and not have to be induced, as is the case of Rh incompatibility.

Consider the very common case, for instance, in which an O woman has an A or B child. Her anti-A or anti-B antibodies, some of which are capable of crossing the placenta, should get into the fetal circulation and agglutinate the fetus's erythrocytes. Although some cases of HDN have been traced to this cause, it is relatively rare. One suggestion of why this is so is that when deaths do occur they may happen so early in pregnancy that fewer medical problems are caused than with later HDN. The amount of embryonic and fetal wastage from this source is speculative; one estimate places it as high as 5 percent of all pregnancies. Theoretically it should be possible to compare directly the frequencies of progeny from matings that are compatible in one direction and incompatible in the other for evidence of this loss. Thus, regardless of the frequencies of homozygous and of heterozygous individuals of group A in a given population, matings of A × O should produce mostly A and some O progeny. From the simplest point of view, these proportions should be the same regardless of which parent is A and which O. However, the A child from an O mother and A father represents an incompatible mother-child combination, since the mother normally has the anti-A antibody that could agglutinate the fetal A cells. The A child from an A mother and O father is a compatible mother-child combination since the mother does not have the anti-A antibody. Of course, in either case, O children would be compatible with the mother.

Therefore, if there exists ABO incompatibility caused by an antigen-antibody reaction, we might detect it by comparing the relative frequencies of A and O progeny from A × O parents, in both directions.

Although the results of several early surveys suggest that there is a reduced production of such incompatible offspring, others have been inconclusive and the problem can still be regarded as unresolved.

Another interesting aspect of ABO incompatibility is its protection against HDN caused by the Rh factors. A number of studies have shown that Rh-caused erythroblastosis is found more frequently when the mother-child combinations are ABO-compatible than when they are incompatible. One explanation for this is that when the child's erythrocytes enter the mother's bloodstream they are destroyed by her natural A and B antibodies before her immune system has a chance to manufacture anti-Rh antibodies.

Other Blood-Group Systems

A dozen more loci responsible for erythrocyte surface antigens are known. Their presence becomes known only when an antibody is produced that cannot be attributed to any previously known group. These can show up when an unexpected case of HDN appears, or when an unexpected cross-reaction occurs in the preliminary test for a transfusion or during one. When such a new group appears, it can be described as "public" if the antigen is carried by the majority of people or "private" if it is carried by relatively few people. If it is public, its discovery usually depends on the formation of antibodies in one of the rare individuals not carrying it, followed by some clinical indication of the existence of that antibody. If private, then individuals carrying it are relatively infrequent and its discovery would depend on detection of the antibody in someone who had been immunized by the rare antigen.

A list of other less important blood groups, along with the designations for their major alleles, is found in Table 20-4.

LECTINS OR PHYTOHEMAGGLUTININS. Substances with human-antibody-like properties may be found in unexpected places. Beans and peas (i.e., the legumes) in particular provide a diverse source of unusual extracts which react with specific antigens; not being antibodies in the usual sense, they are called *lectins* or *phytohemagglutinins*. One such lectin will agglutinate cells of group O, another is specific for anti-A and will differentiate between A_1 and A_2 alleles. Another lectin from the bean *Vicia* reacts with M but very weakly with N. These have the great advantage of being easily prepared, inexpensive, and very potent.

Substances with agglutininlike properties are found in other organisms. An anti-B agglutinin occurs in mushrooms, an anti-H in eels, and an anti-M in the

Blood System	Designations of Alleles
P (= Q of Furuhata)	P^1, P^2, p
Kell	K, k, k^P
Lutheran	Lu^a, Lu^b
Duffy	Fy^a, Fy^b
Kidd	Jk^a, Jk^b
Lewis	Le, le
Diego	Di^a, Di^b
Yt	Yt^a, Yt^b
Dombrock	Do^a, Do
Auberger	Au^a, Au
Stoltzfus	Sf^a, Sf

Table 20-4 Some other blood-group systems and their alleles. (From *Principles of Human Genetics*, Third Edition, by Curt Stern. W. H. Freeman and Company. Copyright © 1973.)

horse and the cow. The existence of substances with these properties from un-expected sources should not be interpreted as an infallible indication of any basic biological identity with the antibodies induced in mammals after exposure to a specific antigen. Rather, they should probably be viewed as compounds with some specific function in those organisms in which they occur that, by coincidence, have a molecular configuration allowing them to combine with human antigens.

Persistence of Allelic Diversity

After all has been said about the characteristics of these erythrocyte cell-surface antigens, some questions still remain unanswered. (1) Why does there appear to be so much allelic variation compared to other loci? (2) What purpose do these loci serve in the first place?

Loci that show allelic forms occurring with moderately high frequencies are said to be polymorphic. It is not known either why there are so many loci producing surface antigens or why so many of them are polymorphic. Possibly the loci have some important function that is not affected by the polymorphic system of alleles. The antigenic properties that we observe unambiguously in transfusion mishaps, or in erythroblastosis, and can readily test for experimentally, may be incidental to some other important but unknown function. Certainly the viability of homozygous null types, *r--/r--*, *Mg/Mg*, and, most commonly, *O/O*, is strong evidence against the overwhelming importance in any specific individual of the antigenically active alleles at these loci.

One of the easiest checks to make involves comparing the distribution of the blood group in the overall population with that in a subgroup selected for some medical reason. Large numbers of such comparisons have been made, and it is now fairly well established that there is an excess of O persons among those suffering from duodenal ulcers (Table 20-5) and an excess of A among those with stomach cancer. While these correlations are interesting and unquestionably valid, one is left with the subjective impression that these may be incidental to the basic functions of the polymorphic loci, functions which remain to be discovered.

We know that the capability of the body to produce antibodies against the erythrocyte cell-surface antigens not only varies from one individual to the next

Table 20-5 The association of duodenal ulcer with blood group O. (R. B. McConnell, in *Selected Topics in Medical Genetics*, ed. C. A. Clarke, Oxford University Press, London, 1969.)

Blood Group	Control Population	Nonbleeding Duodenal Ulcer	Bleeding Duodenal Ulcer
O	3,146	351	329
A	2,648	244	157
Total	5,794	595	486
Percent O	54.3	59.0	67.7

but also shows great variation from one antibody to the next. Possibly the loci (but not necessarily the specific alleles) for the cell-surface antigens are involved in a complex developmental set of interactions of great importance during embryonic or fetal life. They could be part of a system that provides for an extensive immune system capable of identifying and perhaps destroying unusual tissue development, such as cancer, during later life.

References

COHEN, B. H. 1960. ABO-Rh interaction in an Rh-incompatibly mated population. *Am. J. Hum. Genet.*, **12**:180.

FISHER, R. A., R. R. RACE, and G. L. TAYLOR. 1944. Mutation and the rhesus reaction. *Nature*, **153**:106.

GIBLETT, E. R. 1958. Js, a "new" blood group antigen found in Negroes. *Nature*, **181**:1221.

GIBLETT, E. R. 1961. A critique of the theoretical hazard of inter- vs. intra-racial transfusion. *Transfusion*, **1**:233.

GIBLETT, E. R. 1964. Blood group antibodies causing hemolytic disease of the newborn. *Clin. Obstet. Gynecol.*, **7**:1044.

GIBLETT, E. R. 1969. *Genetic Markers in Human Blood*. Oxford: Blackwell.

LANDSTEINER, K., and A. S. WIENER. 1940. An agglutinable factor in human blood recognized by immune sera for rhesus blood. *Proc. Soc. Exp. Biol. Med.*, **43**:223.

LEVINE, P. 1943. Serological factors as possible causes in spontaneous abortions. *J. Hered.*, **34**:71.

McCONNELL, T. D. 1969. Genetics and diseases of the gastro-intestinal tract. In C. A. Clarke, ed., *Selected Topics in Medical Genetics*. London: Oxford University Press.

WATKINS, W. M. 1966. Blood-group substances. *Science*, **152**:172–81.

WIENER, A. S. 1966. The blood groups. Three fundamental problems—serology, genetics and nomenclature. *Blood*, **27**:110.

Questions

Useful terms: glycoprotein, H substance, Bombay phenotype, null allele, phytohemagglutinin, polymorphic.

1. What are the chemical steps resulting from the action of the A and B alleles? How are these related to the H substance?
2. What is meant by the Bombay phenotype? How might a person of this constitution run into serious difficulties?
3. What loci, independent of the ABO locus, modify the manifestation of the antigens produced at the ABO locus?
4. Why was the MN system so delayed in its discovery after that of the ABO system?
5. Is it likely that any new cell-surface antigens will be discovered in the future? If so, how are they likely to be detected?

6. Do the frequencies of the two distinctly different antigenic properties of the MS locus occur randomly in the population?

7. Can the frequencies of the four common genotypes of the MS system be determined from a study of population data?

8. Could the null alleles at any of the cell-surface antigen loci introduce complications in any medicolegal application?

9. What is the basic difference between the sublocus interpretation of the Rh groups and the multiple allelic interpretation?

10. Suppose that an Xg-negative male has developed antibodies against the Xg antigen after several transfusions of blood from his father. Further transfusions from his father are out of the question, and it is desired to get additional blood from one of his family. Who seem the most likely prospects, his brothers, his sisters, or his mother?

11. Where does the nondisjunction probably occur that is responsible for the production of Turner's and Klinefelter's syndromes?

12. How does incompatibility with respect to the ABO groups seem to confer some protection against incompatibility of the Rh groups?

13. What is known about the reason for the relatively high variability in the alleles of the blood-group loci?

21

Variations in Chromosome Structure and Number

The art of chromosome study in humans is of fairly recent origin; as late as 1956 it was generally thought that the diploid chromosome number was 48. The early techniques involved laborious sectioning of material (usually testicular) with a tedious fixing, staining, and sectioning procedure that gave relatively few good preparations. At best, the chromosomes were small, easily misinterpreted, and generally uninformative even in the hands of the most skilled workers. Researchers preferred to work with other organisms such as lilies or amphibians with large, clearly staining, and more manageable chromosomes.

Techniques of Human Chromosome Studies

Early in the 1950's, a number of technical advancements, each minor in itself, were combined to transform this difficult and unrewarding area into an exciting and highly profitable one. In the first place, it was shown that metaphase preparations could be vastly improved by placing the cells in a *hypotonic solution*—one in which the salt concentration is lower than that of a cell. With this treatment the cells swell, separating the chromosomes from each other and making them easier to identify. At about the same time, tissue culture techniques were being developed that allowed proliferation of human cells by rapid mitosis, making it possible to get large numbers of cells in active division. Third, the classical sectioning technique (in which the tissue is cut into a lot of very thin slices to be examined individually) was replaced by an air-drying technique in which whole

347

cells simply flattened themselves on a microscope slide as the liquid fixative evaporated. This not only reduced the time and care necessary for the preparations but also insured that all of the chromosomes of one cell would be found together more frequently. The hazard of the early sectioning technique was that individual cells were sliced at random, sometimes through the nuclei, with a consequent inaccuracy in the chromosome counts.

Several other helpful technical procedures were developed. Lymphocytes, the white cells in the bloodstream, proved to be an easily available source of dividing cells. As mentioned in Chapter 20, phytohemagglutinin, a plant extract, agglutinates the red blood cells to facilitate their removal from the blood sample, at the same time inducing mitosis in the lymphocytes, which might otherwise not divide frequently.

For many years, a drug called colchicine, from the autumn crocus, had been known to inhibit cell division, stopping mitosis at metaphase. (Colchicine is, incidentally, one of the oldest prescriptions for the treatment of gout.) This drug made it possible to obtain much larger numbers of cells in the desired metaphase condition than are found normally.

Finally, the use of many different kinds of chromosome specific chemicals and stains such as orcein, quinacrine, and acridine orange, combined with a variety of recipes for pretreating the chromosomes, made it possible to stain the chromosomes in a variety of distinctive ways with a minimum of fuss. Now, by a combination of some or all of these innovations, a small quantity of blood can be extracted and cultured for several days, and a slide of the cells then made fairly quickly to give a clear, unambiguous karyotype.

EARLY DISCOVERIES.   In 1959, Lejeune, Gauthier, and Turpin showed that children with Down's syndrome had an extra chromosome, now numbered 21. Almost simultaneously it was shown by C. E. Ford in 1959 that females with Turner's syndrome (described clinically in 1938) had only one X-chromosome, and by Jacobs and Strong, also in 1959, that males with Klinefelter's syndrome (described clinically in 1942) had two X-chromosomes as well as a Y.

The XYY constitution was identified in 1961, and 4 years later Jacobs and colleagues published data from institutions suggesting the possible unusual behavior of such males. Since that time, thousands of individuals with many different kinds of abnormal chromosome compositions have been found. Many of these types are sporadic, representing only isolated incidents of specific chromosome anomalies in certain individuals or families. Others fall into more frequent categories which we shall discuss later.

STANDARDIZATION OF TERMINOLOGY.   Because it was difficult to distinguish each of the 23 different chromosomes in the early days of human cytology, workers in different laboratories adopted their own numbering systems. Inevitably, confusion arose. Serious attempts to standardize the nomenclature were made at a series of international conferences in Denver (1960), London (1963), Chicago (1966), and Paris (1971).

The letter grouping from A to G and the numbers 1 to 22, which have been used regularly throughout this text, resulted from those conferences. In addition, a series of standard symbols was adopted to be used in designating the chromosome composition of an individual (Table 21-1). These symbols can be com-

A–G	the chromosome groups	
1–22	the autosome numbers (Denver system)	
X,Y	the sex chromosomes	
Diagonal (/)	separates cell lines in describing mosaicism	
Plus sign (+) or minus sign (−)	when placed immediately after the autosome number or group letter designation indicates that the particular chromosome is extra or missing; when placed immediately after the arm or structural designation indicates that the particular arm or structure is larger or smaller than normal	
Question mark (?)	indicates questionable identification of chromosome or chromosome structure	
Asterisk (*)	designates a chromosome or chromosome structure explained in text or footnote	
ace	acentric	
cen	centromere	
dic	dicentric	
h	secondary constriction or negatively staining region	
i	isochromosome	
inv	inversion	
inv(p+q−) or inv(p−q+)	pericentric inversion	
mat	maternal origin	
p	short arm of chromosome	
pat	paternal origin	
q	long arm of chromosome	
r	ring chromosome	
s	satellite	
t	translocation	
Repeated symbols	duplication of chromosome structure	

Table 21-1 The Table of Nomenclature Symbols adopted at the Chicago Conference on Standardization in Human Genetics in 1966. (From *Birth Defects, Orig. Art. Ser.*, ed. D. Bergsma. Williams & Wilkins, Baltimore, Vol. 11(2).)

bined in standard formats to specify the known or suspected composition of any person (Table 21-2).

AUTORADIOGRAPHY. One of the first techniques developed for separating ambiguous members of the same group of chromosomes was the use of radioactive isotopes. Tritium (^{3}H) is a radioactive isotope of hydrogen, which, when

Table 21-2 The combination of the symbols in Table 21–1 to form descriptions of the karyotypes of specific individuals.

Symbol	Composition
45,X	45 chromosomes, one X chromosome
47,XXY	47 chromosomes, XXY sex chromosomes
45,XX,C—	45 chromosomes, 2X's, one C group missing
48,XXY,G+	48 chromosomes, XXY, an extra member of the G group present
45,X/46,XY	A mosaic with two cell types, one lacking an X and the other XY
46,XY,t(Bp—;Dq+)	A reciprocal translocation between a B and D group chromosome, involving the short (p) arm of the B and the long (q) arm of the D, with a net decrease (—) in length of the B and an increase (+) in length of the D.

it disintegrates, exposes photographic film. For experimental work, tritium is added to *thymidine*, replacing the normal nonradioactive hydrogen in that compound. Thymidine is essential in the cell's synthesis of chromosome material and is incorporated into the new chromatid at the time of replication. If thymidine that contains tritium (called *tritiated thymidine*) is present in the cell at the time the chromosomes are undergoing replication, it becomes part of the new chromosomes, and a photographic film laid over the cells will become exposed near the points of disintegration of the tritium atoms (Figure 21-1).

This technique would be of limited value if all chromosomes incorporated thymidine uniformly. However, not all chromosomes replicate simultaneously during the synthetic phase of mitosis, nor do all parts of one chromosome replicate at the same time. A good example is given by chromosomes 21, 22, and Y, ordinarily difficult to distinguish by the simpler staining techniques. When the chromosomes are presented with tritiated thymidine late during their replication period, chromosome 22 takes up little thymidine, having already replicated, whereas chromosome 21, which is late-replicating, takes up larger amounts and is heavily "labeled." The Y-chromosome, on the other hand, becomes labeled over its entire length when given the thymidine early but is labeled only near the centromere region at a later stage.

It is of considerable interest that the two homologous X-chromosomes present in a female cell behave differently, one replicating early and labeling lightly, the other replicating late and labeling strongly. The conclusion that the late-replicating X is the inactive X (see Chapter 5) comes from a number of arguments, one of the most compelling being that if a female has a grossly defective X-chromosome (one, for instance, with a large deficiency) that X is usually the late-replicating one. Perhaps an even better reason for identifying the late-labeling X as the inactive one is that when there are more than two X-chromosomes present in a cell only one is labeled and all the rest are relatively unlabeled.

Another newly developed staining technique also brings out the difference between the two X-chromosomes. A synthetic chemical called BUDR when

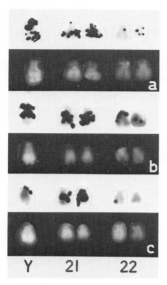

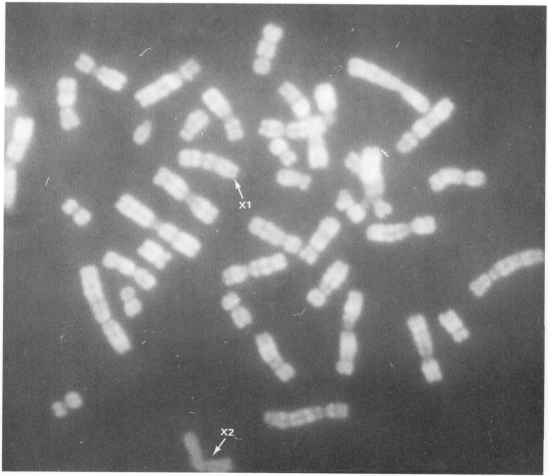

Figure 21-1 Means of distinguishing similar chromosomes. *A*. A comparison of the labeling of chromosomes Y, 21, and 22 by tritiated thymidine, when it is applied in early (*a*), middle (*b*), and late (*c*) stages of replication. The stained chromosomes at the same stages are shown underneath the labeled one. (D. Calderon and W. Schnedl. A comparison between quinacrine fluorescence banding and ³H-thymidine incorporation patterns in human chromosomes, *Humangenetik*, **18:** 63–70, 1973.) *B*. The active (X1) and the inactive (X2) X chromosomes after treating with BUDR which, when applied after replication of the active X but before or during replication of the late-replicating inactive X and therefore incorporated into it, inhibits the normal staining of that inactive X. (Courtesy of P. Pearson, University of Leiden.)

added to cells is incorporated into the chromosome strands during replication in place of one of its natural constituents. When incorporated into the chromosome BUDR has the unusual property of preventing it from appearing stained, when treated with certain dyes. It is possible therefore to add BUDR to a cell after normal replication has occurred, so that none of the chromosomes including the active X contains BUDR, but before the late-replicating X has replicated, so that it alone contains BUDR. The late-replicating X then appears to be very poorly stained (Figure 21-1B).

QUINACRINE STAINING. A significant advance was made in the late 1960's when Caspersson and coworkers showed that different regions of plant chromosomes stained differently, after having been treated with a fluorescent dye called quinacrine mustard. They subsequently showed that with this substance each human chromosome displays its own unique banding pattern, making it possible to identify without question every chromosome of the complement.

On the printed page, karyotypes produced by standard orcein staining often appear to be clearer to the uninstructed eye than those stained by more advanced methods. In actual practice, however, the distinctive banding patterns on each chromosome make the more advanced techniques far superior.

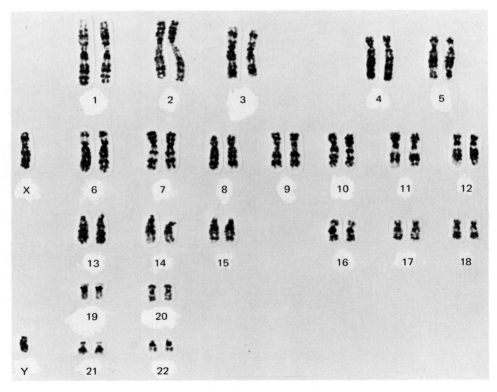

Figure 21-2 Karyotype stained by the Giemsa technique, showing the G-banding. Members of each pair of chromosomes can be identified with ease by their distinctive banding patterns. (Courtesy of W. Schnedl, University of Vienna.)

Because of the great detail they reveal, quinacrine and other staining methods that reveal banding patterns are used whenever possible for the analysis of karyotypes. In fact, quinacrine has revealed that the extra chromosome present in Down's syndrome is the smallest of the complement, and should most properly be called 22, not 21. However, trisomy-21 has become so well established as a synonym for Down's syndrome that a change in numbering would be confusing, and by general consensus the earlier but incorrect numbering has been retained.

Four of the better-known, newer techniques include the quinacrine (Q), Giemsa (G), reversed Giemsa (R), and centric heterochromatin (C) methods. The quinacrine method shows fluorescence from faint to intensely brilliant in the long arm of the Y-chromosome. The G-bands produced by the Giemsa method are located in about the same positions as the Q-bands but are somewhat more detailed (Figure 21-2). The reversed Giemsa method produces a banding that is the opposite of G-banding, and has the advantage of staining the tips of the chromosome arms distinctly. The fourth staining method, C, brings out bands near the centromeres of every chromosome (Figure 21-3), and in a few other places, particularly the tip of the long arm of the Y-chromosome.

The late-replicating X-chromosome of the female, so obvious with autoradiography, is not distinguished from the other X by use of most of these new staining techniques. The long arm of the Y, which ordinarily fluoresces brilliantly after quinacrine staining, can be distinguished at interphase (the resting stage) of mitosis, when it is apparently associated with the nucleolus. The Y-chromo-

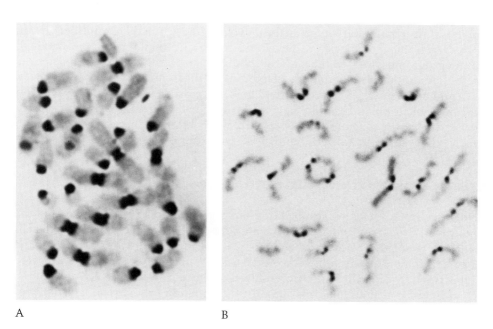

A   B

Figure 21-3 Chromosomes stained by a modified Giemsa technique (C-banding) that specifically reveals the heterochromatin adjacent to the centromere in the tobacco mouse (*A*) and in the common house mouse (*B*). The chromosomes in *B* are from cells in the second meiotic division so that the sister centromeres are still associated. (Courtesy of A. T. Natarajan, University of Leiden [*A*], and C. E. Ford and E. P. Evans, Oxford University [*B*].)

some can even be seen in mature sperm, so that not only can X- and Y-bearing sperm be distinguished but also those carrying more than one Y-chromosome.

Why these different techniques should produce such distinctive banding patterns on the chromosomes has not been fully explained. An early hypothesis was that the highly fluorescent sections contained higher proportions of certain compounds (basic constituents of the chromosomes), which reacted specifically with the dyes used. A more recent suggestion is that the drastic cell pretreatments performed before staining cause the chromosome strands to fall apart; they then reunite to varying degrees in different parts of the chromosome, making a difference in the capacity of the chromosome strands to absorb stain. Since some of the pretreatments in these methods involve a drastic alteration of the protein structure, the differential staining may depend on how much protein is found in different regions of the chromosomes.

These details of chromosome structure have now been incorporated (by a conference in Paris in 1971) into a numbering system specifying the regions of chromosomes involved (Figure 21-4). Although the alternating light and dark areas are called bands, the same term used to describe the patterns of polytene chromosomes of *Drosophila*, they do not refer to the same structural element on the chromosome. The human chromosome bands, totaling a dozen or so per chromosome, must represent gross structural regions, each "band" covering an average of several hundred genes, whereas each stained band of a *Drosophila* polytene chromosome probably marks the location of an actual gene. In rare cases it is possible by special treatments to produce chromosome configurations with an unusual amount of banding detail, but these techniques are not yet practical for regular karyotype analysis (Figure 21-5). It is the unfulfilled dream of every human cytologist to develop a technique that reveals the detail of human chromosomes to the same degree as that of polytene chromosomes, a feat that would have boundless scientific, medical, and social consequences.

Another recently developed method makes it possible to differentiate between the original chromatid and the new replicated copy (Figure 21-6). By this means, it can be shown that the two sister chromatids often exchange segments with each other; such exchange is known as sister-strand crossing over.

Characteristics of Variant Chromosomes

Human chromosomes are grouped in a karyotype according to their sizes and the positions of their centromeres, which are very obvious in colchicine-treated cells since they are the points of attachment of the two chromatids at metaphase. The centromere is the primary constriction. When the centromere is located approximately centrally, the chromosome is metacentric (Figure 2-18); when the centromere is near an end, it is acrocentric. Sometimes there is an additional or secondary constriction along the chromosome length. These are found as well-defined structures in chromosomes 1, 9, 16, and Y, and may occasionally be seen in every other chromosome of the complement.

The D- and G-group chromosomes have very short arms, sometimes appearing to be attached to the centromere by a thin thread. Such short arms are called satellites (Figure 2-18). This chromosome section may not only be variable from person to person, it may also differ between two homologues in a single cell. One

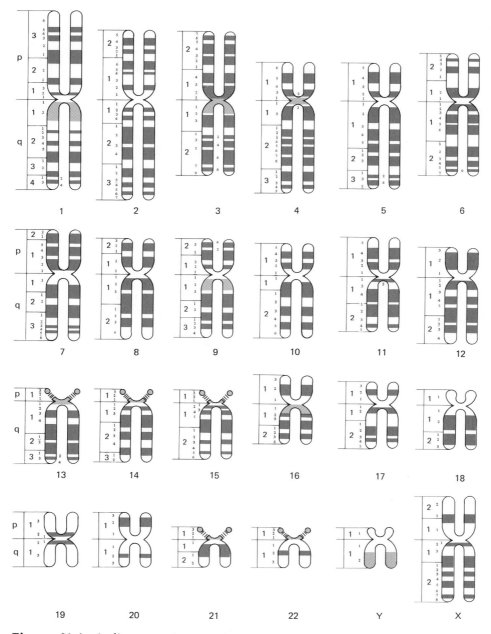

Figure 21-4 A diagrammatic representation of the human chromosome banding patterns revealed by the different kinds of staining techniques. (*Birth Defects: Orig. Art. Ser.*, ed. D. Bergsma. Published by Williams & Wilkins Co., Baltimore, for the National Foundation—March of Dimes, Vol. VIII [7], 1972.)

percent of all persons may have unusually large satellites—on the other hand, satellites are often absent without affecting the phenotypes of the individuals. Nevertheless, there is now evidence that the satellites, or regions very close to

Figure 21-5 An example of an unusually detailed banding pattern present in some chromosomes of a cell stained after treatment with a protein inhibitor. (Courtesy of D. A. Shafer, Georgia Mental Health Institute.)

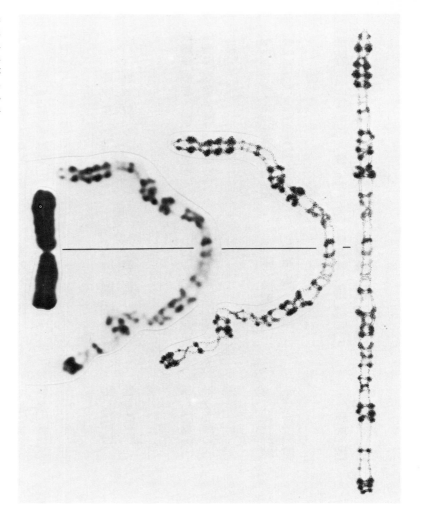

them, are responsible for the organization of the nucleolus in the nucleus of the cell. Since there is a total of ten chromosomes in the D and G groups (five pairs), one might expect a total of ten nucleoli per cell. The maximum number found is six. The reason for this may be that several satellited chromosomes may be associated at interphase and together produce only one nucleolus.

Statistical analyses of the positions of the chromosomes relative to each other at metaphase show that they are not randomly placed, despite the displacement that must occur during the rough treatment of cell preparation. This is particularly true for the satellited chromosomes. Such a metaphase association may be evidence of regular natural groupings of the chromosomes, perhaps at their satellite or centromeric regions, at an earlier stage of mitosis such as prophase or interphase. The predisposition of certain satellited chromosomes to be found closer to each other may be related to the fact that some of them (14 and 21, for instance) show an unusually high incidence of translocation with each other.

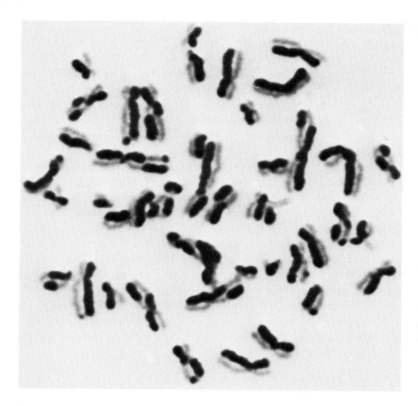

Figure 21-6 Differential staining of the two sister chromatids of each chromosome. The newly replicated chromatid contains more BUDR than the original chromatid and so stains less intensely. Note the frequent exchanges of material, about one per chromosome, between the sister strands. (Courtesy of A. T. Natarajan, University of Leiden.)

The staining patterns of the chromosomes in this figure are different from those of the preceding figure; the technique used here is G-banding and that in Figure 21–10 is R-banding. Still other types of banding techniques are illustrated in other figures in this chapter.

VARIANT AUTOSOMAL TYPES. Morphologically deviant chromosomes which appear to have no phenotypic effect are known. One is an extra long chromosome 1, found in newborns with a frequency of 1 percent (Figure 11-4). This chromosome is sometimes called the "uncoiler" because of an early hypothesis that the extra length was caused by a failure of a section near the centromere to coil tightly. It is now better explained by the duplication of material near the centromere. It is transmitted from a carrier to half the progeny, as would be expected. Following the terminology recommended by the Chicago Conference, the correct description of this anomaly would be 1q+.

In another interesting type of anomaly, a chromosome may have a "fragile" site within it that breaks easily. The broken pieces may then undergo rapid asynchronous replication, producing cells with several small chromosome fragments. This fragility has been found in both chromosomes 2 and 11 and is inherited as a simple dominant.

Many new variants have now been found with quinacrine staining. Chromosomes are particularly variable in the centromeric regions, where some stain brightly and others faintly. Virtually every person has at least one (and usually several) differences between two homologues of a chromosome. These staining differences in chromosomes have already been mentioned as a criterion for judging the zygosity of twins. Two chromosomes that are different in structure are said to be *heteromorphic;* if similar, they are *homomorphic.* If many different forms of chromosomes are found in a population, that population is said to be *polymorphic*.

VARIATIONS IN THE Y-CHROMOSOME.   An argument was presented earlier that the Y-chromosome must be relatively free of essential loci other than those that trigger male sex differentiation. It should not be surprising, then, to find that the Y is one of the most variable of the chromosomes, since changes in structure —additions or subtractions of its genetic material—would involve few genes essential to normal development.

Many cases have been found in which the Y is modified in structure or stainability (Figure 21-7). In some pedigrees there are very long Y-chromosomes, in others, very short ones, with neither type having any particular effect on the phenotype of the males carrying them. One pedigree has been described in which a long Y-chromosome produces much the same symptoms as the extra Y in XYY males, as though some of the limited essential material in the Y had been duplicated. In some instances, the Y-chromosomes have been found that appear to be transmissible dicentrics, i.e., with two centromeres. In one case of a nontransmissible dicentric, no metaphases of the second meiotic division were found in the testis, and the X- and Y-chromosomes did not pair normally. In another case the Y-chromosome was transmitted to one or more generations, and in several other cases dicentric Y-chromosomes were found in females with maldevelopment of the ovaries.

DELETION.   The best-known deletion occurring with a considerable frequency is that of the cat cry (cri du chat) syndrome, first described in France (Figure 21-8). The afflicted infant is characterized by a cry strongly reminiscent of that of a cat (Figure 21-9), a feature which disappears later in life. Although the overall frequency may be as low as one per 100,000, more than 200 cases have now been reported. The affected individuals are invariably mentally retarded, with an array of abnormal clinical features. This syndrome is not lethal, as are most others involving the absence of chromosome material on one of the two homologues, and some of the known patients have survived beyond adolescence.

Figure 21-7 Polymorphism for the Y-chromosome. At the left is a normal Y, at the right a large one with a brightly staining block. (Courtesy of P. Pearson, University of Leiden.)

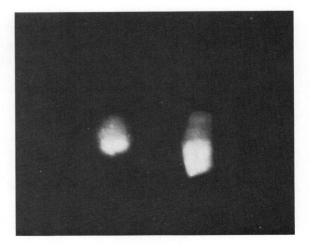

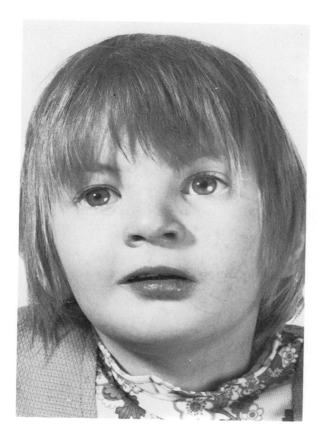

Figure 21-8 A child with the cat's cry (cri du chat) syndrome, caused by a deletion for part of the short arm of chromosome 5. (Courtesy of E. Polani, Guy's Hospital Medical School, London.)

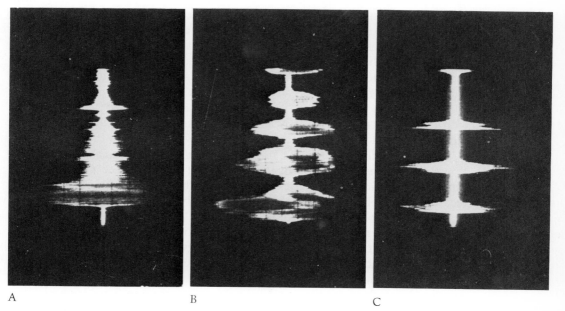

A   B   C

Figure 21-9 An electrical recording of the cry of a normal child (*A*), a cat (*B*), and a child afflicted with the cat's cry syndrome (*C*), showing the strong resemblance of the sound of the cry to that of a cat. (J. Legros and C. Van Michel. Analyse de la voix dans un cas de "maladie du cri du chat," *Ann. Genet.*, **11**:59–61, 1968.)

Figure 21-10 Two number 5 chromosomes (along with two number 4 chromosomes for comparison) from a karyotype of a child with cri du chat, showing one with a deletion in the short arm, responsible for the cat's cry syndrome. (J. Legros and C. Van Michel. Analyse de la voix dans un cas de "maladie du cri du chat," *Ann. Genet.*, **11**:59–61, 1968.)

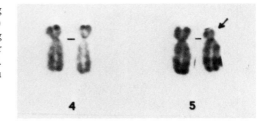

In the vast majority of cases, it can be shown that such individuals have a deficiency for a part of the short arm of chromosome 5 (Figure 21-10). This deficiency (5p−) is variable in size from one case to the next, ranging from less than half to almost all of the short arm.

In several instances a patient with this syndrome has proved to carry a ring chromosome, and upon further investigation the ring is found to involve chromosome 5. It would appear that in these instances, when a ring was formed, the acentric fragment that was immediately lost must have been that section of the short arm of 5 whose deletion is responsible for the cri du chat syndrome.

OTHER DELETIONS. Partial deletions in both the long and short arms of chromosome 18 have been reported, each in several dozen cases. As in other chromosome deficiencies, mental retardation is characteristic, with other developmental abnormalities such as heart defects. In these cases, affected individuals can survive into adulthood.

PARTIAL DELETION OF THE G-CHROMOSOMES. With a very small frequency, patients appear in whom one of the chromosomes of the G group seems to have a deletion. These cases appear to fall into two classes: those in which part of chromosome 21 is deleted, and those in which the affected chromosome is 22. The two classes have some characteristics in common. One of the classes has a number of features that seem to be the antithesis of those found in mongolism, and therefore is referred to as "anti-mongolism."

VARIATIONS IN THE X-CHROMOSOME. Deletions are known for both the short and long arms of the X-chromosome. Women with deletions in the short arm of one (Xp−), the other being normal, appear to have typical Turner's syndrome. Those with deletions in the long arm (Xq−), on the other hand, resemble Turner's syndrome patients except that their stature is normal and their necks are not webbed. One explanation is simply that two short arms of the X are necessary for normal stature.

There are cases in which the X-chromosome is so altered that (as far as can be determined) it consists of two short arms or two long arms (Figure 21-11). In plants, such abnormal chromosomes have been named *isochromosomes* (iso = *equal*), and there is good reason to believe that the two arms so attached are actually identical. In experimental animals true isochromosomes have not been described, and if these abnormal human chromosomes referred to by human cytologists as isochromosomes prove to be true isochromosomes, they will be

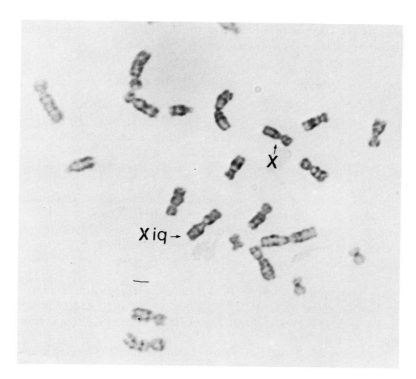

Figure 21-11 Part of a cell of a woman with one normal X-chromosome and an abnormal X consisting of two long arms. The woman has the typical manifestations of Turner's syndrome. (Courtesy of M.-L. Frey, Turku University.)

unique in animal cytology. There is some difference of opinion about the phenotypic effect of deletions and duplications of X-chromosome material on both long and short arms. Perhaps this is related to the problem of identifying all abnormal X-chromosomes and of characterizing their genetic content precisely.

The X-chromatin body formed by these abnormal X-chromosomes is distinctive. In the case of the isochromosome carrying (mostly) two long arms (Xqi), the sex chromatin body is larger than normal; in the others, Xpi, Xp−, and Xq−, it is smaller. This is strong evidence that the abnormal X-chromosomes are the inactivated ones. How does the cell know that the normal X is the one that must remain active? It probably does not. Probably the two X-chromosomes are inactivated at random, and those cells with an abnormal, deficient X remaining active will lack some essential loci and perish, leaving only those cells with the normal X active.

On the other hand, if cells are examined which contain a translocation with a piece of an autosome attached to the X-chromosome, it is this translocated chromosome that appears active, while the normal chromosome is inactive. This is explained by assuming that the inactivation of the piece of an autosome attached to an inactivated X-chromosome creates a pseudodeficiency for autosomal material greater than the cell can tolerate. Such cells are eliminated and those that remain are the cases where the normal X-chromosomes have been inactivated. Thus for the two cases where, on the one hand, it is the normal X-chromosome that is found to be active or, on the other, the normal X-chromosome is inactive, the same explanation holds: those cells with an appreciable deficiency created by the inactivation of one of the X-chromosomes fall by the wayside and the other cells are found exclusively.

References

CASPERSSON, T., G. LOMAKKA, and L. ZECH. 1971. The 24 fluorescence patterns of the human metaphase chromosomes—distinguishing characters and variability. *Hereditas*, **67:**89–102.

HAMERTON, J. L. 1971. *Human Cytogenetics*, Vols. I and II. New York: Academic Press.

HECHT, F. 1974. Autosomal chromosome abnormalities. In *Metabolic, Endocrine and Genetic Disorders of Children*. New York: Harper & Row.

JACOBS, P. A., and A. STRONG. 1959. A case of human intersexuality having a possible XXY sex-determining mechanism. *Nature*, **183:**302–3.

LEGROS, J. 1967. La maladie du cri du chat. *Bull. Soc. Belge Gynecol. Obstet.*, **37:**201–10.

LEJEUNE, J., M. GAUTIER, and R. TURPIN. 1959. Étude des chromosomes somatiques de neuf enfants mongoliens. *C. R. Acad. Sci. (Paris)*, **248:**1721–22.

LEVITAN, M., and M. F. A. MONTAGU. 1971. *Textbook of Human Genetics*. New York: Oxford University Press.

MAGENIS, R. E., F. HECHT, and E. W. LOUVRIEN. 1973. Heritable fragile chromosome No. 16. *Abst. Am. Soc. Hum. Genet.*, 43.

McCAW, B., G. PRESCOTT, R. E. MAGENIS, and F. HECHT. 1972. Familial Turner's syndrome correlated with X-chromosome deletion. *Abst. Am. Soc. Hum. Genet.*, 51a.

Questions

Useful terms: hypotonic solution, colchicine, tritiated thymidine, late-replicating X-chromosome, fluorescence, heteromorphic chromosome, deletion, ring, isochromosome, duplication.

1. List the simple variations in technique which together have made human chromosomes rather favorable objects of investigation.
2. How does the use of radioactive isotopes make it possible to distinguish one chromosome from another?
3. How would you interpret the following chromosome designation? 47,XY,t (Ap+;Cq−)
4. What reason is there to believe that, of the two X-chromosome in a female cell, the late-labeling one is the inactive one?
5. What advantages does quinacrine staining have over orcein staining?
6. How frequent are chromosome polymorphisms in a population?
7. What other kinds of gross chromosomal abnormalities in chromosome structure are known in humans?
8. Discuss the relevance of the cri du chat syndrome to the concepts discussed in this chapter.

22

The Transmission of Abnormal Chromosome Types

What kinds of offspring will a woman with Down's syndrome produce? When a person has an abnormal chromosome, will it be passed on to the progeny? If a couple produces one child with an abnormal chromosome, what are the chances that the next child will be similarly affected? Is it possible for a normal couple to produce chromosomally abnormal children repeatedly?

The science of human cytogenetics has advanced to the stage at which most of these questions can be answered with reasonable assurance, although often the answer given in a genetic counseling session, for instance, would be phrased in terms of *empiric risks*—that is, the estimates of probability based on population studies of similar cases, rather than on theoretical probabilities from mendelian principles.

So many different types of chromosome abnormalities have been described in humans—well into the thousands—that it would be useless to try to describe any appreciable fraction of them here. However, they do fall into characteristic types about which some general statements can be made. These types have been studied in other organisms—corn and the fruitfly primarily, but in many other plants and animals as well. As a rule, human chromosome anomalies behave much the same as similar anomalies in other species. This is the case because meiosis is much the same in all organisms. In fact, it would be difficult to find a meiotic phenomenon in humans that does not have a parallel in other organisms.

If a person carries a chromosome variant as one of two homologues, it will be found in half of that person's gametes and, on the average, in half the progeny.

Such chromosomes may be very useful in assigning loci to specific chromosomes. It was first established that the locus for the blood group known as Duffy is carried on chromosome 1 because, in certain families carrying the so-called uncoiler chromosome 1 (see Figure 11-4) as a heterozygous variant, its pattern of distribution in the pedigrees was matched exactly by that of the alleles of the Duffy blood group.

Transmission in Trisomies

DOWN'S SYNDROME. In rare instances, females with Down's syndrome have produced children. In the slightly more than a dozen known cases when this has occurred, about half of the progeny were normal and the other half affected like the mother. This is the expectation based on a simple model where the gamete has an equal chance of carrying the extra chromosome, or not.

At the first meiotic division, the two homologous chromosomes normally present pair with each other, locus by locus, along their entire lengths. A helpful analogy is that of the zipper: once pairing is initiated at certain homologous loci, the adjacent loci proceed to pair until both homologues are locked together. A third homologue could not pair with either one, any more than a third zipper strand could fasten itself to the two already joined. However, if two homologues are paired in one region but not in another, a third homologue could pair with one of them at the unpaired region to form a combination of three chromosomes (Figure 22-1). Two paired chromosomes and a third unpaired consist of a *bivalent* and a *univalent;* three chromosomes paired two at a time at different regions from a *trivalent.*

Spermatocytes of Down's syndrome males show both kinds of pairing, bivalents along with univalents in some, trivalents in others. Since there are three homologues present but only two directions in which they can go at first anaphase, the separation would be of two of the homologues to one pole and one to the other. We have previously discussed the production of gametes with two homologues by accident of meiosis from normal germ cells; this event is called *primary nondisjunction.* When gametes with two homologues are formed from germ cells with three, the event is called *secondary nondisjunction.* In such a case, the expectation of persons with trisomy-21 is that half of their progeny will be normal and half will have trisomy-21. There is no recorded case of progeny from Down's syndrome males. Although it is known that they can produce

Figure 22-1 The two different kinds of behavior of three homologues at the first meiotic division. *A.* In some cases, two of the three chromosomes will pair for some distance, and then one of the two will pair for the rest of the length with the third chromosome, to form a trivalent. *B.* In other cases, two of the homologues pair completely in a bivalent and the third homologue remains unpaired as a univalent. In both cases, half of the gametes will get one of the homologues and the other half will get two.

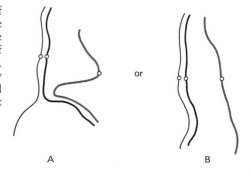

motile sperm, the number is so small that even when found in otherwise normal males, it may be responsible for effective sterility. In addition, trisomy-21 males have reduced sex drive.

THE XXX FEMALE.   In 1959, Jacobs described a female with three X-chromosomes. Our present data indicate that the incidence of such births is about 1 per thousand. XXX women are fertile, produce children, and are reasonably normal, although on the average they show a slight tendency toward decreased intelligence. Sometimes they are characterized by underdeveloped secondary sexual characteristics, sterility, and psychotic behavior. Because they appear normal, no descriptive syndrome classification has been applied to this condition.

A woman with three X-chromosomes would be expected to produce 50 percent eggs with two X's and 50 percent with one, an example of secondary nondisjunction. One might expect, then, that her progeny would be half normal and half with an extra X—either XXX or XXY, depending on whether the XX egg was fertilized by an X- or a Y-bearing sperm. Curiously, this does not appear to be the case; what few children such women have are usually normal. It is possible that the women tend not to produce diplo-X eggs because the XXX germ cells (oogonia) do not readily go through the early germinal divisions. Perhaps a few of these XXX oogonia suffer occasional mitotic nondisjunction or the anaphase loss of the extra X-chromosome to produce XX oogonial clones which then become eggs. Even with her regular XXX oogonia failing to become eggs, an XXX woman could produce enough normal XX oocytes to last her entire reproductive life, since the female infant is born with well over a million reproductive cells. This phenomenon, the production of XX cells from XXX by mitotic nondisjunction, would be consistent with the observed frequency of occasional sterility of triplo-X women—the sterility representing those cases in which an insufficient number of mitotic nondisjunctional events had occurred to produce viable XX oocytes.

Still another possibility is that the three chromosomes form a trivalent at prophase of meiosis, and at anaphase of the first meiotic division the trivalent orients itself in such a way that two of the X's are lost in the polar bodies while one ends up in the ovum. On the other hand, if a bivalent and a univalent are regularly formed, the univalent might be lost by lagging at anaphase, so that each meiotic product would get one chromosome from the separation of the bivalent. Still another possibility is that an XXX embryo is relatively inviable and dies in utero. The apparent high viability of an adult genotype does not necessarily go along with a high embryonic viability. The high viability of individuals with Turner's syndrome after birth compared to the low viability prior to birth gives us strong reason for avoiding such a simple correlation. In both cases, also, these individuals might be mosaics, with the XX oogonia of the XX/XXX mosaic providing the functional eggs.

XYY MALES.   It might be expected that the combination of the three homologues, one X-chromosome and two Y's, during meiosis would give rise to abnormal progeny. The expectations, of course, would depend on the way the chromosomes pair at prophase and separate at anaphase of the first division.

Analysis of spermatogenesis in XYY males gives conflicting and inconsistent results. Although a few cases are known where such males have produced XXY sons, presumably coming from their XY sperm, their progeny are almost always normal XX or XY, without the XXY or XYY types expected of such males if they regularly produced XY or YY sperm. Furthermore, cytology of XYY men shows meiotic configurations involving a single X and a single Y rather than an X and two Y's. The best assumption consistent with most of the data is that there is strong selection against XYY spermatogonial cells (paralleling that presumed to operate against XXX oogonia) and that the progeny produced by XYY males come from the occasional XY cells that appear in the XYY germ line. This would be consistent with the observation that there is wide variability in the fertility of these males which ranges from complete sterility to apparently normal fertility.

Deletion Chromosomes and Sperm Activity

A chromosome that has had a section deleted will lack loci, the number depending on the size of the deletion. The normal homologue present in a heterozygote may allow deletion-bearing individuals to survive, often with little phenotypic effect, if the deletion is small. At first sight it would appear that the expected progeny from a heterozygote would be half with the normal chromosome and half with the deletion chromosome.

However, another question arises here. Are the gametes with a deletion at a disadvantage because some genes normally present in the gamete are completely missing? This is apparently not the case. As far as the egg is concerned, the diploid condition persists until the egg is ovulated, at which time meiosis (which has been stalled at the dictyotene stage since the fetal period) starts again. The haploid stage of the egg lasts for only a short time, for if fertilization is to take place it will happen within a few days. Upon fertilization, any genetic deficiency in the egg is compensated for by normal genes for that region in the sperm. Thus, there is generally no selection against a deficiency-bearing egg from a female heterozygous for that deficiency, and a heterozygous female will produce progeny half of whom are similarly affected. However, if the mother herself has two normal chromosomes, with a deletion occurring anew in her germ line, subsequent heterozygous zygotes with that new deficiency may or may not develop to produce viable progeny, depending on the developmental effect of the new deficiency.

The sperm, on the other hand, spend an appreciable period of time (at least several days) as independent functioning haploid cells, highly motile and therefore with unusual energy requirements. Do their survival and functioning depend on a normal haploid complement? Although we cannot answer this question unequivocally for human sperm, if their requirements for function and fertilization are the same as for other animals, it seems likely that a human sperm is able to function independent of the alleles it does (or does not) carry in the haploid state. We know, in the first place, that half of the sperm lack the X-chromosome, having a Y instead, and these sperm must be as functional as those with an X because the sex ratio is almost 1:1. There is no mutant, no deletion, or no rearrangement in human chromosomes that has been proved to suffer from low sperm transmissibility because of a deletion in the haploid complement. In fact, in *Drosophila* it is possible to produce gametes that lack all of the major chromosomes (i.e., that have virtually no genes at all). These sperm are viable and can fertilize eggs.

One case has been reported in which the sperm's haploid genotype appears to determine its phenotype. Sperm can be agglutinated by antibodies produced against antigens of the histocompatibility locus. If a male is of composition HLA-1/HLA-2, then anti-HLA-1 antibodies will agglutinate almost half the sperm and anti-HLA-2 antibodies will agglutinate the other half. This must mean that the antigens on the surface of the sperm are determined by the alleles present during the stage of differentiation from a spermatocyte into a sperm. This is not quite the same as the genes themselves operating within the sperm, but the end result is similar. Similarly, if a consistent difference were to be shown to exist between X- and Y-bearing sperm, it might not be evidence for a difference in gene activity but rather for a difference in gross structure of the two types of sperm.

The reason for the absence of gene activity in sperm is that the chromosomes are compressed into the tiny sperm head without the elaborate cellular machinery for converting raw material into gene products. The function of the sperm is merely to carry the haploid set of chromosomes from the male to the egg.

There are cases in which heterozygotes of composition AA' (where A and A' represent two alleles of two chromosomes with different structures) produce gametes of the two types with unequal frequencies. Examples have been found in *Drosophila*, the mouse, and corn, and, although the phenomenon is different in each organism, it is similar in that the basis for selecting in favor of one of the homologues at the expense of the other is determined during meiotic (and therefore diploid) stages and not in the haploid gamete. Such unequal production of two types of functional gametes has been discussed earlier (see meiotic drive). A few cases of suspected meiotic drive have been reported in humans, but none has been verified conclusively. This would be an important clinical and counseling consideration, since it could greatly alter the probability of children with the abnormal gene or chromosome being produced by a parent heterozygous for a defect.

Chromosomal Aberrations

There are an unlimited number of different ways in which the 46 chromosomes of the human complement can be "broken" and the segments reattached. The specific instances described in human cells are now in the thousands and that number will continue to grow.

When rearrangements are formed, they appear as isolated cases within simple pedigrees and slowly disappear from the population. However, while they persist, they have the unusual property of causing relatively little damage to most of their carriers but being capable, as a result of meiosis, of causing grossly aneuploid progeny. Furthermore, some (though not all) of the normal children carrying the rearrangement may also produce affected children. The peculiarities of transmission of these are our concern here.

TRANSLOCATIONS. Let us suppose that in a normal oogonial cell two non-homologous chromosomes are broken. Such breakage is a frequent phenomenon. In most cases breakage appears to originate spontaneously—that is, without any prior record of unusual exposure to radiation or any mutagenic or clastogenic compound. When they occur, the broken ends may fail to reunite; they may reunite to re-form the original chromosomes; the two segments with centromeres may join to produce a dicentric, which will be lost in the next cell division or shortly thereafter; or the two may exchange segments to produce a translocation. These possibilities have been discussed in Chapter 12.

When the descendants of the cell, heterozygous for a new translocation, reach the prophase of the first meiotic division, homologous chromosome regions will attempt to pair. For all homologous segments to pair, they would have to achieve the cross-shaped configuration illustrated in Figure 22-2. The strands are numbered in that configuration to indicate the pairing of homologous regions. Whether they succeed in pairing completely may depend on the relative lengths of the homologous segments, with, presumably, the longest uninterrupted sections having the greatest chance of getting together. Whether such a paired configuration, once achieved, will remain that way after the chromosomes have shortened or condensed during prophase of meiosis may depend on whether or not a chiasma, of the sort that regularly occurs between all homologues, occurs between these new pairing segments. If the segment is long, as it is in the case illustrated, the chance of a chiasma is very great; if it is small, the chance is considerably reduced. In any event, in an ideal case of the sort we have illustrated, an exchange might occur in all four arms with the result that these chromosomes would be associated up until the time of metaphase.

The simplest rule governing the manner of segregation at the first division is that segments which have been involved in a chiasma, or exchange, will usually separate from each other at the first meiotic division. If the alternate members of the translocation configuration go to the same pole, it follows that the two normal homologues go to one pole and the two involved in the translocation go to the opposite pole (Figure 22-3A). This kind of segregation is referred to as *alternate segregation*. From such a configuration, two kinds of products may be formed, one with the two unaffected chromosomes and the second with the two translocated chromosomes. The first will give rise to normal progeny with

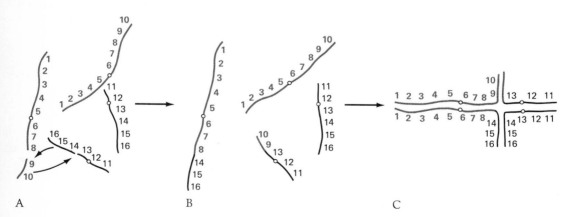

A   B   C

Figure 22-2   Two pairs of homologues, with homologous segments numbered identically, are located at random within the nucleus of a cell. Two breaks occur, at the positions marked with arrows (*A*), and the broken segments reattach to produce a reciprocal translocation (*B*). When homologous regions attempt to pair, the two normal chromosomes pair with the translocated ones in a cross-shaped configuration (*C*).

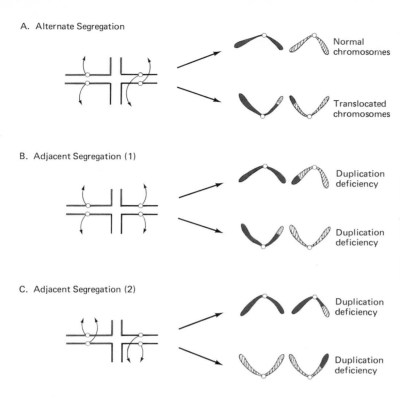

A. Alternate Segregation

Normal chromosomes

Translocated chromosomes

B. Adjacent Segregation (1)

Duplication deficiency

Duplication deficiency

C. Adjacent Segregation (2)

Duplication deficiency

Duplication deficiency

Figure 22-3 The different types of segregation from a translocation heterozygote. *A*. Alternate members of the configuration go to the same pole, resulting in two gamete types, one with the two normal chromosomes and the other with the two translocated chromosomes. *B*. Adjacent segregation, of type 1, where homologous centromere regions go to opposite poles, producing duplication-deficiency gametes. *C*. Adjacent segregation, of type 2, in which homologous centromere regions go to the same poles, resulting in duplication-deficiency gametes.

normal chromosomes, the second to normal progeny heterozygous for the translocation.

It is easily seen, however, that this regular alternate result might not be achieved. Depending on the position of the breaks and configurations formed as a result, the translocation heterozygote may not be paired in all arms, or chiasmata may not hold all four arms together, with the result that segregation would not be of alternate members of the configuration, but rather of two adjacent ones (Figure 22-3*B* and *C*), going to the same pole.

The gametes resulting from adjacent segregation are unbalanced; that is, they are deficient for some loci and duplicated for others. If the amounts of genetic material involved in the duplications and deficiencies, or aneuploidy, are very large, the embryo formed from a gamete of this sort, along with a normal gamete from the other sex, will be aborted at a very early stage in embryonic or fetal life. If the duplications and deficiencies are not extensive, an unbalanced fetus may be born. When translocations are detected in humans, it is usually after a stillbirth or the birth of an abnormal offspring who proves to have an abnormal chromosome make-up that came from the products of adjacent segregation in a normal heterozygous parent. Furthermore, examination of the normal progeny in the sibship will probably show that about half of them have normal chromosomes whereas half carry the balanced translocation as one of the parents did.

Generally speaking, about half of the gametes from a translocation heterozygote of the sort we have described here are balanced, either being normal or having a balanced translocation so that half of the zygotes should develop normally. An embryo having a grossly unbalanced chromosome complement from adjacent segregation will be aborted early or, less frequently, come to term as a quite abnormal child. When there is embryonic or fetal loss that limits the reproductivity of parents one of whom carries a translocation, we would expect the translocations to be eliminated slowly over several generations. Perhaps most of those that are now being observed have been in existence for only a few generations. Curiously, it can be argued that translocations with more drastic aneuploid products causing early embryonic loss may, as a group, persist in the population for a longer period of time than those with lesser effects, because the early (often unnoticed) embryonic losses are more likely to be followed by additional children, half of whom would carry the translocation, whereas births of defective children would tend to limit family size.

COMPOUND CHROMOSOMES.   In 1916, Robertson described some chromosome rearrangements in grasshoppers in which he noted that two rod-shaped chromosomes, acrocentrics, appeared to be attached to each other at the centromeres to form V-shaped, or metacentric, chromosomes. He called these *"compound chromosomes,"* and this name can be applied to translocations that result from breaks near the centromere regions of the two chromosomes involved so that entire chromosome arms are attached to each other. However, such whole-arm translocations are referred to by human cytologists as *centric fusions, fusion chromosomes,* or, more commonly, as *"Robertsonian translocations,"* and when the chromosome arms so attached appear to be either two long or two short arms of the same chromosome number, they are called *isochromosomes* (iso = same) even though, in most cases, the arms must be genetically different from each other.

One compound chromosome found in a large number of pedigrees involves the attachment of the long arms of a 14- and a 21-chromosome. (Figure 22-4A). When such compounds are formed (Figure 22-4B), there is undoubtedly initially a small fragment that can be lost without seriously affecting the genome. When such a compound in a normal cell pairs with a normal 14 and a normal 21, the expectations are relatively simple. In this sort of configuration, the homologues will pair at prophase and separate from each other at anaphase to give a high proportion of balanced alternate gametes producing normal progeny (Figure 22-4C) and a very small proportion of unbalanced gametes (Figure 22-4D).

Of all cases of trisomy-21, about 97 percent arise by primary nondisjunction and are sporadic in nature, being found primarily among children born to older women. The other 3 percent are the products of adjacent segregation in eggs of mothers who are translocation heterozygotes. These are clustered in sibships (i.e., are familial) and occur independently of the age of the mother. Although the trisomy as such is not inherited, the translocation responsible is, and the risks are very much greater for the birth of other children with Down's syndrome to the same mother or to her relatives, both male and female, heterozygous for the translocation (Figure 22-5). The risk for a heterozygous female producing trisomy-21 is 10 percent; the risk for a heterozygous male is 2 percent. The differences in risk for the two heterozygous parents probably reflect the different frequencies of adjacent segregation in meiosis of the two sexes (Figure 22-4C). Note also that other kinds of adjacent segregation will produce inviable types (and sometimes three of the elements go to one pole and one to the other) so

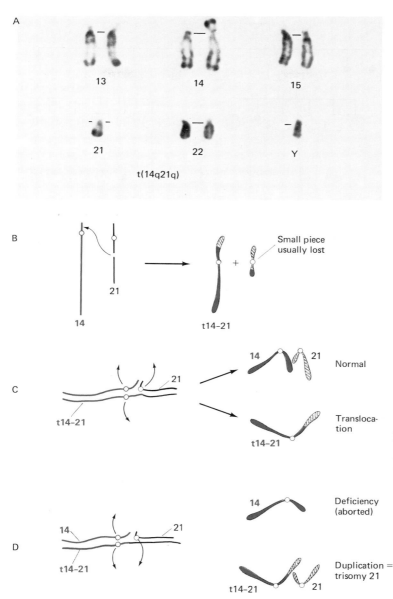

Figure 22-4 *A.* Chromosomes taken from the karyotype of a person heterozygous for a 14-21 translocation. The other unaffected chromosomes of the D and G groups, and the Y, are included to verify the point that the translocation does in fact involve 14 and 21. Note that the long arms of both chromosomes are attached and that the very small product made up of the two short arms of each chromosome is missing. (Courtesy of B. Dutrillaux, University of Paris.) *B.* The formation of a compound chromosome, with most of the material of chromosomes 14 and 21 found on the large piece. The smaller segment may be lost without having any phenotypic effect on the carrier. *C.* Alternate segregation from the heterozygote, yielding one half of the gametes with normal chromosomes, the other half with the translocation. *D.* Adjacent segregation giving rise to deficiency and duplication gametes. Although the zygotes with the deficiency for chromosome 21 are lethal, the complementary class with the duplication will produce a trisomy-21 zygote.

that a pedigree involving a compound, as for other translocations, will show more stillbirths and abortions than normal.

One very unusual compound is the rare 21-21 compound. In this case, the two homologues are attached and the person carrying them is phenotypically normal but can produce only two kinds of gametes: those with the 21-21 compound and those with no chromosome 21 at all. Since the zygotes having only one chromosome 21 (that coming from the normal parent) die in utero, all of the viable offspring of such a person will receive the 21-21 translocation, which, along with

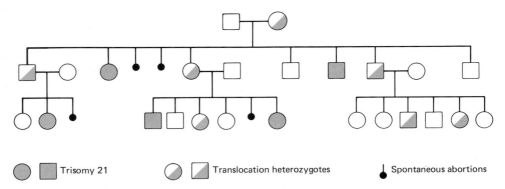

Trisomy 21   Translocation heterozygotes   Spontaneous abortions

Figure 22-5 A pedigree of a family with a 14-21 translocation. The heterozygotes are phenotypically normal but tend to have a high spontaneous abortion rate because of the grossly abnormal gametes they produce. Trisomy-21 is produced with a high frequency, and, among the phenotypically normal children, half carry the heterozygous translocation.

the single 21 from the other parent, will result in Down's syndrome. Here the risk of an affected offspring is 100 percent!

INVERSIONS. Inversions of a section of a chromosome in which a segment is rotated through 180 degrees have vastly different consequences cytologically and genetically, depending on whether or not the centromere is located within the inverted region. If the centromere is to one side of the inverted region (Figure 22-6A), the inversion is referred to as a paracentric inversion and may give no detectable results unless with staining methods such as quinacrine or Giemsa several bands appear to be inverted. Furthermore, the relative position of the centromere with respect to the chromosome ends has remained unchanged so that the overall chromosome form remains the same. On the other hand, when the inverted segment includes the centromere (Figure 22-6B), depending on the

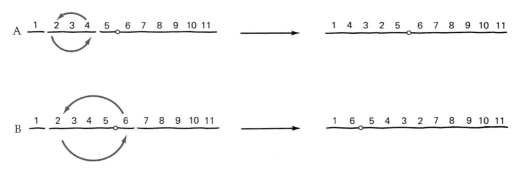

Figure 22-6 The events that lead to inversion formation. *A.* The paracentric inversion produced when both breaks occur in the same chromosome arm. The relative lengths of the two arms remain unchanged, so that the inversion is not cytologically conspicuous. *B.* The pericentric inversion, produced when one break occurs in each of the arms. The centromere region may be shifted in position, giving a new configuration with changes in the relative lengths of the two arms.

relative position of the breaks in the two arms the centromere may be shifted from one position to another to give a distinctly different arm length as well as a different distribution of the bands from the centromere to the tips.

The difficulty that two homologues have in pairing when one has an inversion is quite different from the problem with a translocation heterozygote. In this case the chromosome has to throw itself into a loop to pair with the homologous loci (Figure 22-7). The chromosomes often do not succeed in this. In these cases the frequency of chiasmata and therefore crossing over may be reduced. In fact, early experimental work with such rearrangements in *Drosophila* described their characteristics as "crossover suppressors" rather than as specific chromosome rearrangements. Crossing over within the inversion, however, has very interesting consequences that can be illustrated best by figures in the four-strand stage of the first meiotic division. For the case of the paracentric inversion, shown in Figure 22-8*A*, a single exchange within the inversion causes the formation of a dicentric chromatid as well as an acentric. Both of these are lost immediately since they cannot go through cell division. The two normal chromatids, however, can.

A pericentric inversion, on the other hand, produces two kinds of chromatids (Figure 22-8*B*), both of which are monocentric but differ by carrying duplications and deficiencies. At the first anaphase division, such chromosomes may go off to opposite poles and at the second division can get into the gametes just as the two normal chromatids can. Hence we may expect as a consequence of exchange within pericentric inversions the production of abnormal offspring. When abnormal, but transmissible, chromosome types are manufactured de novo by crossing over within a balanced heterozygote, as in the preceding case, the process is referred to by French cytologists as *aneusomy by recombination*.

An example of this phenomenon is given in Figure 22-9, in which a pericentric inversion (*A*) present in a normal female has, by crossing over with the normal homologue, produced an abnormal chromosome in a son (*B*). The pericentric inversion in (*A*) is obvious because of the shifted centromere posi-

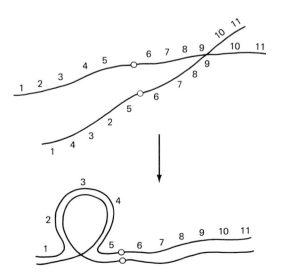

Figure 22-7 Two homologues, different in that one carries a paracentric inversion. In order to synapse completely, the homologues have to throw themselves into an "inversion loop."

Figure 22-8 Crossing over within a heterozygous inversion. The pairing configuration at prophase of meiosis, showing the completely different consequences depending on the type of inversion. In each case, the chromosome is represented as having replicated, so that the two homologues consist of four strands. At first anaphase, the two chromatids making up each chromosome are attached at the centromere regions and so go to the same pole. Separation of the centromere region will occur at the second meiotic division. *A.* When crossing over occurs within the inverted region of a paracentric inversion, a dicentric is formed as well as an acentric fragment. The dicentric and acentric will be lost immediately, so that no abnormal progeny will be produced. *B.* Crossing over in a pericentric inversion loop gives duplication-deficiency products. At mitosis, they can replicate and move to opposite poles, as normal chromosomes do, but, because of their abnormal gene content, they may be responsible for defective embryos.

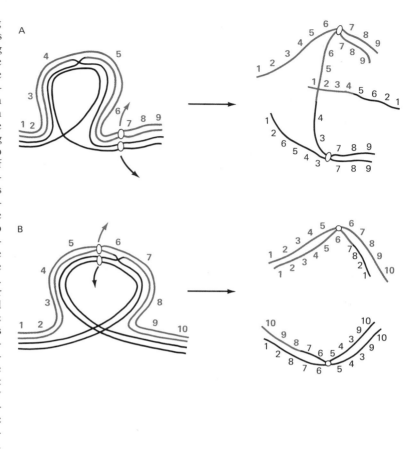

Figure 22-9 *A.* The B group chromosomes of a phenotypically normal mother heterozygous for a pericentric inversion. *B.* A similar set of chromosomes of her son, showing the abnormal chromosome resulting from crossing over between the normal and inverted chromosomes of the mother. (M. Wilson et al. Inherited pericentric inversion of chromosome no. 4, *Am. J. Hum. Genet.,* **22:**679–90, 1970.)

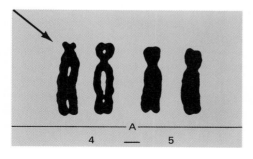
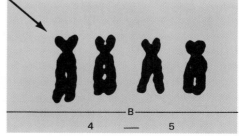

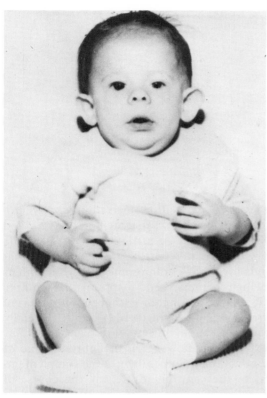

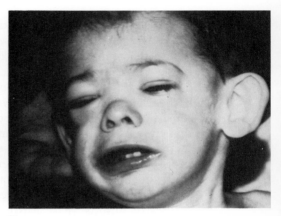

Figure 22-10 The child with the crossover chromosome illustrated in Figure 22-9B, shown at 3 months (A) and in a deteriorated condition at 6 months (B). (M. Wilson et al. Inherited pericentric inversion of chromosome no. 4, *Am. J. Hum. Genet.*, **22**:679–90, 1970.)

A

B

tion, and the crossover product in (B) has an abnormally long arm. The child with the duplication-deficiency product is shown in Figure 22-10.

RING CHROMOSOMES.   As was noted in Chapter 12, rings are interesting because of the anticipated difficulty that their chromatids might have in separating after replication, causing a loss at anaphase. In fact, this difficulty does not seem to be nearly as extreme as one might a priori expect: it is not uncommon to find that a ring-bearing individual has the ring in just about all of his cells. On the other hand, it would be expected that the ring would be lost more often than a normal chromosome and this in fact is found to be the case.

Figure 22-11 shows a ring chromosome separating at metaphase in a *Drosophila* with the rings lagging as though they had somewhat more difficulty in separating than the non-ring chromosomes. This is not an atypical picture for ring chromosomes.

In the formation of a ring chromosome, some genetic material at the ends of the chromosome must be lost and the ring-bearing person will have a deficiency of varying extent, depending on the specific ring. For this reason, the clinical features presented by ring-bearing individuals are of extremely variable sorts, with different degrees of mental deficiencies and other anomalies, and with a life-span that can be quite variable from one individual to the next. A striking case of the effect of a ring for chromosome 1 is shown in Figure 22-12.

Figure 22-11 Two ring chromosomes in *Drosophila* separating at anaphase somewhat later than the others, suggesting some difficulty in separating from each other.

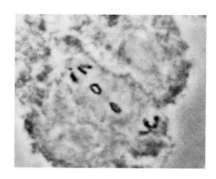

ORIGIN OF CHROMOSOMAL ABERRATIONS.   Many of the aberrations observed in the human populations studied seem to involve breaks that are randomly located within the genome. These breaks cannot usually be attributed to any specific cause. There is, however, a category of regular recurring rearrangements that deserves special mention. Compounds involving chromosome 21 with members of the D group are found with a very high frequency, and it appears that chromosome 14 is preferentially involved. It is possible, of course, that the initial event that gives rise to these whole-arm translocations involves 13, 14, and 15 equally frequently, and that some subsequent selection process eliminates the 13-21 and 15-21 translocations. Another explanation, however, is that chromosome 21 and 14 are more similar to each other than to other members of the complement, particularly in that region adjacent to the centromere. During the course of meiosis or mitosis, these partly homologous regions may engage in a kind of pairing activity. Very often slides of human cells show a nonrandom association of the satellite regions of these chromosomes. Such a physical association of the chromosomes would greatly enhance the frequency with which the initial event giving rise to translocation could take place, or the actual event might be an "illegitimate" meiotic crossover having something of the nature of a crossover occurring in the mitotic cells (a somatic crossover).

RECURRENCE RISKS.   Once a translocation heterozygote has been identified by the production of an affected offspring, the question invariably arises as to the recurrence risk. Because each translocation is usually quite different from the next, with breakpoints at random along the length of the chromosomes, it is difficult to make specific predictions. A simple meiotic diagram, coupled with some guesses about the likelihood of crossing over between the segments, may provide a clue to the probability of unbalanced gametes. Furthermore, the extent of the unbalance deduced from such a diagram may indicate, roughly, the viability of the offspring to be produced. To be conservative, the general rule might be adopted that the translocation heterozygote will produce up to 50 percent unbalanced gametes, with the possible saving feature (from the standpoint of the trauma to the parents) that if the duplications and deficiencies are sufficiently great the embryo will probably be lost spontaneously very early.

On the other hand, for those translocations involving whole chromosome arms, the large homologous segments synapse, and exchange occurs within these segments to insure their segregation to the opposite poles at the first meiotic

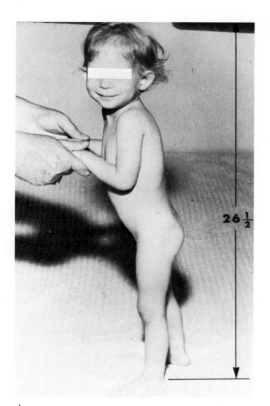

A

Figure 22-12 *A.* A young girl with a ring derived from chromosome 1. At 28 months of age, her total height was slightly over 2 ft and her weight 10 lb 4½ oz. Even at age 3 her clothes were made from the pattern for a 26-inch doll. *B.* The ring chromosome found in the girl above. (C. B. Wolf, J. A. Peterson, G. A. LoGrippo, and L. Weiss. Ring 1 chromosome and dwarfism—a possible syndrome, *J. Pediatr.,* **71:**719, 1967.)

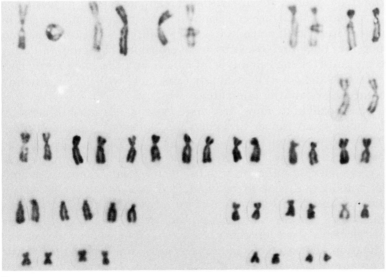

B

Table 22-1 Empiric risks for D-21 translocations: probability of an offspring of one of three different types when the father, or the mother, is a translocation heterozygote. (F. Hecht, Autosome Chromosome Abnormalities, in *Metabolic, Endocrine and Genetic Disorders of Children,* Harper & Row, New York, 1974; M. Mikkelsen, *Humangenetik* **12,** 1, 1971.)

| | Translocation Carrier | |
| | Father | Mother |
Type of Offspring	*Risk (percent)*	*Risk (percent)*
Down's syndrome	2	10
Translocation carrier	49	45
Normal chromosomes	49	45

division. It must be kept in mind in this connection that since male and female meioses are grossly different, with the prolonged dictyotene, stage in the female, there is no reason that the frequencies of the different segregating types should be the same in both sexes.

As an illustration of the above, Table 22-1 gives the observed recurrence risks for the production of offspring with normal chromosomes, translocations, and Down's syndrome from parents having the D-21 translocation. The frequency of observed individuals with Down's syndrome may be somewhat depressed because of their greater inviability prior to birth than normal individuals. However, such tables, based on observation rather than theory, serve as the most realistic basis on which to predict the results from translocation segregation.

Chromosome Abnormalities in Disease

CANCER. It is quite common to find that cancerous cells have a highly aberrant chromosome number, but this aneuploidy probably results from the rapid uncontrolled mitotic activity of such cells, rather than being the cause of it. A number of different theories are in favor at the moment, such as that some chromosomal or mutational change is responsible for initiating the cancer or that a genetic change in the cell increases its susceptibility to some foreign agent, such as a carcinogenic compound or a virus. In all probability, different initiating factors may be the cause of the same type of cancer and the same agent may induce quite different kinds of cancer. In any case, there will almost certainly prove to be no one cause (and no one cure).

Nevertheless, it is possible to show that there exists a relationship between chromosome instability and cancer in some cases. In a rare growth disorder characterized by dwarfism, Bloom's syndrome, the affected individuals who are homozygous for a recessive show a high frequency of spontaneous chromosome breaks in their cells. Such persons have a high risk of developing cancer. Fanconi's anemia, another hereditary disease involving defects not only of the blood but also of skin pigmentation and malformations of the heart and other organs, shows chromosome instability and an increased probability of leukemia. In ataxia telangiectasia, a recessive disorder characterized by progressively decreasing muscular coordination, defect of the immune system, and malignancy of the lymphocytes, chromosome breakage in the lymphocytes is unusually high. It appears in this case, however, that chromosome 14 is specifically involved in translocations with other chromosomes, suggesting that some locus on chromosome 14, specifically responsible for the normal control of lymphocyte proliferation, is released from this control by the new arrangement.

THE PHILADELPHIA CHROMOSOME.   In one form of cancer, chronic myelogenous leukemia, an altered chromosome is found in about 80 percent of all patients (Figure 22-13). The altered chromosome has been shown to be number 22, which is smaller because a piece has been removed and attached almost always near the end of the long arm of chromosome 9. Although marrow cells from these patients sometimes carry other chromosome abnormalities, this is the most consistent change.

The fact that two specific chromosomes, 9 and 22, are preferentially involved in this event, with the necessary "breaks" occurring at approximately the same positions of those chromosomes when the event takes place independently in different patients strongly suggests that they are not selected examples of random chromosome breakage within the cell but that there is some predisposition to association (such as partial homology) responsible for this event. In this case, the rearrangement is found only in somatic cells (in bone marrow cells), and there is no case recorded of its occurring in meiotic cells and being transmitted from parent to offspring.

The exchange of material between chromosomes 9 and 22 appears to involve no net change in the total amount, suggesting that all of the genetic material is still present in the cell. One explanation for this change in the behavior of the cell is that the genes on the altered chromosomes now are no longer able to function normally because, on the one hand, their normal genetic neighbors, with whom they ordinarily cooperate, have been removed to a distant position

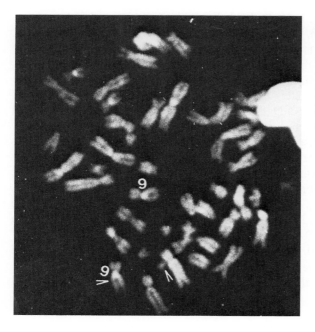

Figure 22-13 Chromosomes from the bone marrow of a patient with chronic myelogenous leukemia, which is generally characterized by a smaller-than-normal chromosome 22, the Philadelphia chromosome, indicated by an arrow. A second arrow points to chromosome 9, to which the missing section of chromosome 22 has become attached by a translocation. (Courtesy of A. T. Natarajan, University of Leiden.)

Figure 22–14   Hypothetical scheme showing the origin of the Philadelphia chromosome. The ends of the long arms of chromosomes 9 and 22 (*J* and *MN*, respectively) are interchanged to produce two new chromosome sequences. Chromosome 9 becomes slightly longer, and 22 is transformed into the smaller Philadelphia chromosome. The breakup of the original gene sequences and/or the new gene associations (either *I* with *M* or *L* with *J*) may then be responsible for a position effect giving rise to the leukemia.

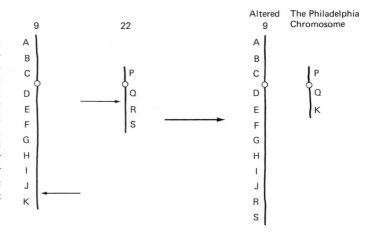

and, on the other, they are now closely associated with different loci that interact to produce new developmental effects. Such a changed behavior of genes, when they are moved to new locations, is called a *position effect* (Figure 22-14).

References

EVANS, E. P., C. E. FORD, R. S. K. CHAGANTI, C. E. BLANK, and H. HUNTER. 1970. XY spermatocytes in an XYY male. *Lancet*, **4:**719–20.

FELLOUS, M., and J. DAUSSET. 1970. Probable haploid expression of HL-A antigens on human spermatozoon. *Nature*, **225:**191–93.

FIALKOW, P. J., S. M. GARTLER, and A. YOSHIDA. 1967. Clonal origin of chronic myelocytic leukemia in man. *Proc. Natl. Acad. Sci. U.S.A.*, **58:**1468–71.

GERMAN, J. 1972. Genes which increase chromosomal instability in somatic cells and predispose to cancer. *Prog. Med. Genet.*, **8:**61–101.

HECHT, F., B. McCAW, and R. D. KOLER. 1973. Ataxia-telangiectasia—clonal growth of translocation lymphocytes. *N. Engl. J. Med.*, **289:**286–91.

HULTEN, M., and P. L. PEARSON. 1971. Fluorescent evidence for spermatocytes with two Y chromosomes in an XYY male. *Ann. Hum. Genet.*, **34:**273.

OKSALA, T., and E. THERMAN. 1974. Mitotic abnormalities and cancer. In J. German, ed., *Chromosomes and Cancer.* New York: Wiley.

ROWLEY, J. D. 1973. A new consistent chromosomal abnormality in chronic myelogenous leukemia identified by quinacrine fluorescence and Giemsa staining. *Nature*, **243:**290–93.

WILSON, M. G., et al. 1970. Inherited pericentric inversion of chromosome No. 4. *Am. J. Hum. Genet.*, **22:**679–90.

WOLF, C. B., et al. Ring 1 chromosome and dwarfism—a possible syndrome. *J. Pediatr.*, **71:**719–22.

Questions

Useful terms: empiric risk, univalent, bivalent, trivalent, alternate segregation, compound chromosome, Robertsonian translocation, familial, aneusomy by recombination, recurrence risk, Philadelphia chromosome, position effect.

1. Make a drawing of the meiotic divisions showing the expectations if trisomy-21 occurs in the cell. Are the theoretical expectations realized in both males and females with trisomy-21?
2. If both sexes with a trisomy were fertile, what kinds of progeny would you expect from a mating of the two?
3. How are the classes of progeny produced by XXX females and by XXY males unusual? Do XYY males produce sons who are also XYY?
4. Is a sperm which is defective in the sense that it lacks a complete chromosome less likely to fertilize an egg than a normal one? Would the same answer be true if the egg were deficient for a chromosome?
5. If a normal couple produces several developmentally abnormal stillborn infants as well as several normal children, what might be your first conclusion?
6. Are the genetic consequences of alternate segregation in a translocation the same as those of adjacent segregation, as far as the well-being of the embryo is concerned?
7. Are the translocations that we observe in a human population at the present likely to be of recent or of ancient origin?
8. How does a compound chromosome differ from a translocation?
9. Can you describe one type of chromosome defect in which it can be predicted that all of the progeny produced by that person with it will be defective?
10. How is a pericentric inversion different from a paracentric inversion?
11. What is meant by aneusomy by recombination?
12. What is the difference between a recurrence risk and an empiric risk?
13. What is meant by the Philadelphia chromosome? How does it originate?
14. Discuss the relevance of the following conditions to the concepts discussed in this chapter: Down's syndrome, XXX female, XXY female, Fanconi's anemia, ataxia telangiectasia, chronic myelogenous leukemia.

23

Genetic Counseling

THE AIM OF GENETIC COUNSELING. When a family is faced with a problem they have reason to believe is genetic in origin, their best course of action is to obtain more information. They must understand the genetic and medical implications of the disorder, and be prepared to make any decisions about alternative courses of action rationally. The condition must be accurately diagnosed and explained to the family. The genetic basis should be determined in order to estimate the likelihood of recurrence. The individuals must be made fully aware of the means by which future problems may be handled or avoided, consistent with their own philosophical, ethical, religious, or moral constraints. In brief, counseling amounts to an intensive course, lasting a few hours at most, in the elements of medicine and genetics as they relate to their specific problem.

Families seek genetic counseling for a wide variety of reasons. According to one study, approximately half of the cases prove to involve genetic defects determined by a single gene. Another 20 percent of cases concern chromosome anomalies, and of these more than 80 percent involve trisomy-21. Another 20 percent are congenital defects with a polygenic or unknown genetic cause. On rare occasions, a family may seek advice because of anticipated problems related to consanguinity or to exposure to mutagens or teratogens.

THE BACKGROUND OF THE COUNSELOR. Until a dozen or so years ago, genetic counseling was handled to some extent by medical practitioners whose knowledge of genetics was minimal and whose main interests may have been focused in other directions, but who tried to communicate the limited knowledge available at that time to the patient. More often, perhaps, counseling was done by academically trained geneticists with an expert knowledge of the basic principles of genetics but with little or no orientation or training or opportunity to deal in depth with the clinical aspects of counseling problems. In both instances, the results were sometimes far from satisfactory.

At present, counseling is carried out more often than not by a group of highly qualified experts, usually functioning in a clinical setting. When such persons are M.D.'s, they are trained primarily in pediatrics, but also in pathology, hematology, internal medicine, obstetrics, gynecology, and neurology. Those with Ph.D. degrees include specialists in genetics, zoology, cytology, cell biology, biophysics, psychology, anthropology, and statistics. In a few cases, counselors have joint degrees in both medicine and genetics. At present, there are no practitioners with advanced degrees in genetic counseling as such, although programs are now in operation at a number of colleges to train both medical and nonmedical personnel to provide specialized services to group counseling clinics.

The Procedures in Genetic Counseling

MEDICAL DIAGNOSIS.   The first step in counseling must be an accurate medical diagnosis of the condition, for if the problem should prove to have no genetic basis, as it will occasionally, genetic counseling will be unnecessary. For instance, more than half of the patients referred to one center for muscular dystrophy proved to have something else. Some genetic disorders can be mimicked by environmental disorders, as is shown by the similarity of genetic phocomelia and the thalidomide-induced phenocopy, both reducing the normal arm length to a fraction of the usual size. Achondroplasia is similar to about half a dozen other developmental anomalies, which in some cases are caused by a recessive rather than by a dominant gene as achondroplasia is. During the diagnostic period, if a chromosome abnormality in any way may be involved, it is essential to karyotype the affected individual and frequently also the parents and sibs. The precise diagnosis may—and probably will—depend on a medical examination, laboratory tests, and x-ray examinations, and for this reason the participation of medically qualified personnel at this stage is mandatory.

THE PEDIGREE.   Making a detailed pedigree is absolutely essential, including all known relatives, with their ages, reproductive history including stillbirths, abortions, and deaths. In some cases a simple inquiry about the country of origin of the ancestors of the two parents, and their surnames, may give a clue as to the likelihood of consanguinity. Sometimes the pedigree will show unambiguously the method of inheritance and, as a result, make a distinction between two similar disorders with different modes of inheritance. Both parents should participate in these sessions from the beginning, when the contribution of family history information from each is important, to the final counseling periods, when both parents must understand the nature of the problem and the range of possible solutions and agree on a final decision.

INFORMATION ABOUT THE DEFECT.   There exists a vast body of medical literature on genetic defects. A catalogue, *Mendelian Inheritance in Man*, issued by
Dr. V. McKusick of Johns Hopkins University and brought up to date every few
years, is useful in providing the basic information about all known genetic
disorders. Another useful source is *Birth Defects: Original Articles Series*, edited
by D. Bergsma.

ESTIMATING THE RISK.   For simple dominant, recessive, or sex-linked genetic
defects, there is no difficulty in making a precise statement of the probability of
recurrence after the birth of an affected child. If one parent is heterozygous for a
dominant defect and the other is homozygous normal, the chance of an affected
child is one half. If two normal parents have produced a child homozygous for a
recessive, it can be assumed that both are heterozygous and the chance of an
affected child is one fourth. If a normal woman has produced a male child with a
sex-linked defect such as muscular dystrophy, the simplest assumption is that
she is heterozygous (unless the affected boy is a new mutation) and that half
of subsequent sons will be affected and half of her daughters will be carriers.
   These are the simple cases. When the problem is caused by a chromosome
anomaly such as a translocation, or a developmental defect such as cleft palate,
there are no specific algebraic expectations for the occurrence of an affected
individual and it is necessary to rely instead on *empiric risks*. These are probabilities based on statistics of similar past cases (Table 23-1).

EDUCATION OF THE PATIENTS IN PROBABILITY.   Probably the greatest difficulty
in counseling is educating the individual in the principles of probability. A
common misunderstanding concerns the independence of the probability of
successive events. It is not unusual for a couple who are both heterozygous for a
simple recessive to be told that there is a one in four chance of having a child
homozygous for that recessive defect, and for them to assume that their already
having had one affected child now insures that the next three children will be
completely normal. Sometimes the use of simple examples like tossing a coin
may seem to get the point across, but if many beginning college students cannot
understand probability, even after hours of patient classroom explanation by
skillful teachers, how can parents taken at random from the population be
expected to master it? This is all the more difficult because the issue is not
hypothetical, but one vitally important to them personally. They may be expected to make a serious decision on the vagaries of chance for which they will
have to accept the consequences for the rest of their lives.

DECIDING ON A COURSE OF ACTION.   Every family with a genetic problem
experiences some emotional trauma. The parents may feel guilty that they were
responsible, angry that they were the ones to suffer the misfortune, and they may
even deny the legitimacy of the child or the existence of the problem. Parents
may recall any of a large number of factors that they consider might have been
responsible: tobacco, alcohol, drugs, weight loss (or gain), coffee, tea, and so on.
It calls for an experienced counselor with an understanding of the intensity of
the trauma and sufficient time and patience to try to educate the parents to the
genetic facts of life—that all persons carry their share of deleterious genes,
independent of status, class, wealth, or ethnic group, and that no one should
assume any personal blame for affected children.

Table 23-1 Empiric risks, for counseling purposes, for cleft lip with or without cleft palate (CL ± CP) and cleft palate (CP) for various family situations. (F. Clarke Fraser, *Am. J. Dis. Child.*, **102**:853, 1961.)

	Proband has	
	CL ± CP	**CP**
	(per hundred)	
I. Frequency of defect in the general population	0.01	0.04
II. My spouse and I are unaffected		
A. We have one affected child		
1. What is the probability that our next baby will have the same condition if:		
a. We have no affected relatives?	4	2
b. There is an affected relative?	4	7
c. Our affected child also has another malformation?	2	2
d. My spouse and I are related?	4	2
2. What is the probability that our next baby will have some other sort of malformation?	same as general population	
B. We have two affected children		
1. What is the probability that our next baby will have the same condition?	9	1
III. I am affected (or my spouse is)		
A. We have no affected children		
1. What is the probability that our next baby will be affected?	4	6
B. We have an affected child		
1. What is the probability that our next baby will be affected?	17	15

After a rational perspective has been assumed, plans can then be made for a future course of action, if any, depending on the nature of the problem and the magnitude of the risk. In some cases, if the parents are so inclined, future pregnancies can be monitored to determine the status of a fetus. In other cases, parents may decide not to have any additional offspring. This is a decision that is left to the parents and can be highly individualistic.

SUBSEQUENT REVIEWS. Many counseling centers attempt to maintain contact with the subjects. This can be helpful in a number of ways. The situation may change with time so that a reevaluation of the original decision may be called for. New clinical methods may develop that make the earlier decision obsolete. Or the initial information given to the parents, as well as their decisions, may become garbled with the passage of time. In fact, studies have shown that many people who appeared to understand the entire situation during the first counseling sessions actually did not, and follow-up studies made it possible to correct any misunderstandings. Furthermore, when the counseling was performed by a family physician (who passed on to the patient the information provided by a genetics clinic), the physician himself often had a limited and somewhat distorted view both of the genetic nature of the disease and of the consequences to the patient. This problem will no doubt disappear as the newer generation of genetically sophisticated medical practitioners replaces the old.

Reviews give clinics and counselors an opportunity to assess their effectiveness. Later studies of counseled parents have revealed the following: that the family's decision is based more on their assessment of the burden involved to the family than on the risk of an affected child, that the amount of the genetic information they retained was minimal, and that the principal barriers to family limitation were religious principles prohibiting contraception, sterilization, or abortion.

Problems of Counseling

Most counselors make a strong effort not to influence the parents with respect to a course of action to be taken but allow the parents maximum freedom of choice in their decisions after supplying them with all available information. However, it may be very difficult to avoid a facial expression, a manner of speech, or other behavior that expresses a point of view. The classic example is the choice forced upon a counselor who must inform the parents that they are both heterozygous for some crippling recessive condition such as Tay-Sachs disease. The parents could be told that "There is a chance of three out of four that the next offspring will be perfectly normal," or, equally well, that "There is a chance of one in four that the next child will be hopelessly doomed to an early death." Each statement is essentially correct, yet the reaction of the parents may be quite different in the two cases. Either statement above would betray the counselor's views of the matter and subtly suggest a course to be followed. The precise way in which this information is communicated may be determined largely by the counselor's intuitive assessment of the psychology of the parents and their likely reaction to the various ways of presentation.

In any case, the background of a counselor may very well determine the attitude with which he or she views human problems, tempering the advice likely to be given. A doctor of medicine whose code of ethics and many years of training have centered almost entirely on the preservation of life is quite likely to give an opinion based primarily on the immediate well-being of the patient, whereas a doctor of philosophy, trained as a population geneticist, may consider that some anticipated future needs of society should take precedence over the welfare of the patient. It would be desirable for each counselor to identify conscious biases and, if possible, to compensate for them when appropriate. This, however, may be an impossible ideal, since most people refuse to admit having biases. Perhaps the most rational counseling is given by several individuals with different backgrounds acting as a team.

HUNTINGTON'S CHOREA.   Some of the problems of genetic counseling are illustrated by the situation that has been described for Huntington's chorea. This is a degenerative disease of the nervous system, determined by a single dominant gene, that develops, on the average, at about age 35. It is characterized by mental deterioration and by purposeless involuntary movements suggesting the waving motions of a dance (hence the name, from the Greek *choreia*, meaning "dance").

On the average, patients with this disease survive only 11 years after the disease has become evident.

Because of the late onset of the disease, most afflicted persons will have married and often will have completed their families before becoming aware of their condition. For this reason, this gene continues to be perpetuated generation after generation. Its overall incidence is one per 25,000 persons, but it tends to be concentrated in a limited number of kinships. On the other hand, as a single dominant with high penetrance, it would appear to be a relatively simple disease to eliminate (in principle). All that is required is that all persons in families in which the disease has appeared not have children.

But is there any justification for suggesting a limitation on the reproductivity of normal persons because they stand a 50 percent chance of carrying an abnormal allele (when there is an equal chance that they will not carry that allele)? Even if it were possible to detect the heterozygotes early in life, would it be acceptable to discourage their reproduction because they stand a 50 percent chance of having a heterozygous child at each birth?

Studies of the attitudes of persons in kinships segregating for Huntington's chorea show that the desire for family size limitation was greatest among the spouses of the affected and among those relatives at low risk, not likely to have the allele, but that those actually affected or young adults with a 50 percent chance of being heterozygotes were less enthusiastic about curtailing family size. This restraint on the part of the unaffected, or those not likely to be affected, and lack thereof on the part of those affected are shown in one study indicating a 27 percent decrease in reproductivity of unaffected sibs compared to a 22 percent increase (over an average sample of the population) for those affected. In fact, a quarter of those in the high-risk category, the possible heterozygotes, indicated that they would probably refuse to take a test, if one were available, to determine whether they actually were heterozygotes.

Prenatal Diagnosis (by Amniocentesis)

Because the possibility of a deformed or mentally defective child haunts virtually all parents at some time, reassurance about the probable well-being of a future child is greatly appreciated by prospective parents. On the other hand, if a fetus can be diagnosed as afflicted with a grossly debilitating disease, some parents may prefer to know and, if their religious and ethical feelings allow, prevent it from coming to term. The decision to decrease the number of births of severely abnormal children may become more prevalent as more and more married couples decide to limit family size anyway, e.g., to two offspring.

Amniocentesis (Figure 23-1), the puncturing of the amniotic membrane to obtain amniotic fluid and fetal cells, was developed originally in the diagnosis and treatment of hemolytic disease of the newborn (HDN). Cytological examination of the cells will reveal the sex of the fetus and possibly any chromosome abnormalities carried by it; in addition, some 30 genetic disorders may be detected biochemically (Figure 23-2).

COMPOSITION OF AMNIOTIC FLUID.   The amniotic fluid is interchanged quite rapidly between the fetus and the mother, being completely replaced about once

Figure 23-1 Amniocentesis, a simple painless procedure often effective in relieving unnecessary anxiety. (Courtesy of Time/Life.)

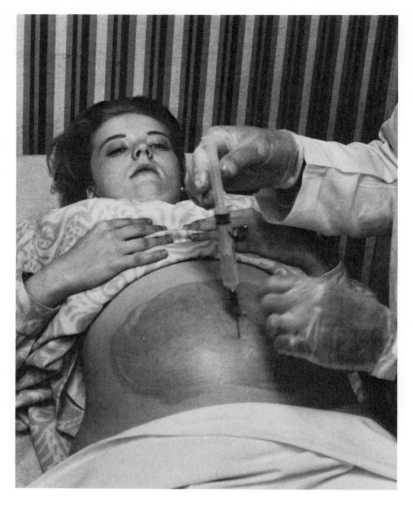

every 3 hrs. The total amount of amniotic fluid is variable, but averages between 500 and 1000 ml at the time of birth. The cellular and biochemical components of the fluid are epithelial cells from the skin, the alimentary tract, and the respiratory tract, as well as cells from the amnion and other tissues. In addition, there is a concentration of protein of about 1 percent. There is a high concentration of some free amino acids in early pregnancy; other chemicals include immunoglobulins and fatty acids, a number of enzymes, and steroids. Some of these components are probably maternal in origin, others are of fetal origin.

THE TECHNIQUE OF AMNIOCENTESIS.   The amniocentesis operation consists of inserting a needle into the uterine cavity and removing from 10 to 20 ml of fluid and associated free cells. The danger to either the mother or fetus is negligible when the procedure is carried out between 14 and 16 weeks of pregnancy. Amniocentesis cannot be accomplished very early because the volume of amniotic fluid may be insufficient. On the other hand, it cannot be postponed too long because amniotic cells become more difficult to culture successfully as

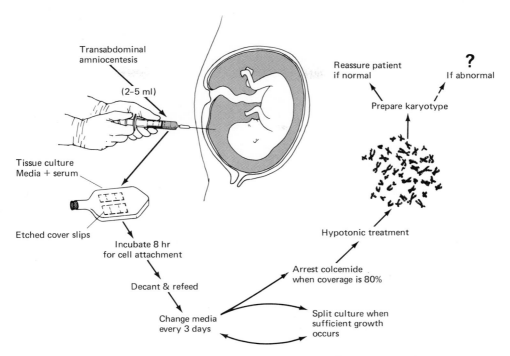

Figure 23-2 Prenatal diagnosis by amniocentesis. The cells present in the amniotic fluid can be cultured and tested for cytological and biochemical properties. (From C. B. Jacobson. Cytogenetic techniques and their clinical uses, in *Genetics in Medical Practice*, Lippincott, Philadelphia, 1968.)

pregnancy progresses, if either biochemical or chromosomal studies are to be made. Culturing the cells takes time—sometimes as long as a month—and the biochemical and cytological studies add another week or two. In the event that serious defects are found in the fetus, or there is some other compelling reason for considering a therapeutic abortion, the parents must be able to make this decision prior to the legal limit if this is being done in one of the half dozen or so states in which it is legal to perform an abortion to prevent the birth of an abnormal child. The law varies from one state to the next, and the limit may be 22 to 24 weeks. If the attempt to culture cells for analysis should fail, or if they should prove refractory to either biochemical or cytological analysis, then a second amniocentesis, with the time necessary for analysis, may extend the required time beyond that allowed for legal abortion.

PROBLEMS OF AMNIOCENTESIS.   Amniocentesis is usually carried out for specific women at risk (Table 23-2), and then only to the extent demanded by the nature of the problem. As a general procedure undertaken for the general population in all pregnancies, it would simply present an intolerable expense.

There are a number of problems with respect to the application of amniocentesis. In the first place, although the risk is very low, the approximate proportion of pregnancies in which complications develop after amniocentesis —including spontaneous abortion, fetal death, or excessive maternal bleeding—is

Table 23-2 Responses to questionnaires from a number of U.S. and foreign medical laboratories indicating the criteria used by them for doing amniocentesis. (From *Birth Defects: Orig. Art. Ser.*, ed. D. Bergsma. Published by Williams & Wilkins Co., Baltimore, for The National Foundation—March of Dimes, White Plains, N.Y.)

Reason	United States		Foreign Countries	
	Yes	No	Yes	No
Family history of gene-transmitted or chromosomal abnormality	77	6	36	7
Previous birth of retarded or malformed child	43	33	19	23
Parent with balanced chromosomal translocation	76	7	37	6
Viral infection first trimester	18	58	8	32
Poor previous pregnancy history	24	52	12	30
History of exposure to mutagen(s)	24	52	12	27
Rh factor	59	19	24	19
Advanced age of mother	69	10	28	14
Sex determination (for X-linked disorders)	51	28	30	10
Research control data	24	44	13	22

of the order of 1 to 2 percent. Second, the attempt to obtain amniotic fluid may be unsuccessful, and, third, the procedures for producing chromosome preparations may fail even in the most skillful hands. With some small frequency, maternal cells are present in the amniotic fluid and could be responsible for a misdiagnosis.

Another possible danger in the procedure is the possible release of fetal erythrocytes into the maternal system in an Rh-incompatible combination, followed by immunization of the mother. In general, amniocentesis will not be performed unless there is a substantial medical reason for it. At the present time, for instance, most physicians would object to performing amniocentesis for the sake of ascertaining the sex of the forthcoming child in the absence of any compelling medical justification.

RISK CATEGORIES. One proposed scheme for classifying patients eligible for amniocentesis is based on the relative risk involved:

1. A *high-risk* group. When the mother is known to be heterozygous for a serious X-linked recessive disorder, half of the male progeny can be expected to carry this disorder. If a woman has had a child with muscular dystrophy of the Duchenne type, she might prefer not to undergo a similar experience a second time. Since this particular disease cannot be detected by any prenatal tests, she may elect simply not to allow any male fetuses to come to term. With other sex-linked diseases, such as the Lesch-Nyhan syndrome, where it is possible to make an accurate diagnosis prenatally, a woman need be concerned only if the male

fetus is actually affected. Patients who have produced one abnormal child, e.g., one with Tay-Sachs disease, have a 25 percent chance that each additional child will be similarly affected. Tay-Sachs homozygotes can be identified in the fetal stage.

Another group in the high-risk category includes those where one parent is heterozygous for a chromosomal rearrangement as a result of which there is a sizable chance (5 percent or more) that the fetus will carry an abnormal chromosome complement. Often this can be detected by karyotyping.

2. A *moderate-risk* category. Here we include women who are of age 38 or older, since the probability of a child with a chromosome aberration such as one of the trisomies may run as high as 4 percent.

3. A *low-risk* group. In this group are women aged between 35 and 38, and those who have normal chromosomes but who have previously had a chromosomally abnormal offspring. In these cases, the likelihood of an affected child is greater than that for the population at large, but less than that for the moderate-risk category.

Of course these are only rough guidelines for determining the desirability of a course of action. In practice the final decision would have to depend to a great extent on the patient's perception of the magnitude of the risk and the likelihood that any positive action would follow the discovery of an abnormal fetus.

Table 23-3 shows the outcome of pregnancy in more than 300 cases in which amniocentesis was performed. Of 331 cases, mostly of high- and moderate-risk mothers, 41 abnormal fetuses were detected and 35 of these were aborted.

Screening Programs

With our advanced genetic knowledge and sophisticated clinical methods, what better use to put them to than in the prevention of abnormal births? Screening programs, in which very large numbers of people (often members of an ethnic group at high risk) are routinely checked for the presence of some defective

Table 23-3 The outcome of pregnancy in 331 cases of high- and moderate-risk mothers. (H. L. Nadler, Indications for Amniocentesis in the Early Prenatal Detection of Genetic Disorders. In *Birth Defects: Orig. Art. Ser.*, ed. D. Bergsma. *Intrauterine Diagnosis.* Published by Williams & Wilkins Co., Baltimore, for The National Foundation—March of Dimes, Vol. VII(5):5, 1971.)

	Studied	*Affected Fetuses*	*Aborted*	*Delivered*
Chromosomal translocation carrier	41	10	9	32
Maternal age > 40	119	4	4	118
Previous trisomic Down's syndrome	67	2	1	66
Other	28	5	5	23
X-linked recessive	27	12 males	8	19
Metabolic	49	8	8	42
Total	331	41	35	300

allele, or newborns are checked for an abnormality, have been set up for several genetic defects.

One might argue that the more information a person has about his or her genetic constitution, the more intelligent can be the choice made about the production of offspring, and the better off the individual and society will be. To accomplish this ideal, a screening program that will identify those with genetic defects for which we presently have tests is a first step.

If the preceding paragraph seems idealistic, self-evident, and noncontroversial, it is only superficially so. In point of fact, when screening programs have been instituted they have created a great deal of dissension, and, although there is no doubt that they have accomplished much good, they have also done some harm. In fact, in some cases the harm may outweigh the good. To illustrate this point, we will examine four different kinds of screening programs: for Tay-Sachs disease, sickle-cell anemia, phenylketonuria, and chromosome abnormalities.

Tay-Sachs Disease

Tay-Sachs disease is a simple recessive disease in which the nervous system degenerates. It is obvious after a few months of age and always leads to death in early childhood. Usually the child does not learn to walk; muscular control degenerates rapidly. At age 18 months, the affected child is usually deaf and blind, with complete rigidity of the muscles, and will die before the fourth or fifth year. The affected child lacks an enzyme, and this results in the accumulation of fatty tissue in the cells. Accumulation in the brain cells, in particular, leads to the loss of coordination, seizures, blindness, and finally death.

This disease occurs in approximately one in 6,000 Ashkenazi Jewish births, but only in two in 1 million non-Jewish births. From these figures we can calculate that about one in 40 Ashkenazi Jews and about one in 400 non-Jews are carriers for this recessive.

The absence of the enzyme hexosaminidase in the homozygote can be detected by amniocentesis, thus making it possible to monitor pregnancies at risk, with the choice of abortion being left to the parents. Furthermore, the enzyme is known to be present in only about 50 percent of the normal amount in known heterozygotes, so that heterozygotes can be detected (Figure 23-3), but parents are ordinarily not aware of their heterozygosity until the birth of the first affected child. Since the overall frequency of normal children from heterozygotes is $3/4$, there may be several normal children before the affected child—and the family may limit its size prior to the conception of a second affected child. For these reasons, only 18 percent of all homozygotes are diagnosed prenatally, even when amniocentesis is available.

A screening program was set up in 1971 in the Baltimore-Washington area, where it was estimated that about 8,000 persons are heterozygotes. Both husband and wife of several hundred couples are heterozygotes and their progeny are therefore at risk. The propriety of therapeutic abortion of homozygotes detected in utero is a matter of considerable disagreement among the Jewish religious leaders. If elective abortion were decided upon by all carriers for this disease when the fetus appears to be a homozygote, then clearly the incidence of affected newborns would diminish rapidly. On the other hand, since such early

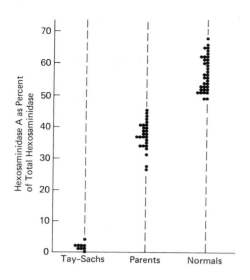

Figure 23-3 The detection of heterozygotes for the recessive allele causing Tay-Sachs disease. The percentage of hexosaminidase A is very low in affected children, and lower in their parents (who must be heterozygotes) than in normal individuals. (Reprinted by permission of J. S. O'Brien and the *New England Journal of Medicine,* Tay-Sachs disease—detection of heterozygotes by serum hexosaminidase assay, **283:**15–20, 1970.)

losses of the homozygous fetus would probably be compensated for by the parents deciding to have an additional offspring—a phenotypically normal offspring, but heterozygous two-thirds of the time—the net effect of amniocentesis would be to increase the frequency of the defective gene in that population.

SICKLE-CELL ANEMIA.   Sickle-cell anemia, caused by homozygosity for the hemoglobin-S allele, affects one in every 500 Black children in the United States (about 50,000 of this nation's 22 million Blacks). Homozygotes suffer from considerable pain as a result of chronic anemia, strokes, and other disasters of the circulatory system. Such homozygotes very often die before 10 years of age and the majority of them by 30. More than 2 million Blacks are heterozygous, described as having the "sickle-cell trait." Under ordinary conditions, heterozygotes with sickle-cell trait are indistinguishable from normal, although they may at times be slightly anemic, and a drastic reduction in oxygen pressure may cause their otherwise normal cells to sickle. Deaths have been reported for individuals with the sickle-cell trait under anesthesia, during severe physical stress, and after blood transfusions, but these are most unusual, and heterozygotes should be considered quite normal. It is possible to detect heterozygotes by a relatively simple test (one of the least expensive and best-tested systems for detecting them costs only 2 cents per test!), so mass screening programs are technically feasible.

Amniocentesis has not been effective for the simple reason that the hemoglobin present in the fetus is of a completely different sort from that present in the adult, and fetal hemoglobin does not show evidence of the sickling phenomenon. In fact, the adult hemoglobin that reveals the sickling characteristic may not appear until 5 or 6 months after birth. Therefore, the only effective countermeasure against the production of homozygotes has been to warn the heterozygous parents of the 25 percent probability of their having a homozygous offspring. Some recent work has indicated that some very small amount of adult hemoglobin may be present in the fetus. This may make it possible eventually to detect sickle-cell homozygotes prior to birth.

ETHICAL PROBLEMS.   Few issues of genetic concern have given rise to such widespread controversy as the sickle-cell screening program, and some of its severest critics have come from the Black population itself. Some criticisms are directed toward the manner of carrying out the programs, sometimes by untrained and insensitive personnel. Others are directed at the racial overtones of this kind of screening program, particularly when it is made compulsory by law, as it is in ten states.

It is certainly a legitimate concern to wonder if any ethnic group should be subjected to a screening program that is aimed specifically at them as a population group. Still another question that arises is whether individuals who do not wish to be exposed to a medical (or quasimedical) examination should be forced to do so.

From the standpoint of the Blacks, there has been some skepticism as to whether the millions of dollars that have been legislated for the alleviation of this disease were not given so much for humanitarian purposes as for political gain. Counseling has been interpreted as pressure for the heterozygous Black person to have fewer children, and the word "genocide" has been used to describe counseling that might limit productivity. Some have questioned the motives of those in the medical profession and in hospitals who have accepted large sums of money from the government to do research in this area.

There has been some confusion of the homozygote with the heterozygote in the public eye: the first is afflicted with sickle-cell anemia, whereas the second is a carrier and is said to have the sickle-cell trait. The distinction may not be clear. The original Massachusetts law refers to "the disease known as sickle-cell trait or sickle-cell anemia." The legislation passed by the Congress of the United States includes the statement that "the Congress finds and declares (1) that sickle-cell anemia is a debilitating inheritable disease that afflicts approximately two million American citizens." Surely this refers to the number of heterozygotes with the trait, not the homozygotes with the anemia. The New York State law requiring testing prior to marriage (along with tests for venereal disease, a most unfortunate association) states that it is for the purpose of "discovering the existence of sickle-cell anemia," when it is clearly meant for detecting the trait. In many cases, even medical personnel have confused the sickle-cell trait and sickle-cell anemia, and this unfortunate terminology is the source of a great deal of the problems that arise. In some cases, individuals who have the sickle-cell trait (i.e., who are heterozygous) have been fired from their jobs or otherwise discriminated against on the grounds that they have a serious disease, which is simply not the case.

Phenylketonuria (PKU)

Large numbers of studies have been made on PKU incidence in different populations and it appears to occur with a frequency ranging from three to ten per 100,000 in different populations of Northern European origin (Table 23-4). It has also been found in Japan and in the Middle East, but it is very uncommon among both Jews and Blacks. Most of the homozygotes are found in mental institutions, where they constitute about 1 percent of the mentally defective population. The age distribution of these patients shows a lesser than expected frequency under

Country	Estimated Number Screened	Cases Found
Australia	479	0
Belgium	20,000	5
Canada	65,898	6
Denmark	12,500	0
Germany	67,309	10
Ireland	25,000	4
Israel	65,000	5
New Zealand	1,840	0
Poland	130,912	21
Scotland	16,500	2
Sweden	21,505	1
Switzerland	7,002	1
United States	2,238,634	205
Wales	1,536	0
Yugoslavia	23,690	3
	2,697,805	263

Table 23-4   The results of screening newborn infants for phenylketonuria in 15 countries. (From R. Guthrie, Screening of Inborn Errors of Metabolism in the Newborn Infant. In *Birth Defects: Orig. Art. Ser.*, ed. D. Bergsma. Published by Williams & Wilkins Co., Baltimore, for The National Foundation—March of Dimes, Vol. IV [6]:93, 1968.)

5 years and over 35 years. The former is undoubtedly a reflection of a late diagnosis or a late development of the phenotype. The lack of affected individuals at older ages results from the relatively early deaths of homozygotes, usually from infectious diseases that are common in institutions.

DIAGNOSIS.   The homozygotes can be detected by the presence of phenylpyruvic acid, one of the breakdown products of phenylalanine when not converted into tyrosine, with the use of ferric chloride ($FeCl_3$). When phenylpyruvic acid is present in a test sample of urine, the ferric chloride turns the color to olive green. The most common and practical test is called the Guthrie test, in which a drop of the infant's blood is checked to see if it inhibits the growth of a special mutant strain of bacteria. This happens when the blood carries minute quantities of the phenylpyruvic acid present in higher quantities in PKU children than normal. It is a very sensitive test and may be used shortly after birth.

It is interesting that the original diagnosis was made because the mother of two of the first patients insisted that biochemical tests be made because she was surprised by the physical similarities of her two offspring, and by their characteristically smelly urine. It is from the presence of the metabolites of phenylpyruvic acid, the phenylketones in the urine, that the abbreviation PKU is derived.

TREATMENT.   The basic principle in treating this disease is to feed the child a diet with a low concentration of phenylalanine. It was attempted originally by replacing the protein in diets by a mixture of pure amino acids. This, however, is very costly. A more practical method is to remove the phenylalanine from a diet containing otherwise normal proteins. The problem that arises is to include sufficient protein along with phenylalanine to maintain body protein, and to allow for normal growth, at the same time not exceeding the phenylalanine requirement. A diet too deficient in phenylalanine may cause severe complications, including anemia, with eventual fatal consequences. After a proper phenylalanine treatment, the biochemical abnormalities disappear. The peculiar odor of the urine vanishes, skin blemishes clear up, hair and skin pigmentation darkens, and the patient becomes more manageable, with an increase in attention-span and improvement of motor performance. Even PKU adults who are beyond the period when specific nerve development can be

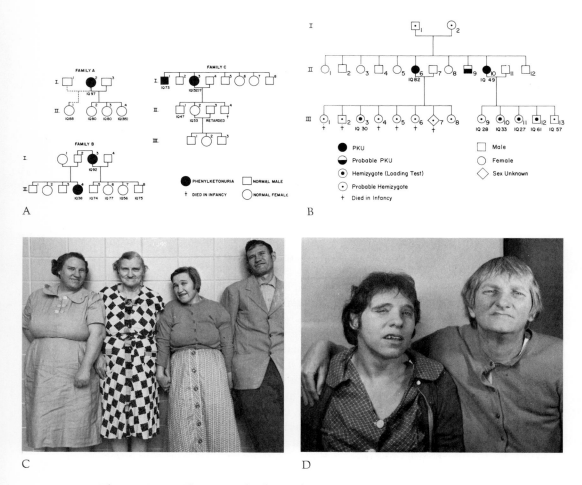

Figure 23-4 The unusual effects of maternal PKU. *A.* Three families in which the mother was homozygous and all children (heterozygotes) suffered as a result. *B.* A pedigree with high mortality in one sibship, and striking effects on all progeny, four of whom (*III,3, 10, 11,* and *12*) could be shown to be heterozygotes. *C.* Individuals *III,10, II,10, III, 11,* and *III,12* of pedigree *B. D.* Individual *III,3* and her mother, *II,6* of pedigree *B.* (Reprinted by permission of C. C. Mabry and the *New England Journal of Medicine,* Maternal phenylketonuria: Cause of mental retardation in children without metabolic defect, **269:**1404–1408, 1963.)

improved by a low-phenylalanine diet can have their behavior patterns improved considerably by such a diet.

The most important aspect of the treatment is that the low-phenylalanine diet will improve the intelligence of the homozygote when it is applied at an early age, and may even reverse previously established impairment during the first few years of life. The homozygotes are apparently born normal and degeneration begins within the first few weeks. The enzyme responsible, phenylalanine hydroxylase, does not, however, become functional in the liver until shortly after birth. Since the treatment must be applied within a few weeks of birth before intellectual impairment sets in, it becomes a major problem to diagnose the infants in time to start treatment. The more the treatment is delayed, the less the improvement, as a rule.

INFLUENCE OF MATERNAL PKU ON OFFSPRING.   One of the curious aspects of PKU retardation is that when homozygous females marry and have children (Figure 23-4), which happens fairly frequently for those without severe mental deficiency, their non-PKU children may nevertheless be abnormal. From a total of 14 homozygous mothers there were produced 34 non phenylketonuric children, 30 of whom were retarded. In fact, the degree of mental retardation was greater for the children than for the mothers. In addition, the children had a low average birth weight, and sometimes microcephaly. It is reasonable to conclude that elevated levels of phenylalanine (or its products) in the mother cross the placenta and concentrate in the blood of the baby. These compounds are then harmful to the brain development of the fetus. These cases can be handled, apparently, by prescribing a low-phenylalanine diet during pregnancy. One difficulty with this is that the pharmaceutical dietary preparations with a low phenylalanine content may be unpalatable or even nauseating to the patient. Another problem is that the phenylalanine concentration in the blood must be constantly monitored to make sure that the mother is getting neither too much nor too little.

As the treatment of PKU infants becomes more common, an increasing number of homozygous women will appear normal, get married, and have children, without anyone realizing that the fetuses are at grave risk. Reagents have been developed for widespread screening of pregnant women to uncover any unsuspected homozygotes.

Chromosome Abnormalities

Screening programs that center on adults have provided important cytological information about the distribution of chromosome defects in the population. Not only have relative frequencies been determined, both inside and outside of institutions, but information on the range of physical and mental normality provides a basis for predicting the likely course of new cases as they arise. There can be no doubt that without the kind of information that comes from population surveys there would be a tendency to exaggerate both the frequency and the effects of chromosome abnormalities found in adult populations.

Several large-scale studies have been made of the chromosomes of newborn infants (in Boston, Denver, and New Haven) and of young children (Baltimore).

In principle, this seems like a desirable procedure, making it possible to determine those cases in which a behavioral condition may stem from some previously undetected chromosome abnormality. Some of the problems that arise in this connection are those common to all screening programs. A unique one is whether such determinations ever become self-fulfilling prophecies. In the case of an XYY boy who, according to the law of averages, would probably be indistinguishable from XY in height, intelligence, and behavior, will his parents and teachers be able to treat him as a normal boy and will their attitudes of solicitude or apprehension divert his behavior into an abnormal direction? Is there any likelihood of his developing those traits as a consequence of his being informed that he is XYY and that such persons often behave in an antisocial way? Questions such as these appeared so important to a group of citizens in the Boston area that they forced the suspension of a well-established and medically approved program of screening the chromosome compositions of newborns.

Problems Arising During Counseling and Screening

MISTAKEN PATERNITY.   Almost 5 percent of the typical U.S. middle-class population of children have mistaken paternity; that is, their mother's husband (and their legal father) is not their biological father, a fact not generally known to either husband or child. It is understandable that the mother might be unwilling to transmit this information during a counseling session, particularly if her husband is present. Mistaken paternity may be revealed if a child is homozygous for an allele that only the mother carries (Tay-Sachs disease or PKU) or if the child carries an allele that neither the mother nor her husband has (sickle-cell trait). It can also be uncovered accidentally in the course of routine blood-group testing of the family. When evidence of mistaken paternity is uncovered, the confidentiality of this information must be maintained with utmost discretion.

THE PROBLEM OF CONFIDENTIALITY.   Who is to have access to the information revealed during counseling procedures? Suppose that a pedigree analysis uncovers some serious defect. Who, if anyone, is to inform relatives who might be similarly affected, or who might be carriers? It is possible that the patients themselves will prefer not to discuss this with close relatives, and may even decide that this essential information should not be communicated to them by anyone. The ethics of confidentiality are involved in a case like this, since the physician cannot take it upon herself or himself to discuss the nature of the defect with other people. This is reinforced in 34 states of the United States that have laws governing the confidentiality of privileged medical communication.

ETHICAL ISSUES.   All screening programs are subject to many fundamental ethical questions. Do parents have a fundamental right to produce children under any circumstances? Does this right exist even if it is known that there is a high probability, or even certainty, of a defective offspring? Does the woman have an absolute right over the embryo and fetus developing within her? If the fetus is considered to be a living being, does it have the inalienable right to be born free of physical and/or mental defects? Does the fetus have rights in society;

if so, when in gestation do they begin? If it were possible to examine all fetuses for physical and mental defects prior to birth, should this be done? If a woman has had one trisomic offspring, should she consider not having any additional offspring? Should a woman known to be heterozygous for a serious sex-linked recessive gene by the birth of an affected son, feel free to abort all subsequent male fetuses?

Does society have the right or responsibility to lay down rules for parents at risk of producing abnormal offspring? Or "too many" offspring? Is it fair to develop a screening program in which the focus is entirely on "bad genes" found predominantly in one ethnic group? Is it within the jurisdiction of religions to stipulate the conditions for having (or not having) offspring? If a physician fails to inform a woman of a high probability, or certainty, of a defective child, can he be held legally responsible? Does a human have a right to a certain tranquility of spirit, peace of mind, without being badgered by unsolicited, unwanted, sometimes depressing genetic information about which he can do nothing anyway? Would it be desirable to have a national repository of information containing what is known about each person's genetic constitution?

These are only a few of the many questions that have become acute during the past few years. Since there can be no hard and fast answers to these questions, the controversy surrounding them can be expected to continue for years to come.

References

BERGSMA, D., ed. 1973. Contemporary genetic counseling. Proc. Symp. Am. Soc. Hum. Genet. In *Birth Defects: Orig. Art. Ser.*, Vol. IX. Baltimore: Williams & Wilkins.

BODMER, W., and A. JONES. 1974. Genetic screening—the social dilemma. *New Sci.*, **63**:596–97.

COHEN, M. M., and A. D. BLOOM. 1971. Monitoring for chromosomal abnormality in man. In *Monitoring, Birth Defects and Environment*. New York: Academic Press, pp. 249–72.

EMERY, A. E. H., ed. 1973. *Antenatal Diagnosis of Genetic Disease*. Baltimore: Williams & Wilkins.

GORDON, H. 1971. Genetic counseling. Considerations for talking to parents and prospective parents. *JAMA*, **217**:1215–25.

GRAHAM, J. B. 1972. Genetic effects of family planning. *Quart. J. Family Planning*, **21**:27–29.

HECHT, F., and L. B. HOLMES. 1972. What we don't know about genetic counseling. *N. Engl. J. Med.*, **287**:464–65.

HOOK, E. B., D. T. Janerich, and I. H. Porter, eds. 1971. *Monitoring, Birth Defects and Environment. The Problem of Surveillance*. New York: Academic Press.

MABRY, C. C., J. C. DENNISTON, and J. G. COLDWELL. 1966. Mental retardation in children of phenylketonuric mothers. *N. Engl. J. Med.*, **275**:1331–36.

MILUNSKY, A. 1974. *The Prenatal Diagnosis of Hereditary Disorders*. Springfield, Ill.: Thomas.

MOTULSKY, A. G., ed. 1970. *Counseling and Prognosis in Medical Genetics*. New York: Hoeber.

NADLER, H. L. 1971. Indications for amniocentesis in the early prenatal detection of genetic disorders. In D. BERGSMA, ed., *Birth Defects: Original Article Series*, Vol. VII. Baltimore: Williams & Wilkins.

O'BRIEN, J. S. 1972. Ganglioside storage diseases. In H. Harris and K. Hirschhorn, eds., *Advances in Human Genetics*. New York: Plenum.

REISMAN, L. E., and A. P. MATHENY. 1969. *Genetics and Counseling in Medical Practice*. S. Louis: Mosby.

STEVENSON, A. C., et al. 1970. *Genetic Counseling*. Philadelphia: Lippincott.

WERTZ, R. W., ed. 1973. *Readings on Ethical and Social Issues in Biomedicine*. Englewood Cliffs, N.J.: Prentice-Hall.

Questions

Useful terms: genetic counseling, empiric risk, amniocentesis, high-, moderate-, and low-risk categories, screening program.

1. For what reasons do families seek genetic counseling? What does the genetic counselor try to accomplish?
2. Explain why the participation of medically qualified personnel is essential in the process of counseling.
3. What kinds of questions are usually asked of the family in making out a pedigree?
4. Can you explain the difficulties that those being counseled may have in understanding the simple rules of probability?
5. How successful has genetic counseling proved to be, based on later studies of counseled parents? Have subsequent studies of counseled persons shown counseling to be (in general) an effective procedure?
6. Can you imagine a situation where counseling by an untrained person could lead to a harmful result?
7. If you were to act as a genetic counselor, what personal biases or convictions that you have would inevitably show up in the kind of advice that you would give? How could such an expression of personal bias be avoided?
8. Does the experience with Huntington's chorea suggest that people are likely to curtail their reproductivity when there is a likelihood of some abnormal consequence?
9. Describe how the procedure of amniocentesis is accomplished. Of what use is it?
10. Which general classes of genetic disorders belong to the various risk categories?
11. What is meant by a screening program? For which diseases have these been instituted in one part of the country or another?
12. What is meant by "sickle-cell trait"? How has this terminology given rise to unnecessary problems?
13. Do you think that screening programs serve a useful purpose?
14. How is a PKU treated? What problems may result from these treatments? When do heterozygotes for PKU stand a higher probability of being abnormal?

15. Describe some of the problems that might arise when large segments of the population are screened for chromosomal abnormalities.
16. Discuss the relevance of the following conditions to the concepts discussed in this chapter: muscular dystrophy, phocomelia, achondroplasia, cleft palate, Tay-Sachs disease, Huntington's chorea, HDN, sickle-cell anemia, phenylketonuria (PKU), microcephaly, XYY male.

24

Genetic Variation in Human Populations

Darwinism and Neo-Darwinism

When Darwin proposed his theory of natural selection by the survival of the fittest in the 1860's and showed how this could account for the process of evolution, one essential step was missing from his otherwise convincing argument. Darwin knew nothing about the nature of the hereditary material or how it might be transmitted from one generation to the next. This was a serious conceptual gap, because it was not at all clear how a new beneficial change could maintain itself in the population. On the contrary, it seemed reasonable to imagine that it would be diluted by half in the immediate progeny, by another half in the next generation, and so on, disappearing rather quickly. His own ideas about the mechanism of inheritance, based on a biological communication system, whereby the different cells of the body send messages to the germ cells was essentially Lamarckian; that is, it was fundamentally based on the assumption of the inheritance of acquired characters. Although this idea may appeal to the biologically unsophisticated who are willing to accept it as a valid explanation for such peculiarities as the length of the neck of the giraffe (who must reach to eat the leaves of a tree), it is a hypothesis for which there is no scientific support despite serious attempts by many conscientious scientists to confirm it.

DE VRIES "MUTATION THEORY." It was not until about 1900 that Hugo de Vries, working in the Netherlands, pointed out the possibility that the sharp, clear-cut morphological changes he saw in the evening primrose might provide the variations on which natural selection could act. In retrospect it appears that de Vries was quite incorrect in his interpretation of the details of this phenomenon. In fact, almost all of the apparent mutations he found were gross chromosomal abnormalities—trisomies, duplications, and deficiencies—but he did redefine the word "mutation" and correctly interpreted its importance in evolution.

The work with *Drosophila* started by T. H. Morgan in 1910 led to the accumulation of large numbers of new spontaneous mutant types. Although these new mutations affected virtually every aspect of the flies' morphology, in some respects they were similar. In the first place, these new mutations were almost without exception recessive, and, second, from the developmental standpoint the mutational change in fact had a relatively slight effect on the total animal. No mutational change caused the sudden conversion of one species into a new one, or, for that matter, from one type in a species into another very dissimilar type. These two characteristics, recessivity and relatively slight developmental effects, were fashioned into a modified theory of evolution known as neo-Darwinism, according to which the developmentally small, usually recessive, mutational changes arising spontaneously and affecting parts of the animal at random were transmitted in a mendelian fashion and constituted the variability upon which natural selection could then act. Both of these properties of mutation were consistent with the observation that evolution must be a very slow process, for, although evidence for it could be unambiguously derived from a large body of biological and geological data, the instances in which speciation appeared to have been noted within recorded time were so few as to appear to be exceptional textbook cases rather than illustrations of a universal phenomenon.

THE PROBLEM OF RECESSIVITY. The fact that most changes appear to be recessive created a dilemma in evolutionary thought. If virtually all mutations of evolutionary significance were recessive, then natural selection would have to act on the homozygotes, a class of very low frequency. Thus, if one mutational event occurred in a diploid population of size 500,000, the gametic frequency of the allele would be one in a million, or 10^{-6}; the expected frequency of homozygotes for that specific allele would, in later generations, be the square of that allele frequency or 10^{-12}, i.e., once in every million million individuals, or, if each generation of that species consisted of 500,000 individuals, once every 2 million generations, since $5 \times 10^5 \times 2 \times 10^6 = 10^{12}$. It should be noted that for most species, other than large mammals, 500,000 must be a gross underestimate of population size and for each increase in population size by a factor of 10 the incidence of homozygotes for a specific mutated allele must decrease by a factor of 100. Further, regardless of how much better this new hypothetical homozygote should be, it would disappear into the gene pool to reappear some thousands or more generations later with an imperceptibly increased frequency of the homozygote.

Several extensions to the general theory took care of these points. For one thing, allele frequencies might gradually change by mutation of one allele to another, or *mutation pressure*. Furthermore, populations might be broken up for geographical or other reasons into small inbreeding units. These smaller populations would then give a better opportunity for homozygotes for recessive mutations to appear and would serve as pockets of variability within the overall population on which natural selection could operate. This division of the total species into smaller subpopulations would represent the optimal condition for survival of the species under conditions of unpredictable variation in environmental conditions in which some of the subpopulations would be better fitted for survival than others. With the expansion of experimental work on factors of evolutionary significance in many plants and animals, and the development of many new techniques for analysis, this generalization has been extended and sharpened over the years, particularly with new data on chromosome and allelic variability within natural populations, but in principle it has remained unchanged.

Factors Affecting Allele Frequencies

MUTATION.   Mutation, broadly defined to include chromosomal changes such as inversions and duplications, is the ultimate source of all genetic variability in a population. Genes and chromosomes are quite stable and remarkably effective in maintaining their precise individuality over a large number of replications. Thus, in humans, where there may be from 30 to 50 successive mitotic divisions in the germ cells in each generation, only one gene in 100,000 to 10 million, roughly, will suffer a detectable change after that many replications. The fact that most random changes in the gene are found to be deleterious has been discussed in Chapter 10.

The reason that they are recessive in the diploid cell deserves a comment. Whatever the function of the gene is—making a protein, sending out a signal to other genes, making other nucleic acids—it is generally the case that the cell (and organism) can get along reasonably well if only one, and not both, of the homologous genes at a given locus are active in this way. Therefore, a mutational change that inactivates, or lessens, the activity of a gene will appear to be recessive.

What is the frequency of potentially beneficial mutations? Most workers in the field make a rough guess of something like one out of every thousand mutational changes, but this is pure speculation. It may be argued that such mutations must be rare because mutation as a recurrent phenomenon would have produced such beneficial alleles many times in the previous history of the species, allowing for its earlier incorporation into the genotype of that species. Of course, mutations that at one time may have been deleterious or neutral may be beneficial at a later time superimposed on a new genetic background, or under new environmental conditions. Even so, the overall frequency of beneficial mutations must be very low.

The problem of the time of action of new mutational changes is an awkward one. In a single-cell organism, with no differentiation of cellular structure, a new mutant need function only under a limited set of cellular conditions. However, in multicellular organisms with a vast variety of tissue types, any beneficial mutational change would have to operate in precisely the right tissues. In other words, in addition to a potentially beneficial mutational change, an event of low frequency in itself, that change must also become operational in the right place at the right time. For instance, consider the case of a uniquely new mutational change which results in the synthesis of a hitherto unknown digestive enzyme in an animal. This enzyme would have to be produced in cells specialized in the digestive tract, where it would confer an advantage to the animal. However, if its action were not limited to these cells, but also extended to other cells in the system, the net result could be disastrous to the animal, which might suffer from having its own cells digested away. Thus, the problem of location and timing in highly organized multicellular systems adds an additional probability of low value that will decrease still further the chance of incorporation of a potentially beneficial change.

SELECTION.   The all-powerful force involved in evolution is natural selection, to which every individual of every species is exposed every second of its lifetime and by which, operating through the phenotypes of the individuals, specific genotypes are either discriminated against or allowed to produce more than their fair share of progeny. This leads to a change in gene frequencies in favor of those that are "best fitted." Although the force of selection is directly effective at the level of the organism and population groups, and indirectly on the geno-type and even more indirectly on specific genes, our understanding of how selection operates is sharpened if we assume that selection is operating directly on specific loci and momentarily overlook the obvious complication that an organism is more than the sum of its individual genes.

SELECTION AGAINST RECESSIVES.   The most common form of mutational change is that of the deleterious recessive. That all diploids normally carry such dele-terious recessives has been shown repeatedly in experimental organisms. For humans, these can be demonstrated by comparing viability of the progeny of first-cousin marriage with that of the general population. It is very important to understand that all people of all ethnic groups carry their fair share of poten-tially damaging genes. These are commonly referred to as the genetic load, i.e., the burden placed on future generations by our deleterious genes. This has been elaborated on in Chapter 10.

Because most recessive alleles of low frequency are found in heterozygotes, selection against homozygotes reduces allele frequencies slowly, the speed decreasing as the allele drops in frequency (Figure 24-1). This slowness serves as a strong argument against any program, involving any expense or effort, that would attempt to eliminate recessive alleles in a population by curtailing the reproductivity of the homozygotes.

Natural selection will constantly operate to reduce the frequency of deleterious alleles. They would be eventually eliminated completely except that mutation replaces the lost ones. As we have seen in Chapter 10, when the rate of loss of recessives by homozygosity is equal to the rate of gain by mutation, an equilib-rium condition for that allele is reached, whereupon the frequency remains stable. This is most dramatically illustrated by the sex-linked recessives with a severe phenotypic effect where the frequency of affected males is exactly equal to the mutation rate.

SELECTION FOR A RECESSIVE.   We have already noted that selection in favor of beneficial recessives is relatively ineffective because selection for a recessive depends on the production of the homozygote, which appears only very rarely for alleles of low frequency. Figure 24-2 shows the extremely low rate at which a gene with a hypothetical advantage of 1 percent would be incorporated into the

Figure 24-1 Graph showing the extremely slow decrease in frequency of a deleterious recessive that, when homozygous, completely wipes out the reproductivity of the affected individual. Starting with an initial allele frequency of one per thousand, it will take 1,000 generations to reduce it by 50 percent and 10,000 generations to reduce it to about 10 percent of its original frequency.

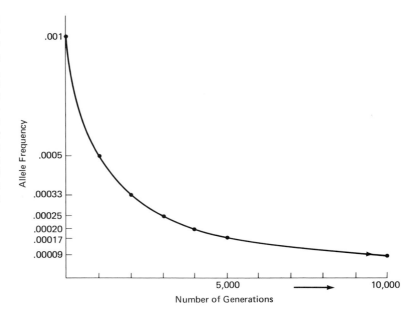

population. It may be wondered why, if such selection is going on at a large number of loci, we do not see evidences of it in the population at the present time. Possibly our analytic tools are not adequate to identify cases in which a specific allele confers some definite advantage to the homozygous individuals.

SELECTION AGAINST A DOMINANT. Selection against a deleterious dominant is immediate and decisive. If its effect is drastic, then selection eliminates the dominant allele immediately either by the loss of the individual carrying it or by his failure to produce progeny carrying it. Some unknown proportion of early abortions must be caused by new dominant lethal mutations.

SELECTION IN FAVOR OF A DOMINANT. On rare occasions a beneficial dominant mutation may arise. When it does, its frequency in the population will rise rapidly, depending on the extent of the greater reproductivity of the individuals carrying it (Figure 24-3). Incorporation into the population will be even swifter if the homozygote for the new dominant allele has a higher fitness than the heterozygote. It should come as no great surprise that the only case of an apparently beneficial dominant for which the population is not heterozygous is one in which the homozygote is actually less fit (i.e., sickle-cell anemia), this latter detrimental feature having prevented the complete incorporation of that allele into those populations.

Other Allele Interactions

HETEROSIS. When two inbred lines of plants or animals are crossed with each other, the F_1 will often exceed either of the two parents in physical vigor. This is known as *hybrid vigor* or *heterosis*. It is best known in its commercial applica-

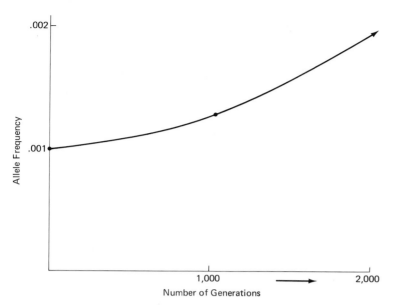

Figure 24-2 Selection for a recessive that endows the homozygote with 50 percent greater reproductivity than normal. Even with this high positive selective value, it will take approximately 2,000 generations to increase the allele frequency from .001 to .002.

tion in corn, where by trial and error with various lines of highly inbred corn it was found that certain crosses could produce hybrids with very high yields. (As a matter of practical procedure, four inbred lines may be used to get two hybrids which then produce further hybrids.) It has been suggested from time to time that hybrid vigor might play a role in human crosses. A few studies comparing physical characteristics (i.e., height) of previous generations with the present one suggest an increase not entirely attributable to improved nutrition and therefore possibly the result of the recent breakdown of isolated breeding

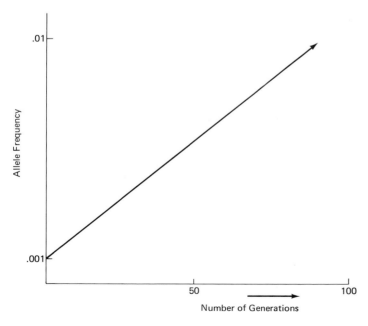

Figure 24-3 The immediate and rapid rise in frequency of a beneficial dominant. In the case chosen, the dominant allele is assumed to have a reproductivity only 10 percent greater than normal. In a few (about 100) generations, it will increase in frequency in the population tenfold.

groups. Some students of history interested in breeding patterns in different societies try to relate the blossoming and subsequent decline of civilizations to a change in heterozygosity, but there is little to support the hypothesis that mankind might now be improved, in physique, mentality, or any other way, if outbreeding increased, with a consequent increase in heterosis. Neither is there any reason to believe that an increase in inbreeding which would decrease heterozygosity and reveal more recessive traits would benefit mankind. Perhaps the amount of genetic variability within each human population group is so great that whatever advantage might be promoted by heterosis is already found within those populations.

POLYMORPHISMS.   Variation with respect to blood-group and histocompatibility antigens has been mentioned earlier. Polymorphisms exist for other gene products and are well known in those cases in which enzymes have been subjected to structural analysis, usually by electrophoretic mobility. More often than not, it can be shown that within a class of active enzyme there are some with demonstrably different mobilities; such equivalent enzymes are referred to as isozymes (iso = *equal*). Isozyme polymorphism has been found in virtually every organism investigated for its presence and has provided new insights into the relationship of organisms to each other and of different population groups within species. It is tempting to speculate that these polymorphisms represent one kind of genic variability involved in evolutionary progress. Another point of view is that these polymorphisms are neutral changes which do not modify the activity of the enzymes in question to the extent that they would be of any great evolutionary importance. In any case, these allelic polymorphisms do vary in frequency from one population group to another, and, since related groups are usually more similar to each other than unrelated ones, they provide a powerful tool for the anthropologist interested in determining the interrelations of various groups of people.

POLYPLOIDY AND GENETIC REDUNDANCY.   In the course of evolution there must occur from time to time an event that increases the number of loci in the genome: man must have more genes than a bacterium. A simple and effective means for accomplishing this is found in plants, where a doubling of the diploid chromosome number may produce a viable and fertile tetraploid individual. The establishment of a stable tetraploid line in plants is aided by three conditions not found in the higher animals. First, a single chromosome-doubling event in premeiotic cells of a plant can give rise to both pollen and eggs with that doubled chromosome number, and these gametes, being located on the same plant, can produce tetraploid individuals which, again producing gametes of both sexes, may perpetuate the polyploid line indefinitely. Second, a sex chromosome mechanism such as the X-Y combination in man is usually absent. When sex chromosomes are present in varying numbers in polyploids, they cause the development of intersexes and other sterile or poorly fertile types which make it more difficult for an isolated chance incidence of polyploidy to establish itself as a viable vigorous polyploid line. Third, vegetative reproduction in plants by runners, tubers, and so forth allows a single individual with an unusual chromosome number to proliferate without running either the hazards of meiosis or of the reproduction of deleterious sterile types produced when a gamete for a polyploid fuses with one from a diploid.

In animals such as man, in which the two sexes are represented in different individuals and in which sex is genetically determined, polyploidy is not a suitable means of adding to the gene pool. In any case, as we have seen earlier, polyploid fetuses in man are aborted early. Additions to the total number of loci must come primarily from the occasional accident of duplication of loci on single chromosomes. Such duplication makes it possible for one set of loci to mutate to undertake a completely new function, at the same time that the second set retains the original function, which itself may be quite essential for normal development (Figure 24-4).

In summary, we can be sure that the genetic structure of the human species is changing, although at a rate not perceptible to us in any obvious fashion. Although it is certain that the gene pool is not static, that some alleles are increasing in frequency and others are decreasing, for all practical purposes it may be considered to be in equilibrium. The major factor in altering gene frequencies is selection. Mutation is constantly introducing new alleles into the population in each generation. In some populations that are small or consist of very small isolates, chance may determine whether certain alleles will persist or be lost.

Primate Variability

TASTE SENSITIVITY.   It is interesting that primates other than man show gene diversity not unlike that of man. About 70 percent of American Whites have a strong revulsion to the bitter taste of a substance called phenylthiocarbamide

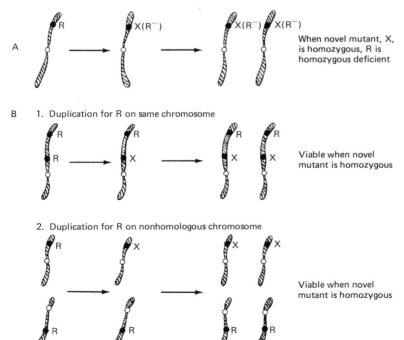

Figure 24-4 The value of duplications in evolution. *A.* Under ordinary circumstances, if a gene *R*, which is essential to the organism, mutates to another form *X*, with a completely different function, the new mutant cannot be incorporated because the homozygote, now deficient for *R*, will be lethal. *B.* However, if the locus is duplicated elsewhere, alleles at one of the two loci may mutate to a novel allele *X*, with a new function because of the presence of another set of *R* alleles.

(PTC); the rest find it relatively innocuous. The latter appear to be homozygous for a simple recessive allele found with a frequency of about .55. A large number of different primates have been tested for their ability to taste PTC. Chimpanzees appear to show a clear bimodal distribution of the ability to taste the substance, just as humans do, and in one pedigree two nontaster parents produced only nontaster offspring, similar to the situation in humans. In a small sample of spider monkeys (fewer than 12), no tasters were found, whereas all marmosets were tasters. It is not known why this polymorphism should be present in the primates, but it has been suggested that it is connected with their diet. Certain plants, relatives of cabbage and cauliflower, have a particular taste related to that of PTC and are known to affect thyroid disease in humans. One suggestion is that the effect of certain nutritional factors on the thyroid is the important element in maintaining this taster polymorphism in various members of the primates.

BLOOD GROUPS. Large-scale tests have been made on the primates for their blood-group antigens. In most cases, the common human antigens A, B, M, and N can be found in some if not all of the different primates investigated. It should be kept in mind, of course, that the tests here involve a test for agglutination of the erythrocytes of the lower primates when tested by the specific antibody active in man and this is no proof of the absolute identity of the antigens in both cases. However, for purposes of argument, we will assume that they are the same.

A large number of primates have been tested for the blood groups; the results are summarized in Table 24-1. Different primate groups are quite different and no single antigen of the ABO group is present in all primates. The gorillas are rather interesting in that their erythrocytes have neither the A, B, nor H antigens, although the saliva shows a presence of B and H-group substance, but not A. The MN antigens appear to be present in most of the primates (including chimpanzees and gorillas) but absent in others (some gibbons and orangutans), and one allele is present and the other absent in still other forms.

TRANSPLANTATION ANTIGENS. The rhesus monkey and chimpanzee both have been shown to have a transplantation locus of much the same sort as humans; in fact, chimpanzee serum contains antibody of specificity against HL-A cells

Table 24-1 The presence of the ABO antigens among primates other than man. The numbers tested are given for the first three listed; only the presence or absence is indicated for the remaining five.

Primate Tested	*Number Tested (when available)*	*O*	*A*	*B*	*AB*
Chimpanzee	250	35	215	0	0
Gibbon	52	0	10	20	22
Orangutan	26	0	22	1	3
Gorilla	—	+	—	—	—
Baboon	—	+	+	+	+
Macaque monkey	—	+	+	+	+
Rhesus monkey	—	—	—	+	—
Patas monkey	—	—	+	—	—

when tested against humans and two of the antigens present in the chimpanzee appear to be alternative alleles just as they are in humans. In one study involving 62 chimpanzees, there was evidence to indicate the likely presence of seven of the known human antigens and the possible presence of six others.

COMPARISON OF PRIMATE KARYOTYPES. Chromosome preparations of some of the primates show very strong resemblances to those of the human, and banding patterns make it possible to homologize the different chromosomes. As an example, in the case of the chimpanzee it is clear that there are few differences between their chromosomes and those of man. There appear to be at least six pericentric inversion differences, and one translocation, the latter probably being involved in the reduction of chromosome number from 48 to 46. In any case, except for the difference in total numbers, the chromosomes appear so similar that it would be difficult for anyone but an expert human cytologist to tell the difference (Figure 24-5).

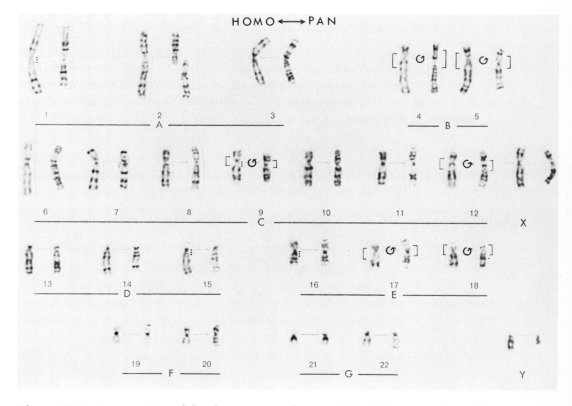

Figure 24-5   A comparison of the chromosomes of man and the chimpanzee, those of man being the left member of each pair, with those of the chimpanzee on the right. The only striking difference between the chromosomes is that the chimpanzee has two smaller chromosomes that correspond to the larger number 2 in man. The circular arrows indicate hypothetical inversion differences that have occurred as man and the chimpanzee both evolved from some more primitive primate. (From C. Turleau and J. de Grouchy. New observations on the human and chimpanzee karyotypes, *Humangenetik*, **20**:151–57, 1973.)

Human Population Differences

THE MEANING OF RACE. It is convenient to use the word "race" to refer to human population groups occurring in different parts of the world. Traditionally this word has referred to physical differences by which one group of people distinguish themselves from another group, independent of the extent of the genetic basis for these differences. We shall use the more precise definition of race provided by Curt Stern: a race is a group, more or less isolated geographically or culturally, who share a common gene pool and who, statistically speaking, are somewhat different at some loci from other populations.

Of the total number of loci possessed by the human species, somewhere between 10,000 and 100,000, all but a trivial number of their alleles must be common to all groups. There are a small number of alleles found primarily in one population and rarely in another. The skin color differences among African, Oriental, and European populations are obvious examples, but a list of such allelic differences would represent a minute fraction of the total loci. There is no case known in which any gene of significance is present in one racial group and not in others.

The word "race" has a distinct usefulness in the discussion of human populations; it is unfortunate that over the years it has developed connotations of prejudice and intolerance to the extent that some workers (particularly in the social sciences) are reluctant to use the word at all. However, avoiding the use of the word would require either the substitution of a synonym or the omission of any discussion of the genetic variability that characterizes different segments of mankind.

Table 24-2 The frequencies of the O, A, B, and AB blood groups in a selected sample of populations, some with greatly different frequencies.

| Population Tested | Total | Phenotypic Frequencies | | | | Allele Frequencies | | |
		O	A	B	AB	A	B	O
Iowa City, United States	49,979	22,392	21,144	4,695	1,748	.264	.067	.669
Nord, France	103,242	45,524	44,878	9,303	3,537	.271	.064	.664
Dublin, Ireland	36,879	19,981	11,919	3,926	1,053	.195	.070	.735
Armenians, U.S.S.R.	44,632	12,913	22,272	5,883	3,564	.351	.112	.537
Moscow, U.S.S.R.	63,549	21,598	23,932	13,105	4,914	.261	.154	.585
Uzbekistan, Asiatic U.S.S.R.	2,400	735	770	643	252	.242	.208	.550
Ibadan, Nigeria	26,027	13,411	5,559	6,038	1,019	.136	.146	.718
Bihar, India	9,257	2,909	1,995	3,551	802	.165	.273	.562
Eastern Islanders, Oceania	754	247	492	8	6	.418	.009	.573
Pure Cherokee Indians, North Carolina,	166	157	6	3	0	.018	.009	.973
Miyagi, Japan	240,204	32.3%	36.4%	22.8%	8.5%	Allele frequencies not calculated		

Table 24-3 The frequencies of the MN blood types in a sample of populations, some with widely different allele frequencies.

Population Tested	Total	Phenotypic Frequencies			Allele Frequencies	
		M	MN	N	M	N
New York City, U.S.A.	3,268	1,037	1,623	608	.566	.434
Danes, Copenhagen	4,319	1,270	2,147	902	.543	.457
Armenians, U.S.S.R.	5,728	1,743	3,222	763	.585	.415
Eskimos, Greenland	569	475	89	5	.913	.087
Papuans, Oceania	240	5	43	192	.110	.889

BLOOD-GROUP DIFFERENCES.   All population groups show some allelic diversity and the alleles they carry may have quite different frequencies. Tables 24-2, 24-3, and 24-4 give the frequencies in various populations of the ABO, MN, and Rh blood groups. Some of these differences are quite interesting. The Basques, an isolated population living in the Pyrenees, the mountain range between France and Spain, have the Rh^- allele of the Rh blood-group system with a high frequency. It is the only population known to have this allele with a frequency over 50 percent, whereas the frequency of that same allele in the American Indians and Australian Aborigines is zero. On the other hand, a blood-group allele known as Di^a is found only in Asians and American Indians. A wide array of other differences can be seen by examining the blood-group frequency tables.

DISTRIBUTION OF GENETIC DISEASE.   Some genetic diseases are very highly limited in range, such as Tay-Sachs disease, found among the Ashkenazi Jews from Central Europe and virtually not found in other groups of Jews or Caucasoids. Tuberculosis is less frequent among Jews than among other peoples, and scarlet fever is relatively rare among Asians and Blacks; this is evidence, along with the concordance rates in twin studies, that there exist genetic factors conferring resistance to these diseases.

Table 24-4 The frequencies of the Rh phenotypes in various populations.

Population Tested	Total	Phenotypic Frequencies		Allele Frequencies	
		Rh+	Rh−	D	d
Massachusetts, U.S.	91,817	76,415	15,402	.591	.409
Bohemians, Czechoslovakia	80,518	66,811	13,707	.587	.413
Spanish Basques, Pyrenees	480	343	137	.466	.534
Eskimos, Western Alaska	2,522	2,521	1	.980	.020
Javanese, Jakarta	48,964	48,964	0	1.00	0
Aborigines, Australia	281	281	0	1.00	0

Figure 24-6 The distribution of the allele for hemoglobin S, responsible in the homozygote for sickle-cell anemia, found in Africa mostly south of the Sahara and north of the River Zambezi, in all the major Mediterranean countries, in the Middle East, and in South India and Assam. (E. M. Edington and H. Lehmann. The distribution of hemoglobin C in West Africa, *Man*, **36**:34–36, 1956.)

By far the best known of the genetic diseases with an unusual distribution is sickle-cell anemia, caused by homozygosity for an allele producing abnormal hemoglobin. Although it occurs primarily in the Black population from Africa, it is also found in Italy and Greece and, with a lower frequency, in other parts of the world (Figure 24-6). Around the Mediterranean Sea there is a high frequency of an allele producing still another abnormal hemoglobin; this allele, like that for sickle-cell anemia, causes anemia (thalassemia) when homozygous. In each case it is thought that the heterozygote for these alleles has increased resistance to malaria (Table 24-5), thereby conferring an advantage to populations in malaria-infested regions of the world, despite the serious effects on the homozygotes (Figure 24-7).

CHROMOSOME POLYMORPHISMS. The discovery of the bands on mammalian chromosomes has made it possible to compare, in detail, the chromosomes of one form with another. The karyotypes of some of the other primates are quite similar to that of man (Figure 24-5). If the variability of chromosome structure among primates is quite limited, one might guess that there would be little difference in chromosome structure among different races of men, and this is precisely the case. Some chromosome polymorphisms are found with a low frequency, the level of which may be somewhat different from one group to the next. One study has shown that the Y-chromosome may be somewhat larger in Japanese than in other groups, but it should be kept in mind that this chromo-

Table 24-5 Apparent correlation of the presence of the sickle-cell allele, present in heterozygotes, with the severity of malaria in that region. (From A. C. Allison, in *Genetic Polymorphisms and Geographic Variations in Disease*, ed. B. S. Blumberg, Grune & Stratton, New York, 1961. By permission.)

Tribe	Locality	Number Tested	Percent Heterozygotes	Malaria Severity
Ganda	Kampala	334	19.5	+
Amba	Bundibugyo	220	39.1	+
Chiga	Kabale	206	1.0	−
Suba	Rusinga Island	173	27.7	+
Simbiti	Musoma, Kanesi	126	40.5	+
Sukuma	Mwanza	175	26.9	+
Kituyu	Nairobe	227	.4	−
Kamba	Machakos	213	0	−

some must be largely inert genetically. It is possible, even quite likely, that as cytological techniques improve some minor chromosome differences may be discovered between different races, but the failure to uncover any now with the methods presently available denies the existence of differences of any great extent.

OTHER INTERESTING DIFFERENCES. Japanese, Taiwanese, and Koreans have a much lower threshold for the effect of alcohol; after drinking a small amount that has no detectable effect on Whites, they may show marked flushing of the face and evidences of minor intoxication. This may be a genetic characteristic common to all populations of Oriental origin. The susceptibility of the American Indian (also of Oriental origin) to "firewater" has become legendary through innumerable written and filmed versions of the Wild West. A study of 3,000

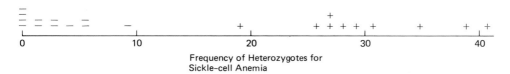

Figure 24-7 The frequencies of heterozygotes for the sickle-cell alleles, which give a direct measure of the allele frequency, compared with the relative severity of malaria, against which the allele may provide some protection. Twelve populations have low frequencies of heterozygotes, below 10 percent, and are found in regions of low (−) malaria incidence. Ten populations have frequencies of 19 percent or higher and are all found in regions with severe malaria problems. (Data from A. C. Allison, in *Genetic Polymorphisms and Geographic Variations in Disease*, 1961, ed., B. S. Blumberg. By permission of Grune & Stratton, Inc., New York.)

Indians from 16 tribal groups demonstrates that the dry cerumen (ear wax) that is characteristic of Orientals is also found among the American Indians. A study of 835 longshoremen from New York City showed that the Black workers had better hearing than individuals of Italian, Irish, or Jugoslav background. Some individuals are unable to digest ordinary cow's milk because they lack the enzyme lactase that attacks the sugar lactose and breaks it down. This problem becomes particularly acute for children between the ages of 2 and 4. These people are usually of Asian or African ancestry.

Human erythrocytes have an enzyme called pyridoxine kinase. In one test for this enzyme all Black subjects had an average activity of almost exactly half that of the White subjects, as though this particular locus were present twice in Whites. It has been suggested that perhaps this difference is responsible for the virtually complete immunity of Blacks to the type of malaria called vivax.

At the Kikuyu community outside of Nairobi, Kenya, 65 infants were tested for mental and motor skills at 2-month intervals up to the age of 16 months. The African infants surpassed the performance of U.S. children, both White and Black, on 38 items of the mental test and on 20 items of the motor test and were behind on only seven items of the mental test and two of the motor test. This difference was temporary, however. After a year or so, all of the children were about equal in performance.

Pigmentation Differences

Differences in skin color in the various races undoubtedly have an explanation in the different selection pressures that are imposed on the various phenotypes and therefore change the frequency of the genes responsible for the production of the pigment melanin. The most generally accepted theory is that the lighter skin colors represent an adaptation to conditions of scarce ultraviolet light where greater efficiency in converting and synthesizing vitamin D is made possible by the greater absorption of that light. The darker skins are more common in the tropical zones, serving as protection against the harmful effects of ultraviolet light (i.e., carcinogenesis) while still permitting the necessary level of vitamin D synthesis. Another theory suggests that the pigmentation provides a protective mechanism for the eye in those regions where solar radiation is at a maximum; still another that the presence of the pigments is correlated with changes in the immune system, providing greater protection against infection.

Different studies on the skin color of progeny of matings of individuals of different color have led to estimates of the number of loci involved from two to half a dozen, with the figure of three to four as the preferred number. If we assume that there are three loci, each with two alleles, the Black genotype would be represented by $AABBCC$ and the White by $A'A'B'B'C'C'$. These alleles are postulated to be additive and equal in effect on pigmentation, so that an F_1 of composition $A'AB'BC'C$ would be intermediate in color. Genotypes with any combination of three alleles from one race and three from the other, such as $A'A'B'BCC$ or $AAB'B'C'C$, possible from subsequent intermarriages, would have a similar phenotype.

Nothing is known of the genetic basis for the color differences between the Oriental and other races.

MISCEGENATION OR RACIAL MIXTURE. Up until several centuries ago, groups of humans maintained their individuality over long periods of time simply because of limitations of transportation and communication. At present, isolation is virtually impossible, and small isolates are becoming dispersed at a continually increasing rate.

There is no evidence that racial admixture or *miscegenation* (*misce* = mixture) is accompanied by any particularly positive or negative effects from the standpoint of genetics. We do not know of any instances in which miscegenation has been either advantageous or disadvantageous except possibly for those rare cases of crossing of groups that are grossly different in size. A pigmy woman, for instance, might have great difficulty bearing a child whose father is from a neighboring tribe with larger stature, because of the inadequacy of the maternal birth canal to allow passage of the large fetus; such a serious medical problem might limit the degree of race crossing. This represents an exceptional case, however, and in general it appears that race crossing is without serious consequences.

At first glance it might seem that with sufficient crossing between the races, over a long period of time, racial distinctions would disappear and the human population would be relatively homogeneous. This might be the case if inheritance were of a blending nature, but it is not. For instance, this would almost certainly not be true for differences in pigmentation.

Consider the consequences if two populations of equal size, one *AABBCC* and the other *A'A'B'B'C'C'*, married completely at random for several generations. Using our simple rules of probability, we can calculate that the chance that a person will have all six alleles of a specific type is $(\frac{1}{2})^6$, or $\frac{1}{64}$. Thus slightly less than 2 percent of the population would be of each of the original genotypes. Any assortative mating of similar phenotypes would increase the proportion of types at the extremes of the distribution.

These simple considerations make it clear that mankind cannot be molded into one phenotypically similar conglomerate; genetic diversity is a permanent feature of the human species.

THE DEGREE OF WHITE-BLACK ADMIXTURE IN THE UNITED STATES. In this connection, it is of interest to consider the amount of racial mixing that has occurred historically between the Whites and Blacks in the United States. This can be done by comparing the blood-group frequencies of these two groups along with the frequencies of alleles in the region of Africa from which the ancestors of most U.S. Blacks came during the days of slave trading in the New World. The method is simple. Suppose that one group has a frequency of an allele *L* of 100 percent and the other a frequency of 0 percent. Then if a hybrid population has a frequency of 25 percent it can be hypothesized that the population was made up of one quarter of the population of 100 percent and three quarters of the population of 0 percent. No other set of values will give the right proportion.

It is clear that we can directly calculate the proportions of two populations that must go into a hybrid population in order to give the gene frequencies found in it. An allele for a variant of the immunoglobulin molecule has a frequency of 100 percent in the African and 0 percent in Caucasian groups. In one Black population in the United States it has a frequency of 73 percent. This must mean that 27 percent of the allelic content of that population comes from the Whites. The alleles in the Duffy blood group can be used similarly. Here the frequency is

43 percent in Whites and 0 percent in the Blacks in Africa. The American Black population has an allele frequency of 9.4 percent, suggesting a 22 percent contribution from the White population to the Black. These figures may vary, of course, depending on the population being studied. A group of Blacks from South Carolina, studied by this method, showed only 6 or 7 percent White alleles. Information from several studies of this type may be combined graphically to illustrate the relative "distance" between different population groups (Figure 24-8).

It is not possible to make any corresponding calculations with any great validity on the proportion of Black genes in the White population of the United States because of several complicating factors. Much of the miscegenation occurred during the early days of slavery, whereas the population of the United States is derived to a great extent from the massive immigration from Europe that occurred in the nineteenth and early part of the twentieth centuries. The fact of pervasive racial discrimination in the United States insured that the extent of introduction of genes from the Black population into the White would depend to some extent on the frequency with which individuals with Black ancestors crossed the color line to become part of the White population.

The Intelligence Controversy

In recent years a number of psychologists have suggested that intellectual capability differs among various races and that this should be a matter of some concern; these pronouncements have been severely criticized by others in the field as evidence of racism, and a heated debate has developed. To clarify the basis for the controversy, it should be pointed out that when Black children take IQ tests they will, on the average, make lower scores than the corresponding age groups of White children. Goldsby, a well-known Black educator, has suggested that this difference may be about fifteen points. This can be interpreted in a number of ways. One of the common ways is to consider those environmental factors that modify individual performance (see Chapter 17) and to explain this difference on the basis of such factors.

There is no doubt that Black families are on the average more economically deprived, culturally disadvantaged, and generally discriminated against, with the result that the academic orientation of children during the early years of life will not be as great as in White families. Furthermore, one can argue that IQ tests are not in fact "culture-fair" but rather reflect the educational ideals of the middle-class White academic community.

A different point of view has been taken by an educational psychologist named Jensen, who holds that these differences arise in part from a basic genetic difference between Blacks and Whites and that early childhood schooling should be redesigned in such a way as to conform to the child's aptitudes instead of forcing children of both groups into the same mold. In principle, this is an idealistic argument suggesting that compensation in educational advantages be made for those with capabilities different from the average. In practice, however, this assumption would probably work in just the reverse fashion. Instead of providing more funds for the education of those with individual needs and providing curricula centered on the development of the maximal potential of each human, the

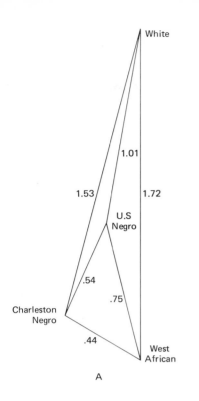

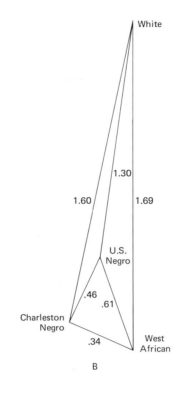

Figure 24-8 A diagrammatic scheme showing the relationship of four populations to each other, based on morphological data (*A*) and on allele frequencies for blood groups and hemoglobin types (*B*). (W. S. Pollitzer. The Negroes of Charleston [S.C.]; A study of hemoglobin types, serology and morphology, *Am. J. Phys. Anthropol.,* **16:** 241, 1958.)

predictable consequence would be to decrease money, facilities, and manpower available to the disadvantaged groups on the grounds that such resources should be preferentially provided to the most capable.

Another scientist (Shockley) upholds the validity of a difference of IQ performance as a basic measure of the innate difference in intelligence between Blacks and Whites and argues that a greater reproductive rate of the less capable in the population will inevitably have an effect on the average genetic endowment of the population. According to this argument, since the lower economic groups of the Black population whom he assumes to be less well endowed genetically have a higher reproductivity than any other groups in the country, the intelligence of the Black population will inevitably decrease. Furthermore, because of the greater reproductivity of Blacks than Whites (about 40 percent greater in the 1960's) the Black population will become a larger proportion of the total U.S. population, which therefore will become, on the average, less capable genetically.

The arguments against this point of view are several. We do not know to what extent in any population the "genes for intelligence" are stratified according to economic level. In fact, the intelligentsia of the United States at the present time are more often than not the children or grandchildren of rather poor, uneducated immigrants of 50 or 100 years ago. In the second place, it must be kept in mind that even though the average performance on an IQ test differs for the two races, there is considerable overlapping of the specific scores considered individually. Therefore, to single out individuals and subject them to different educational

Table 24-6 The relative change in fertility of women from 1966–1970 to 1971–1972, broken down according to the income of the family and color. (Reprinted with permission from F. S. Jaffe, *Family Planning Perspectives*, Vol. 6, No. 2, 1974.)

Income Status	Drop in Fertility in Percentages	
	White	Non-White
Very low income	−8.1	−14.0
Marginal income	−13.0	−14.3
Higher income	−8.7	−13.4

programs on the basis of race would be considered by most people unreasonable in principle and repugnant in practice.

Finally, reproductive trends are highly variable, changing markedly from one period to another for reasons that are not yet understood. During the last few years the birth rate of the Black population has declined considerably and, in fact, on a percentage basis has declined to a greater extent than that of the White (Table 24-6). If higher than average reproductivity is a general feature of the lower economic classes and if high reproductivity is considered a detriment to society at the present time, then it would seem clear that the humane approach would be to increase vocational or professional opportunities available to these low-income groups.

In general, the reaction of most biologists to these arguments that there exist important genetic differences in innate abilities between different races is that the evidence is not convincing, that the concerns are premature in view of changing social, educational, and reproductive patterns, and that, in any case, such statements aggravate social tensions in a troubled society, which is likely to react by denying minority groups precisely those advantages necessary for eliminating the gap between them and the more privileged.

References

ALLISON, A. C. 1964. Polymorphism and natural selection in human populations. *Cold Spring Harbor Symp. Quant. Biol.*, **29**:137–49.

BODMER, J. G., and W. F. BODMER. 1970. Studies on African Pygmies. IV. A comparative study of the HL-A polymorphism in the Babinga Pygmies and other African and Caucasian populations. *Am. J. Hum. Genet.*, **22**:396–411.

BODMER, W. F. 1970. Intelligence and race. *Sci. Am.*, **223**:19–29.

BOYD, W. C. 1952. *Genetics and the Races of Man.* Boston: Little, Brown.

CAVALLI-SFORZA, L. L., and M. W. FELDMAN. 1973. Cultural vs. biological inheritance. *Am. J. Hum. Genet.*, **25**:618–37.

CHIARELLI, A. B., ed. 1971. Comparative genetics in monkeys, apes and man. *Proceedings of a Symposium of Comparative Genetics in Primates and Human Heredity, Ernice, Sicily.* London: Academic Press.

CORLISS, C. 1974. Review: Masatoshi Nei and Arun K. Raychoudhury. 1972. Gene differences between Caucasians, American Negro, and Japanese populations. *Science,* **177**:434–39.

GOLDSBY, R. A. 1971. *Race and Races.* Springfield, Ill.: Thomas.

JENSEN, A. 1971. Can we and should we study race differences? Race and intelligence. *Harvard Education Review.* (See subsequent articles in same journal.)

KING, J. C. 1971. *The Biology of Race.* New York: Harcourt Brace Jovanovich.

MORTON, N. E. 1969. Birth defects in racial crosses. In *Proceedings of the Third International Conference on Congenital Malformations,* pp. 264–74.

MORTON, N. E., C. S. CHUNG, and M. P. MI. 1967. *Genetics of Interracial Crosses in Hawaii.* New York: Karger.

POLLITZER, W. S., E. BOYLE, JR., J. CORNONI, and K. NAMBOODIRI. 1970. Physical anthropology of the Negroes of Charleston, S.C. *Hum. Biol.,* **42:**265–79.

RACE, R. R., and R. SANGER. 1968. *Blood Groups in Man.* Philadelphia: F. A. Davis.

REED, T. E. 1969. Critical tests of hypotheses for race mixture using Gm data on American Caucasians and Negroes. *Am. J. Hum. Genet.,* **21:**71–83.

REED, T. E. 1969. Caucasian genes in American Negroes. *Science,* **165:**762–68.

Questions

Useful terms: Darwinism, natural selection, Lamarckian, mutation pressure, genetic load, heterosis, isozyme, genetic redundancy, race, miscegnation.

1. Describe the Darwinian and Neo-Darwinian views of the process of evolution and mention some criticisms of these ideas.
2. Do you consider the theory of evolution to be a valid biological explanation for the diversity of living things?
3. Why are new mutations more likely to be deleterious than beneficial?
4. Explain why selection for (and against) a new recessive allele is relatively ineffective, particularly when the allele has a low frequency. How does this differ from selection for or against a dominant allele?
5. Would the speed of evolution in a polyploid species differ depending on whether the effective mutations are of a dominant or of a recessive sort?
6. Do man and other primates show similar genetic variability?
7. Do you think that you would be able to distinguish a metaphase cell of a man from that of a chimpanzee?
8. What is the evidence that sickle-cell anemia is a penalty that some populations pay in return for an increased resistance against malaria?
9. If the human population should be completely panmictic over several centuries, would racial differences disappear?
10. How is it possible to calculate the extent to which the U.S. Black population has genes from the White population?
11. Do you think that anything is gained by a free discussion of the possibility that different racial or ethnic groups have different innate capabilities? Give arguments on both sides.
12. Discuss the relevance of the following conditions to the concepts discussed in this chapter: taster, Tay-Sachs disease, rubella, thalassemia.

25

The Chemistry of the Gene

Up to this point we have been concerned with the transmission of genetic traits in human populations from one generation to the next and their overall effects on human development. Now we will discuss the structure of the gene, cellular gene products, sequences of some mutant proteins, and the mitochondrial genes.

Historical Background

In 1913, Sturtevant showed that *Drosophila* genes were arranged in a linear sequence on the chromosome, and for years there was speculation about the kind of molecule that might make up the genes in that linear order. For about 30 years the most popular idea was that, of all the large molecules found in the cell, only proteins were complex enough to store the wide variety of genetic information that would be necessary to produce a complicated organism. Then in 1944 Avery and his colleagues devised an important experiment. They purified long molecules of DNA from normal bacteria and mixed this DNA with bacteria from a very stable mutant strain (one that had never reverted spontaneously to normal). They obtained normal bacteria from the mixture, demonstrating that the information for the normal characteristic was contained in the purified DNA, which had been able to enter and function within the mutant cell.

An understanding of how DNA could contain genetic information did not come until 1953, when, at Cambridge University, Watson and Crick discovered the structure of DNA, using new high-resolution x-ray diffraction patterns they borrowed from Wilkins, a worker at King's College, University of London.

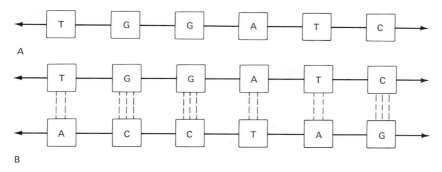

Figure 25-1 *A.* A small section of a single strand of DNA showing the linear attachment of molecules of the four subunits A, T, G, and C, apparently in no special order.

B. Diagrammatic representation of the rule that an A subunit on one strand will pair only with a T on an adjacent strand (and vice versa) and that a G on one strand will pair only with a C on the other (and vice versa).

The Structure and Function of DNA

Since DNA contains all of the information on how to construct an entire organism, it might be expected to be too complex a molecule to understand. On the contrary, its simplicity is striking.

A single strand of a DNA molecule is a long chain of just four subunits occurring repeatedly in what, at first sight, might appear to be a random order. We will use the abbreviations A, T, G, and C* (Figure 25-1A). In nature, however, DNA is found as two long DNA strands held tightly together by many very weak hydrogen bonds according to the following simple rule. An A on one strand binds weakly to a T on the other, and a G on either one of the two strands binds with a C on the other (Figure 25-1B). Any variations of this that occur can be considered to be mistakes. In addition, the strands can be considered to be directional, and we will use the number 5' at the beginning and 3' at the end of a sequence to indicate one direction, 3' at the beginning and 5' at the end for the reverse.

It is clear, then, that if we know the sequence on one strand as 5'---AATGCTAC ---3' we can state categorically the sequence in the parallel strand, in the case given, 3'---TTACGATG---5'. Since the sequence of the subunits on one strand predetermines the sequence on the other, they are said to be complementary to each other.

REPLICATION. Double-stranded DNA found in chromosomes in the nucleus can direct its own *replication*. If the two strands of a double-stranded DNA are separated and a new complementary strand is manufactured for each one, the result is two identical double-stranded DNA's (Figure 25-2). Description of the

---

*For the student with a better-than-average background in chemistry, A, T, G, and C stand for the DNA subunits adenine, thymine, guanine, and cytosine. More precisely, however, the subunits of DNA are deoxyadenylic acid (A), deoxyguanylic acid (G), deoxythymidylic acid (T), and deoxycytidylic acid (C). They are covalently linked through phosphodiester bonds between the 3'-carbon of one subunit and the 5'-carbon of the next.

Figure 25-2 Replication of a section of a DNA strand. When two complementary strands (*A*) separate (*B*), each can then form a new complementary strand (*C*). The end result consists of two double strands with exactly the same composition as the original.

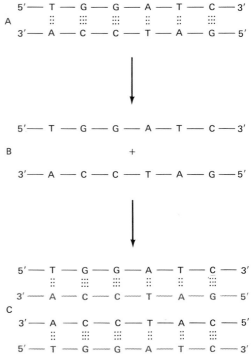

specific enzymes and the exact chemical steps involved in DNA replication is currently a subject of intense research.

Watson and Crick showed that the two strands of double-stranded DNA are not straight like a ladder but form a *double helix* resembling a spiral staircase (Figure 25-3*A*). The point on the DNA at which replication is occurring is called a *growing fork* (Figure 25-3*B*). The problem of exactly how the cell unwinds its long double-stranded DNA during replication is still a puzzling one.

RNA SYNTHESIS.   There is another kind of molecule, *ribonucleic acid (RNA)*, which is similar to DNA in that it is composed of four subunits. RNA is normally single-stranded and all of the RNA subunits are different from those of DNA. We will designate them A, U, G, and C.* It is synthesized in the nucleus enzymatically from one of the two strands of DNA such that it is complementary to that strand, according to the base-pairing rules in Table 25-1. The production of an RNA molecule complementary to a length of DNA is known as *transcription*. Some of the RNA synthesized in the nucleus is transported out to the cytoplasm, where it is utilized primarily in the production of proteins. Two structurally different types of RNA perform this function: *messenger RNA (mRNA)*, which constitutes the template from which the protein is manufactured, and *transfer RNA (tRNA)*, which performs the unusual task of determining which amino

*The subunits of RNA are adenylic acid (A), uridylic acid (U), guanylic acid (G), and cytidylic acid (C), not to be confused with DNA subunits. They are covalently linked through phosphodiester bonds between the 3′-carbon of one subunit and the 5′-carbon of the next.

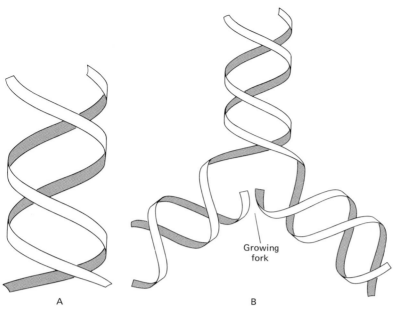

Figure 25-3 The basic spiral structure of the DNA molecule (*A*) showing how the spiral unwinds and replicates at the growing fork (*B*). (Redrawn from D. Fraser, *Viruses and Molecular Biology*, Macmillan, 1967. By permission.)

Growing fork

A

B

acids must be joined together in accordance with the sequence information of subunits on the mRNA. This conversion of the information of the sequence of the mRNA subunits into a specific chain of amino acid subunits is known as *translation*.

PROTEIN SYNTHESIS. There are 20 different amino acids in the long chains that make up proteins (Table 25-2). The total number, the proportion of each type, and their sequential order give the protein its specificity. This is determined by the order of the subunits A, U, G, C in the mRNA, synthesized on the DNA strand. However, since there are many more amino acids (20) than RNA subunits (4), some sort of system for specifying relationships, or *code*, is needed.

Let us think about how many RNA subunits would be needed in the code to specify each amino acid. A one-to-one relationship of one RNA subunit to one amino acid clearly would not be adequate, since then only four different amino acids could be specified, one for each RNA subunit. Pairs of RNA subunits would also not be enough to uniquely determine each amino acid because there are only 16 possible pairs to code for 20 different amino acids. On the other hand,

DNA Subunit	Complementary RNA Subunit
A	U
C	G
G	C
T	A

Table 25-1 Base-pairing rules for DNA-RNA complementarity.

Table 25-2 List of the amino acids found in proteins, along with the three-letter abbreviation of each.

Abbreviations for Amino Acids:

ala	alanine
arg	arginine
asn	asparagine
asp	aspartic acid
cys	cysteine
gln	glutamine
glu	glutamic acid
gly	glycine
his	histidine
ile	isoleucine
leu	leucine
lys	lysine
met	methionine
phe	phenylalanine
pro	proline
ser	serine
thr	threonine
try	tryptophan
tyr	tyrosine
val	valine

Other Abbreviation:

stop	termination of a gene

three RNA subunits would allow 64 (4 × 4 × 4) combinations, more than enough. In fact, it is the set of three or triplet of RNA subunits (called a *codon*) that codes for an amino acid. The correspondence is called the *genetic code* (Table 25-3).

The code is contained in the tRNA molecules, in cooperation with associated enzymes. One section of the tRNA recognizes a triplet of mRNA subunits and to another section an amino acid is enzymatically attached. The recognition of the RNA codon at the one end is readily accomplished by the presence of the complementary codon on the mRNA.

A chain of amino acids, or a *polypeptide*, is constructed with the aid of a structure called a *ribosome*. A ribosome is composed of many proteins and several RNA's called *ribosomal RNA* (*rRNA*). A mRNA molecule becomes attached to a ribosome and moves across it, its codons being "read" in a manner analogous to a tape being read by a tape head. In this way successive triplets are recognized by the recognition end of the appropriate tRNA and the amino acid at the other end is attached to the growing amino acid chain (Figure 25-4). This process continues until the synthesis of the chain of amino acids is terminated. Thus, the sequence of amino acids in a protein is exactly determined by the sequence of the subunits in DNA.

Table 25-3 The genetic code. The bases are those found on the messenger RNA—the DNA bases would be the complements of these. The amino acid specified by each group of three bases, or codons, is given in small letters to the right.

First Base	Second Base U		Second Base C		Second Base A		Second Base G		Third Base
U	UUU	phe	UCU	ser	UAU	tyr	UGU	cys	U
	UUC	phe	UCC	ser	UAC	tyr	UGC	cys	C
	UUA	leu	UCA	ser	UAA	stop	UGA	stop	A
	UUG	leu	UCG	ser	UAG	stop	UGG	try	G
C	CUU	leu	CCU	pro	CAU	his	CGU	arg	U
	CUC	leu	CCC	pro	CAC	his	CGC	arg	C
	CUA	leu	CCA	pro	CAA	gln	CGA	arg	A
	CUG	leu	CCG	pro	CAG	gln	CGG	arg	G
A	AUU	ile	ACU	thr	AAU	asn	AGU	ser	U
	AUC	ile	ACC	thr	AAC	asn	AGC	ser	C
	AUA	ile	ACA	thr	AAA	lys	AGA	arg	A
	AUG	met	ACG	thr	AAG	lys	AGG	arg	G
G	GUU	val	GCU	ala	GAU	asp	GGU	gly	U
	GUC	val	GCC	ala	GAC	asp	GGC	gly	C
	GUA	val	GCA	ala	GAA	glu	GGA	gly	A
	GUG	val	GCG	ala	GAG	glu	GGG	gly	G

PROTEIN STRUCTURE.   The specificity of proteins is a function of their three-dimensional structure, and this can be broken down into several orders of complexity. In the first place, the primary structure of the protein depends on the specific sequence of amino acids that make it up. On top of this is the secondary structure that comes about from a helical coiling (not to be confused with the helical structure of DNA) or bending of the polypeptide chain. This, then, is folded into a three-dimensional complex, the tertiary structure that exposes certain chemically active sites and makes the protein reactive, as, for instance, when a protein catalyzes a chemical reaction and is therefore an enzyme. Finally, several different polypeptide chains may come together to form a more complex protein. This association is called the *quaternary* structure. This three-dimensional structure of the protein is further determined by electrical charges, hydrogen bonds, and bonds between the sulfur atoms of different amino acids. Finally, in forming its complex, the protein molecule may pick up an external compound or atom that may then be responsible for the unique property of that particular protein. Thus, the protein molecule that makes up

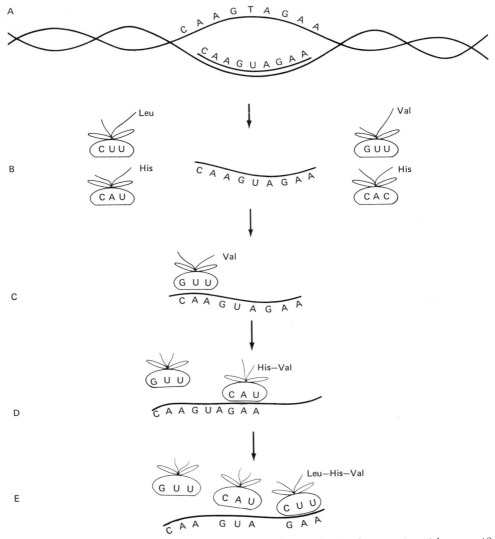

Figure 25-4 An idealized sketch showing the synthesis of a protein with a specific amino acid sequence determined by the coding on the DNA molecule. *A.* One DNA strand, at a region of separation from the other, determines the synthesis of a complementary messenger RNA (mRNA) strand (with the T of DNA replaced by U in RNA). *B.* The mRNA leaves the nucleus. Transfer RNA's (tRNA), of which only three types are shown, have a set of three bases at one end, while the corresponding amino acid is at the other. *C.* The first three bases of the mRNA (CAA) are complementary to GUU; its specific tRNA pairs with it and thereby brings in a molecule of *valine* for the protein. *D.* The next codon on the mRNA (GUA) is complementary to CAU, which characterizes the tRNA for *histidine*. Histidine is thus brought into position to join with valine. *E.* The process is repeated for the third codon (GAA) whose complementary tRNA (CUU) specifies *leucine*, which is then added to the growing polypeptide. Not shown is the ribosome, a much larger body in which these events take place. These three amino acids are found at the beginning of the β chain of the hemoglobin molecule.

hemoglobin will pick up heme, which carries iron, and this becomes an integral part of that molecule and is responsible for its oxygen-carrying capacity.

Chromosomes

We can see stained chromosomes at the metaphase of mitosis because at that time they are tightly packaged in a condensed form. Those chromosomes are made up primarily of DNA and proteins. In the normal human diploid cell, there are about 6 ft of double-stranded DNA, distributed into 46 chromosomes. The proteins include those that are responsible for the complex folding of DNA and the enzymes that are involved in replication and transcription. Some of the proteins binding to DNA probably determine which genes are turned off or are transcribed at any particular time. The geometrical and functional relationships of the DNA and proteins making up the chromosome are the subject of intensive research at the present time.

The Mitochondrial Genes

Virtually all of the genes in the human cell are in the chromosomes found in the nucleus, and, in fact, no abnormal human phenotypes have been shown to be due to genes in the cytoplasm. It has been shown, however, using biochemical methods, that there are more than 19 genes in the *mitochondria.*

Mitochondria are cytoplasmic organelles in which energy-yielding reactions occur. In each mitochondrion there is at least one circular DNA molecule (5 μm long in contrast to the 2 meters of DNA in the nucleus) along with the enzymes required for the replication of the DNA, transcription, and protein synthesis.

No genetic difference in man has yet been found which can be attributed to mutational changes in the DNA located outside the chromosomes, although such changes are known in other organisms. If such changes should be found, we might expect that they would appear to be maternally inherited, since the mitochondria are transmitted from one generation to the next through the egg cell.

Mutations

Several kinds of spontaneous changes in DNA sequence are observed with a low frequency. One DNA subunit may be replaced by another subunit. Consider a length of DNA of which one strand is transcribed to yield a mRNA, which in turn is translated into an amino acid chain. If there is a triplet AAA in the transcribed strand of DNA, then the mRNA contains the codon UUU. During DNA replication, if a mistake occurs to change the DNA triplet to AAG, the RNA triplet will become CUU, which codes for the amino acid leucine (see Figure 25-5; note that triplets in text are written 5′ to 3′). Leucine will replace phenylalanine at that position in the protein.

Furthermore, this mistake in DNA replication will be perpetuated, because when the strand containing the G subunit reproduces, the complementary strand

Figure 25-5 Without the mutation, the AAA DNA triplet is transcribed as the RNA codon UUU coding for phenylalanine in the protein. A mistake occurs during replication, replacing AAA with AAG in the transcribed strand. Note that the sequence is read from 5' to 3', in this case from right to left. During a second replication, the complementary CTT is synthesized off the AAG. Transcription of AAG yields CUU and hence leucine is put into the protein.

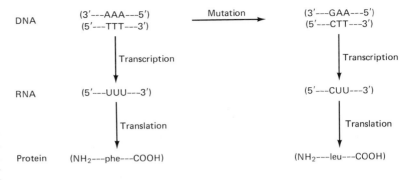

DNA (3'---AAA---5')
(5'---TTT---3')

Mutation →

(3'---GAA---5')
(5'---CTT---3')

Transcription

Transcription

RNA (5'---UUU---3')

(5'---CUU---3')

Translation

Translation

Protein (NH₂---phe---COOH)

(NH₂---leu---COOH)

will carry C instead of T, and all subsequent double strands derived from the original mistake will have a G–C pair instead of the original A–T pair.

The effect that such a substitution will have on the overall molecular structure of the protein is somewhat unpredictable. Some such changes may destroy the function of the protein; others will have almost no effect. Several hundred genetically determined variant hemoglobin molecules are known in which such a simple change has taken place.

If a DNA subunit is deleted, then the ribosome, translating sets of three subunits, gets out of phase after the deletion and the rest of the amino acid sequence is garbled (Figure 25-6). The presence of an extra DNA subunit would also put the triplet translation out of phase. Since the reading of the DNA subunits three at a time is analogous to viewing them through a frame three units long which is moved in jumps of three units along the DNA molecule, a change that throws this precise reading out of phase is called a *frame-shift mutation*. Deletions or insertions of larger numbers of DNA subunits could similarly cause gross changes in the polypeptide produced (unless a deletion or addition involved three, or a multiple of three, subunits, in which case one or more amino acids would be missing from the polypeptide chain, but the translation would continue accurately after the deletion or insertion). When the reading is thrown out of register

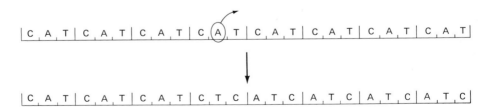

Figure 25-6 The effect of a base deletion on protein synthesis. In this simplified model, when the A is removed, the sequences, when counted off by threes from left to right, become ATC, and the new coding has completely different consequences.

in this way, the new triplets specify a new set of amino acids. The polypeptide chain may show no similarity to the original whatsoever, at least with respect to the section occurring after the change. Such a change, particularly if it occurs near the beginning rather than near the end of the gene, may produce an inactive product. No doubt many simple recessives originate in this way. On the other hand, the newly changed gene might, in rare instances, prove to have some unexpected usefulness to the cell or to the organism, in which case it would act as a beneficial dominant, if only one such new allele would be necessary to produce the new product. Whether such a new allele could replace the old in the population might very well depend on whether the old was dispensable (which most alleles are not) or whether there were other loci in the genome with similar functions (Figure 24-4). The speed of incorporation of such beneficial dominants would be very high, and we would not be likely to catch one in process of being incorporated unless there were some additional reason why it could not reach a frequency of 100 percent, as is the case with the gene for sickle-cell anemia, which happens to be quite deleterious in the homozygous state.

From this point of view, organisms with a large amount of genetic material present in duplicate (i.e., with *redundancy*) would be at an advantage in evolution. This redundancy could take the form of duplications of genes inevitably found in polyploids or in the duplicated segments within a chromosome. In this connection, it is worth pointing out that a calculation of the amount of genetic material in a cell of man compared to an estimate of the number of genes leads to the conclusion that only 5 percent or so consists of active genes with the other 95 percent consisting of superfluous DNA.

Another point of interest comes out of the consideration of base changes, particularly deletions, leading to mutational changes. It is clear that a single base loss can lead to the production of a protein, e.g., an enzyme, with partial or complete inactivity. This loss can occur at any one of a large number of positions along the DNA molecule, and the frequency of mutation from this event would depend on the total of such losses leading to a demonstrably different allele. However, a reverse mutation that would restore the initial condition of the DNA precisely would require a specific change that would essentially restore the original series of bases. For this reason, we might expect the frequency of true reverse mutations to be very much lower than that of forward mutations—perhaps so low that their effect in modifying gene frequencies in the human population would be insignificant.

XERODERMA PIGMENTOSUM. One defect that commonly occurs in DNA is the chemical change in two T's adjacent on a single strand, forming what is called a *dimer*. Such a dimer is produced when ultraviolet light hits the DNA. Clearly when two adjacent T units are linked to each other (Figure 25-7), reproduction of the complementary strand is impossible at this point and a defective strand is produced. There is an enzyme, normally found in all cells, which recognizes

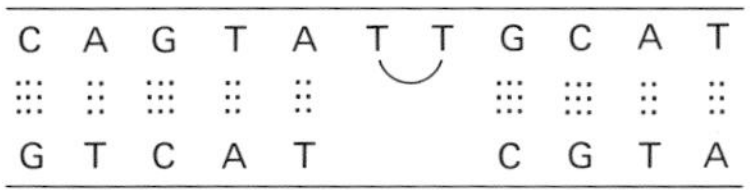

```
C A G T A T T G C A T
::: :: ::: :: ::   ::: ::: :: ::
G T C A T       C G T A
```

Figure 25-7 The linkage of two T bases to form a dimer. The strand produced by replication will be defective at this point.

such dimers in the DNA molecule and excises them from the affected strand of the double-stranded DNA. Then other enzymes resynthesize the correct sequence from the normal complementary sequence on the other strand.

Some individuals lack one or more of the enzymes needed to repair the dimer. After exposure to ultraviolet light, these people develop severe lesions, which eventually become cancerous and lead to death. This condition is called xeroderma pigmentosum (Figure 25-8). It is possible that a good part of the known carcinogenic effect of ultraviolet light may originate in the production of dimers in strands that happen not to get repaired and then produce defective DNA strands in the progeny cells.

The Hemoglobins

The hemoglobins found in human erythrocytes have been subjected to more detailed analysis than any other protein in any organism. At present, well over a hundred different hemoglobins have been found and more are constantly turning up. The most common variants include the one responsible for the dis-

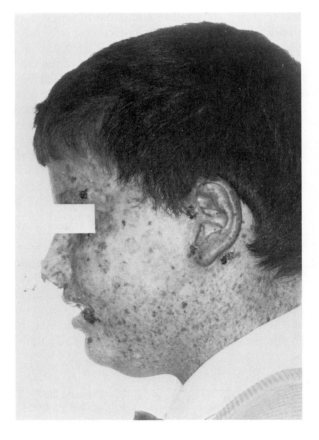

Figure 25-8 Xeroderma pigmentosum, caused in most cases by a demonstrable defect in a DNA repair system. In the early life of homozygotes, it is manifest as unusually heavy freckling in parts of the skin exposed to light. Heterozygotes are also heavily freckled, although not so severely as the homozygote, and have a normal life expectancy. (Courtesy of P. E. Polani, Guy's Hospital, London.)

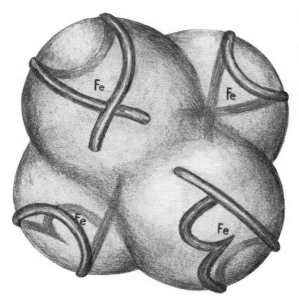

Figure 25-9 A simplified representation of the hemoglobin molecule. An alpha and a beta chain are combined in one direction and a similar chain is found at right angles behind the first. The heme groups which carry the molecules of oxygen are indicated in each chain by the symbol for iron, Fe. (Courtesy of L. Bernini, University of Leiden.)

ease sickle-cell anemia and another for thalassemia, a variant found primarily in people inhabiting the coasts of the Mediterranean.

NORMAL HEMOGLOBIN STRUCTURE.   Each hemoglobin molecule is composed of four chains, two of a type called *alpha* (α) and two *beta* (β) (Figure 25-9), and each of these four chains has an iron-containing segment called a *heme*. The heme is constant in all forms of hemoglobin; the variants found in the polypeptide, or globin, chains. Normal *adult hemoglobin, HbA*, consists of two identical alpha and two identical beta chains, with a structural formula $\alpha_2\beta_2$. These four chains in their final globular form, along with the hemes, have a molecular weight of approximately 70,000. The length of the alpha polypeptide is about 141 amino acids, the beta 146. A different form of hemoglobin, *fetal hemoglobin (HbF)*, is found in the fetus and is replaced by adult hemoglobin shortly after birth. Its alpha chains are like those of HbA but the beta chains are different and are designated as gamma (γ), hence the formula $\alpha_2\gamma_2$. The gamma chain of the fetal hemoglobin differs from the beta chain of adult hemoglobin by 39 different amino acids. It can be readily surmised from our knowledge of the production of polypeptides by codon sequences in the chromosome that the alpha, beta, and gamma polypeptide chains are synthesized by distinctly different gene loci.

ABNORMAL HEMOGLOBINS.   The study of abnormal hemoglobins began in 1949 when Pauling and coworkers showed that hemoglobin A and the variant found in sickle-cell anemia, hemoglobin S, could be distinguished from each other by their speed of migration in an electrical field. At about the same time, the clinical effect of this particular abnormal hemoglobin in producing serious anemia in homozygotes was pointed out by Neel.

The enzyme trypsin is used to cleave the polypeptide chains of hemoglobin, which it does at points where lysine or arginine is found, and the polypeptide is broken up into smaller fragments that can then be separated by allowing them to migrate at their own characteristic rates in a suitable medium. An electrical field is applied at right angles to give a two-dimensional separation. Figure 25-10 shows the fragment present in normal adult hemoglobin, but not in sickle-cell hemoglobin, and one present in sickle-cell and not in normal adult hemoglobin. This fragment comes from the beta chain of the hemoglobin and can be shown to result from the substitution of valine for glutamic acid at the sixth amino acid position in that chain. This is the only difference between the two hemoglobins, and all of the medical symptoms that come from the presence of sickle-cell hemoglobin are the result of what appears to be at first sight a trivial substitution of one amino acid for another. Under conditions of a low oxygen pressure, the HbS molecules aggregate and form long rodlike conglomerations that distort the normal spherical shape of the erythrocytes and cause them to adopt elongated or sickle shapes. An individual who is heterozygous for the gene for sickling produces two kinds of hemoglobin, HbA and HbS. Those homozygous for the allele produce only HbS.

For a reason that is not clear, the vast majority of the abnormal hemoglobins have their amino acid substitutions in the beta sequence. However, alpha chain variants are known and in one case a single family showed both an alpha chain (Hb-Hopkins-2) and a beta chain (HbS) abnormality. The independent segregation of these two abnormalities in the sibship showed that the genes responsible for the alpha and beta chains must be at separate loci, and either some distance from each other on the same chromosome or on separate chromosomes. This is confirmed independently by the allocation, by other methods, of the locus of the alpha gene to chromosome 4 and the beta gene to chromosome 2.

Proteins of the Human Blood Serum

When the erythrocytes and other cells form a clot with the help of fibrinogen, the liquid that remains contains a large number of different proteins and is called blood serum. These proteins may be separated by a number of techniques. One of the most effective is to place them in solution in an electrical field, where they move through a medium such as a starch gel with a speed that depends on the charge they carry. In this way, proteins with different charges may be physically separated from each other. Figure 25-11 shows a schematic representation of the kinds of proteins that may be separated from each other by a difference in charge at the two ends of such a solution. Of particular interest to us are the *haptoglobins* (alpha globulins), the *transferrins* (beta globulins), and the *immunoglobins* (gamma globulins).

HAPTOGLOBINS.   Haptoglobins are globulin proteins with the apparent function of binding hemoglobin that is released by disintegrating erythrocytes, thereby preventing the excessive loss of this valuable protein by excretion. Two of the haptoglobin types differ only in that one migrates somewhat faster than the other in a starch gel; therefore, one is called HP^{1F} and the second HP^{1S}. HP^{1F} and HP^{1S} are alleles of the *HP* locus. Analysis of the polypeptide sequence of HP^{1F} and HP^{1S} shows that they differ at only one amino acid.

Figure 25-10   The separation of the components of normal hemoglobin (*A*) and hemoglobin S (*B*), showing that they differ by only a single group. (C. Baglioni. An improved method for the fingerprinting of human hemoglobin, *Biochim. Biophys. Acta*, **48:**392, 1961.)

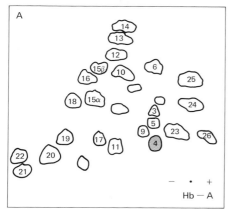 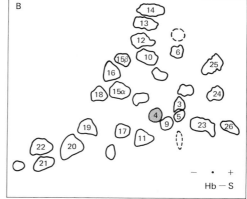

HP^2 is a distinctly different allele, producing a polypeptide almost double the length of HP^1 and apparently, consisting of sequences of both HP^{1F} and HP^{1S}. It has been suggested that a breakage and fusion of strands of two homologous chromosomes could give rise to a gene of the composition that HP^2 has. The HP^2 allele has reached a very high frequency in some populations, notably Orientals and Caucasians. If the allele arose only once, as its probable manner of origin suggests, then it is curious that its frequency is high. There is little evidence that such an allele could have any kind of selective advantage, in view of the fact that populations which appear to be almost or totally devoid of haptoglobins in their serum, the case for a few Black populations, do not show any deleterious effects.

TRANSFERRINS.   Transferrins are proteins (Figure 25-11) that function in binding iron and transporting it from the plasma to the cells of the bone marrow, where it is used in manufacturing hemoglobin. There are a large number of variants. One of them (TfC) is very common and the others (TfB, TfD, etc.) are relatively rare. Because of the rarity of the alleles other than *TfC*, virtually all phenotypes other than homozygous *TfC* are heterozygotes with a *TfC* allele along with one of 18 other alleles.

   There is one case reported of an individual who lacked transferrins entirely. She suffered from severe anemia at the age of 3 months, required frequent blood transfusions, and died in early childhood with severely damaged internal organs from a massive accumulation of iron. Her mother had three other pregnancies that produced inviable fetuses. It is possible that the two parents were both heterozygous for a rare recessive gene that failed to produce normal transferrins.

IMMUNOGLOBULINS.   The immunoglobulins are the serum proteins with antibody properties. They are found in five different classes—IgG, IgM, IgA, IgD, and IgE. The last two occur at a very low level in the blood plasma and not much is

Figure 25-11 The separation of a heterogeneous mixture of blood serum proteins by electrophoresis, showing the main categories identifiable in this way.

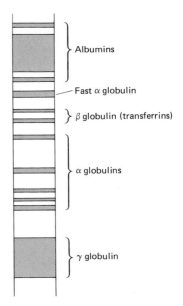

Albumins

Fast α globulin

β globulin (transferrins)

α globulins

γ globulin

known of their function. Of the other three, IgG is the smallest and is the only type that regularly crosses the placenta. Hemolytic disease of the newborn is the responsibility of the IgG antibodies, whereas hypersensitivity (allergies) involves the IgE antibodies.

In all immunoglobulins, two heavy and two light polypeptide chains form the basic structural unit (Figure 25-12). There exist three different classes of heavy chains, and the light chains come in two different varieties that are found associated with all three heavy classes. IgM has a high molecular weight because five of these units are held together by common sulfur bonds; IgG consists of two structural units hooked together. IgA, the major antibody found in secretions, may consist of a single unit, or of two, three, or four units joined with sulfur bonds.

In amino acid sequence studies of light or of heavy chains, part of the polypeptide appears to have a relatively constant sequence from one molecule to another within the same individual, whereas the other part has a relatively variable sequence. It is the variable portion of that molecule (Figure 25-12) that gives it its characteristic antibody specificity.

How the numerous immunoglobulin chain sequences arise is still not understood. Those theories which have gained wide acceptance at the present time

Figure 25-12 A schematic for the gross structure of the immunoglobulin molecule showing the two heavy chains, the two light chains, and the portion of the molecule with variable structure responsible for the antibody specificity of the molecule.

Heavy chain

Light chain

Variable half

involve the assumption either (1) that all of the chain sequences are inherited, the diversity of specificity being derived from the manner in which the peptide sequence is determined (constant and variable regions might be spliced together in some fashion) and from the particular combination of light and heavy chains in each antibody, or (2) that only a few chain sequences are inherited and diversity arises from somatic mutation or from recombination of similar sequences.

References

ALLFREY, V. G., and A. E. MIRSKY. 1961. How cells make molecules. *Sci. Am.*, September.

BROCK, D. J. H., and O. MAYO. 1972. *The Biochemical Genetics of Man.* New York: Academic Press.

CLEAVER, J. E. 1968. Defective repair replication of DNA in xeroderma pigmentosum. *Nature,* **218:**652–56.

CRICK, F. H. C. 1954. The structure of the hereditary material. *Sci. Am.*, October.

CRICK, F. H. C. 1962. The genetic code. *Sci. Am.*, October.

EDLIN, G. 1972. Molecular genetics—a survey of highlights. *BioScience,* **22:**77–81.

FRUTON, J. S. 1950. Proteins. *Sci. Am.*, June.

GIBLETT, E. R. 1969. *Genetic Markers in Human Blood.* Philadelphia: F. A. Davis.

HARRIS, H. 1970. *The Principles of Human Biochemical Genetics.* Amsterdam: North-Holland.

HURWITZ, J., and J. J. FURTH. 1962. Messenger RNA. *Sci. Am.*, February.

KENDREW, J. C. 1961. The three-dimensional structure of a protein molecule. *Sci. Am.*, December.

STENT, G. S. 1971. *Molecular Genetics.* San Francisco: Freeman.

WATSON, J. D. 1976. *Molecular Biology of the Gene,* 3rd ed. New York: Benjamin.

Questions

Useful terms: double helix, growing fork, transcription, messenger RNA (mRNA), transfer RNA (tRNA), ribosomal RNA (rRNA), translation, genetic code, codon, complementary codon, polypeptide, mitochondrial gene, dimer, haptoglobins, transferrins, immunoglobins.

1. Why was it thought in the earlier days of genetics that the genetic material must be similar to protein? What was the first conclusive evidence that it is DNA?

2. With a simple diagram illustrate the complementary nature of the two chains that make up the DNA molecule and show how this complementary nature makes the precise replication of the DNA thread possible.

3. How does RNA differ from DNA? What are the three major types of RNA and what function do they perform?

4. What is meant by the genetic code?

5. Do proteins have a structure similar to that of DNA?

6. Show very simply how changes in the DNA molecule may lead to a new mutation. What are the various kinds of changes called?

7. What kind of change in the DNA molecule can ultraviolet light produce leading to skin cancer? What is the name of the disease that afflicts people unable to repair this damage?

8. How does the abnormal hemoglobin present in sickle-cell anemia differ from normal hemoglobin?

9. Discuss the pros and cons of the possibility of making specific changes in the DNA molecule in order to repair defective genes in mammals.

10. What is the relation of the chromosome and the DNA contained in it?

11. Would the chromosomal theory of inheritance be disproved if it could be shown that some basic characteristic were transmitted from one generation to the next entirely by way of the female sex?

12. Discuss the relevance of the following conditions to the concepts discussed in this chapter: xeroderma pigmentosum, sickle cell anemia.

26

The Manipulation of Man's Genetic System

Biological Manipulation in General

Within recent years much attention has been focused on the possibility of altering the genetic components of the cell to achieve some desirable end. To be sure, the manipulation of the external environment to improve other species has always occupied man's attention—in agriculture, for example. But to reach within the cell and, in one way or another, to alter the genetic mechanism so that it is able to fulfill some function not previously possible vastly expands man's potentialities to control his destiny, from the practical viewpoint, and provides a new outlet for scientific creativity, from the esthetic viewpoint. In popular terms, such experimentation is generally referred to as "genetic engineering."

ANIMAL AND PLANT BREEDING IN THE PAST.   Genetic engineering has been going on since prehistoric time. The domestication and selection of breeds of animals have been carried on for thousands of years. Babylonian records show that artificial pollination of the date palm occurred several thousand years B.C., but it may not have been immediately obvious to the primitive agriculturist that the production of the new generation depends on a contribution from each of the two parents.

It seems intuitively obvious that two individuals will produce progeny more similar to themselves than to other individuals taken at random from the population. Simple selection for specific traits is the means by which large numbers of

439

varieties of cultivated plants and animals came into existence. It is only within the last century that our knowledge of genetics has advanced to the point at which more efficient mating schemes have been developed. Professional plant and animal breeders have developed efficient and intricate mating systems for manipulating the genetic constitution of living things to produce what is, in their opinion, some desired end result. Perhaps the best known case is that of hybrid corn, in which highly inbred strains, relatively weak and poor-yielding plants, are mated to each other to produce hybrids that far surpass the parental types in yield. The point here is not so much that this has been and is being done, with great success, but rather that this biological experimentation is socially acceptable and, in fact, highly valued by society. No one has seriously criticized the production of high-yielding grain crops, for instance, on the grounds that such genetic manipulation interferes with God's or Nature's original design of living things.

THE SPECIAL CASE OF HOMO SAPIENS. However, these comments do not apply to humans who occupy a special place in their own concept of the universe. There are many reasons for why it would be difficult, even impossible, to apply simple genetic principles to the breeding of people, but even if it were possible there are restraints of many kinds (including persuasive ethical considerations) that limit the extent to which we may feel free to tinker with our own constitution. Nevertheless, with the present high level of sophistication in the biological sciences, some possibilities for genetic engineering to benefit the human population are seriously being considered. In this chapter we shall discuss some of these, as well as some of the many problems that must be faced before any of these procedures could be considered to be a practicable possibility.

One popular conception of the ultimate goal of the geneticist is that of drastically altering the human species, a view promoted by multitudes of low-cost Hollywood films. On the one hand, it is supposed that the scientist's ideal is to make all individuals absolutely identical, or, on the other, that it is to create multitudinous human freaks.

The fact is that geneticists, being human themselves, feel that the human race as it is presently composed would not be improved much by tampering with its phenotype. Little would be gained by adding another eye, a couple of antennae, or a pair of wings. Technological advances have made many of the improvements of science fiction obsolete. No pair of wings could compete with a jet aircraft; telepathy is accomplished far better by the selective communication made possible by the telephone, radio, and television than by ESP. We can survive at temperature extremes, at the bottom of the ocean, and in outer space; changes in the genome that might promote adaptations to such conditions would be superfluous. At the present time, in fact, most thoughts about the alteration of human genetic make up are directed toward the alleviation or elimination of genetic defects that disturb our well-being.

PRESENT PROBLEMS AND FUTURE PROSPECTS. A list of the new approaches that are either with us now or will undoubtedly be available to mankind in the near future—almost certainly within the next century—would include the control of sex of offspring, the use of sperm banks to provide for fertilization of eggs in special cases, the transplantation of organs from one person to another (already discussed in Chapter 9), the implantation of fertilized eggs into a foster mother, the modification of prenatal and postnatal development to provide for "better" individuals (*euphenics*), the production of new individuals with the same genetic composition as that of a presently existing one (*cloning*), and the manipulation of the genetic material either to repair defective genes or to improve the performance of presently existing ones.

As we consider each of these aspects of possible future applications, we will be faced constantly with a number of questions. Who will decide what is a desirable or an undesirable trait? How is it possible to guard against an "incorrect" determination when there is no hard and fast basis for that judgment in the first place? When does life of the human start? Is it permissible to destroy a sperm and an egg but not a fertilized egg, an embryo, or a fetus? Where does one draw the line? Would a program to improve the state of humanity likely diminish our sense of dignity as human beings by putting us at the same level as economically useful plants or animals? Would any plan for the "improvement" of the human race simply reflect the fears, biases, and prejudices of the individuals making the decisions? If a program of therapy must be limited to a fraction of those who would probably benefit from it, who will decide which individuals will receive the treatment and which will not? Are there any foreseeable long-range problems that might develop as the consequence of a certain line of research, so potentially hazardous that that particular line of investigation should be prohibited? Might the proposed procedure conflict with the religious or ethical beliefs of a sizable fraction of the population and if so to what extent would society have the right to curtail the freedoms of the individual for the purpose of "improving" the human race?

These, and many other questions like them, are neither trivial nor rhetorical, nor do they have a specific "right" answer. The response to any of them will depend on the society at that time, the prevailing religion and philosophy of life, and, at a more commonplace level, the age and politics of the individual.

Selection at the Level of the Gamete

CONTROL OF SEX. In principle, choosing an offspring of a desired sex is possible at the present time, since the sex of a fetus may be ascertained as early as amniocentesis can be performed. However, few persons would agree that the sex of a fetus is, in itself, a matter of such importance as to justify amniocentesis, and even fewer would agree that the "wrong" sex of a fetus would be sufficient cause for a "therapeutic" abortion.

With astonishing regularity, reports are published both in the daily press and in reputable scientific journals suggesting that the sex of an offspring may be predetermined by alkaline or acid treatments of sperm or by physical separation of X- and Y-bearing sperm prior to fertilization of the egg. Even if these reports should not be verified by more extensive and strictly controlled experiments, it is likely that a procedure will be developed in the near future that will change the present approximately equal probabilities of a boy and a girl. Whereas it was necessary previously to undertake actual breeding experiments to test schemes for the differential selection of one kind of sperm over the other, the discovery of the fluorescence of the Y-chromosome within the sperm after quinacrine staining makes it possible to evaluate new techniques for separation of X and Y sperm much more effectively. There is one foreseeable problem. If the separation depends on physical differences between the X and Y sperm, such as the size or volume of the sperm head, then aneuploid sperm, those with an extra autosome, or with one missing, would necessarily be selected at the same time that the X and Y sperm were being selected, since they would have similar size and volume differences. For this reason, we might expect that such a separation procedure used to increase the probability of fertilization by either an X- or a Y-bearing sperm would, at the same time, increase the likelihood of a grossly abnormal embryo.

It is interesting that a procedure like the choice of sex of offspring, which was considered to be a bold spectacular advance not too long ago, is now theoretically possible by amniocentesis, but raises little more than subdued yawns from the population at large, reminiscent of the reaction of the world to the fifth walk on the moon. Perhaps there is a lesson to be learned here—that some of the anticipated advances will turn out to be not quite so exciting in the future as we now imagine they will be.

SPERM BANKS. In the 1930's, H. J. Muller suggested that some benefit to society could result from the creation of sperm banks where the sperm of distinguished males could be preserved for later use. (In theory a similar system might be set up for the preservation of eggs of distinguished women, but the techniques for doing this have not yet been developed.) At the present time, in the United States, there are tens of thousands of cases of artificial insemination per year, applied in most cases to women whose husbands are sterile. In such cases the sperm may be taken from males at random in the population or, even worse (as Muller slyly remarked), from medical students who are remunerated for their contribution. Although it is true that these anonymous donors may be chosen so that their phenotypes bear some resemblance to those of the wives' husbands, in order to insure that the F_1 at least not be strikingly different from the expectation of a natural child, it is certainly the case that the donors are virtually never checked for the presence of alleles of even the most common genetic diseases that can be readily detected in the homozygous or heterozygous state. Would it not be wise, asked Muller, to select as sperm donors individuals who have been shown to be free of genetic defects and who, further, are not run-of-the-mill, as most of us are, but are outstanding in some respect that might be attributable in part to an unusual genotype? Thus one might preferentially choose the sperm of individuals who are distinguished as composers, humanitarians, or authors, rather than mental incompetents, bums, the chronically unproductive, and so on, although, to be sure, these two lists overlap to a good extent.

It would, of course, be simple-minded to imagine that such germinal selection would produce the characteristics of the individual from whom the sperm were obtained. In the first place, the sperm would be a random haploid gene complement of the diploid condition that made up the individual. In the second place, virtually all of the characteristics that make one human seem superior to another are multifactorial in nature, to whatever extent that "superiority" is genetically determined. Nevertheless, it would seem a simple matter of common sense to suggest that if artificial insemination is to be a widespread practice the donor sperm should come from individuals who have been checked for deleterious alleles, insofar as that is possible. And, for those whose philosophy includes the belief that there is a strong genetic component to ability and accomplishment, the selection of "superior" donors holds out the additional hope that perhaps sometime in the future a new unusual genotype, including some genes perpetuated by the selection of donors, will appear to make some extraordinary contributions to the human race.

Asexual Reproduction

PARTHENOGENESIS.   The production of a new individual from a single cell, without the combination of genes from two parents, is known as parthenogenesis. This may take several forms.

It might be possible to induce an unfertilized egg to initiate cleavage divisions. An embryo with this origin would be haploid; from what we know of the large number of deleterious genes present in every human, we can expect the haploid to be inviable, and, in fact, no haploid humans, or even mosaics partly haploid, have ever been seen. In another type of parthenogenesis, unfertilized eggs fuse with other meiotic products (i.e., a polar body or another egg) to restore the diploid condition. Such a zygote would have a greater chance of being viable, depending on the specific meiotic products so fused. Still another possibility would involve the induction of an oogonial (and therefore diploid) cell to behave like a fertilized egg and to commence cleavage. However produced, the cells so induced to undergo parthenogenesis could then be returned to the mother, or to a foster mother.

Any embryo developing from these events could have only maternal genes and any resulting individual would probably bear an unusually strong physical resemblance to the woman providing the egg. On the other hand, depending on the source of the cells involved in the fusion event, there would be a high likelihood of homozygosity for a good fraction of the total genome, since there would be no way of insuring that the two cells did not carry identical alleles at a given locus. The hazard of producing a defective embryo is so great that any efforts to induce artificial parthenogenesis in humans would be very risky, indeed.

CLONING.   In cloning, the nucleus is removed from an ordinary somatic cell and is implanted into an egg, which develops to produce an individual with precisely the same genes as those in the cell from which the nucleus was taken. The individual produced would then be the "identical twin" of the person from whom the cell was removed. This has been accomplished in the frog, where nuclei taken from cells in the intestine have replaced the nucleus of an egg, and the egg has then developed to produce a normal mature animal. It is relatively simple in the frog because its eggs are very large, as is known by anyone who has looked at the gelatinous masses, dotted with frog eggs, in brooks and pools during the spring of the year. The operation of removing the nucleus of the frog egg is technically difficult but not impossible. Furthermore, frog eggs ordinarily develop to maturity in a pond, where there are wide fluctuations in conditions such as temperature and surrounding media, so that there is no problem in persuading them to develop in the laboratory. On the other hand, the mammalian egg, as we have noted before, is quite small, demands very special conditions, and must be implanted into a human hormonally receptive to the development of an embryo.

These problems, however, are matters of technique. Although it may be difficult to insert a needle into the small mammalian egg to remove the nucleus and to replace it with one from another, there is no reason, in principle, that the nucleus cannot be effectively removed by destroying it, say by an intense beam of radiation, with a foreign nucleus being subsequently introduced by the cell fusion technique described earlier. Thus such an egg, induced to undergo cleavage and put into a suitable environment (i.e., a foster mother), could theoretically give rise to a newborn genetically identical to the adult from whom the nucleus came. Even though full success has not yet been achieved in this area because of the problems of technique, it seems reasonable to predict that within 10 or 20 years this will be a feasible procedure.

Besides any philosophical, ethical, or religious objections that might arise from the possibility of humans being produced by unorthodox methods, some apprehensions stem from the thought that it might become possible to populate the world with specific genotypes. Thus, one unpleasant picture is that of aggressive nations creating multitudes of physically capable, somewhat unintelligent, and psychologically abnormal individuals as professional armies. Equally repelling is the thought that powerful egotists might decide to create many individuals patterned exactly after themselves. Because of such potential problems, many concerned scientists have suggested that measures should be taken now to prevent any further development of such techniques.

Probably some of the distaste for the application of techniques like cloning is unwarranted, at least to the extent that it is imagined that a group of persons with identical genotypes would be exact carbon copies of each other. There is also the implication that such a group would somehow form a psychic bond, not just among themselves, but particularly with their donor, whose motive might be to achieve some measure of immortality. From our study of data from identical twins, even those raised together, the concordance of psychological traits is imperfect at best. It is quite likely that if a group of cloned individuals were ever to be produced they would, as they diverged psychologically during development, consist of a wide assortment of types and not be simple replicas of each other.

At first sight, it might appear that techniques that would increase the population size should be discouraged. Why should science and society make an effort to develop procedures that would inevitably accentuate population problems to the detriment of humanity when child production is already carried out very efficiently and without great expense? The best answer to this question is that techniques like cloning have a greater usefulness in other ways. The greatest problem facing the biologist at the present time is that of differentiation—why certain cells become muscle and others nerve for instance, but also why some cells become cancerous and others fail to differentiate normally at all. The understanding (and possible cure) of a large number of diseases depends on mastering such basic biological phenomena. It seems certain that the experimentation on nuclear transplantation, with the knowledge about differentiation that will come from it, will eventually be responsible for saving millions of human lives and untold suffering. The potential problems presented by misuse of these techniques seem inconsequential in comparison.

Modifications of Normal Prenatal Development

Since the publication of Aldous Huxley's *Brave New World*, the specter of the mass production of embryos and fetuses, carefully designed for specific purposes, has caught the popular fancy. This area of biological engineering has several different aspects.

EXTRAUTERINE DEVELOPMENT.   Strictly speaking, the phrase "test-tube baby" should be applied to development outside of the mother from the stage of the fertilized egg to that of the newborn. The contribution of the mother to the embryo and fetus during development is so substantial and complex that it may very well be impossible to duplicate these conditions outside the womb. Not only are simple nutrients provided by the mother (which might easily be supplied), but also more complex proteins, such as antibodies (which might be provided with some difficulty), and a set of physical conditions appropriate for development are found in the womb that may simply not exist elsewhere. It is not just a question of the warmth and oxygen supply, which could be duplicated easily, but also of the precise cellular structure of the uterus that is responsible for specific physical and chemical signals during development. If the embryologist should reach the level of sophistication that would make it possible to reproduce these exactly (at a cost of at least millions of dollars per embryo), the question would immediately arise as to whether the effort was worthwhile, since the entire procedure is being accomplished quite efficiently naturally at relatively little expense. It is difficult to imagine circumstances, in civilization as we know it or as we project its course in the near future, that would justify developing this line of research as an end in itself.

EXTRAUTERINE FERTILIZATION.   The fertilization of an egg by a sperm in vitro, with the return of the zygote to a female for further development, after a few cleavage divisions, will be much less difficult to accomplish. The fertilized egg may be expected to survive the trauma of the test tube for a few days, provided that it is returned reasonably promptly to its natural abode. It might, however, be implanted into a foster mother rather than into the woman from whom the egg came.

   Such a procedure is not farfetched, and is not beyond the capacity of science at present. It has, in fact, been accomplished experimentally in mice and rabbits. In cattle, the production of large numbers of progeny quickly from a single pair using such foster mothers would be of enormous economic value. In the case of humans, the procedure might have some usefulness in cases in which an otherwise fertile woman was unable to bring a child to term.

EUPHENICS.   Whereas *eugenics* is aimed at improving the genotype, *euphenics* is the improvement of the phenotype. This might be done, for instance, by modifying the course of development of the embryo in utero to produce somewhat "better" individuals. It has been suggested, for example, that the cells giving rise to the neurons of the brain might somehow be induced to undergo (by some as yet unknown treatment) just one more mitosis during early development. The number of brain cells could then be doubled, and, it might be anticipated, the effect on intelligence would be striking. Some people, but probably not many, would object to some simple treatment of the embryo of fetus in utero that would promote the development of a much higher intelligence of the newborn.

An increase in apparent intelligence brought about by treatment of the fetus in utero may have been inadvertently accomplished already. The mechanism by which this comes about is not clear, but women who are subject to miscarriage may be treated with male sex hormones to suppress the physiological reactions responsible; when girls are born, they sometimes suffer from evidence of masculinization. Such girls also show, as a group, an increased ability to perform well on IQ tests, a large proportion being in the class of the quite superior (this result being astonishing because of the well-known fact that girls do better on the whole than boys on tests).

Selection at the Level of the Organism

EUGENICS.   Programs of negative eugenics that would alter the genetic composition of humans by interfering with the right of individuals to choose mates or to have offspring, or with the right of every offspring to be allowed to live, are often regarded as unacceptable because of the conflict with the ideal of freedom of choice. To be sure, these rights are limited at the present time. Congenitally feeble-minded individuals are generally put into institutions where they are prevented from marrying and producing offspring. However, this may be done from concern over the welfare of the children born under such conditions rather than from a wish to prevent the perpetuation of any deleterious genes. A dramatic demonstration of the existence of genetic factors—fortunately fairly rare—responsible for subnormal mentality is given by the case, found in almost any mental institution, of two parents clearly in the moderately retarded category, along with their similar children.

Although it might seem self-evident at first glance that negative eugenics might reasonably be practiced only in extreme cases, there is still cause for concern. That concern centers on the possibility that once criteria are set up that allow individuals in certain classes to be deprived of their reproductivity the criteria can then be adjusted to fit the prejudices of those who happen to hold power at any given time. Positive eugenics, which involves the promotion of more than average numbers of offspring from individuals considered to be superior, is faced with an equal criticism, since the definition of superior can be adjusted to suit the fancies of those who happen to be making the rules at the time.

CARRIER DETECTION.   The *detection of homozygotes* in utero provides a means whereby grossly defective embryos may be aborted, if this is acceptable to the parents. Unfortunately, heterozygous parents are usually not identified as such until after an abnormal offspring has been produced, and if an aborted embryo is later replaced by a normal child, that child has a two thirds chance of being heterozygous. Thus, although monitoring for homozygotes by amniocentesis may decrease the frequency of homozygotes born, it will increase the frequency of the defective allele in the population because in two thirds of the cases the poorly reproductive homozygotes will be replaced by normally reproductive heterozygotes.

Clearly, if all heterozygotes could be identified, and if they either chose not to have any children or chose not to have heterozygous children, any defective gene could be eliminated in one generation (except for rare new mutations). In theory this is possible, in practice not. Even if all heterozygotes, e.g., for Tay-Sachs disease, could be identified, it would be difficult to persuade all of them that they should not have children because half of their offspring, who would be quite normal, would be heterozygous for this defective gene. It would be even more difficult to persuade such heterozygotes of the advisability of a therapeutic abortion of a phenotypically normal embryo because it was heterozygous for a defective allele. The destruction of phenotypically normal embryos heterozygous for an allele that, in the homozygous state, produces a serious defect is an ethical question that has not yet been faced explicitly by society; it would probably be considered an unacceptable procedure at the present time. Be that as it may, it is a curious contradiction that large sums are spent looking for cures for many genetic diseases—which, when eventually found, will almost certainly turn out to be incomplete solutions—when in fact a theoretical solution is now at hand in the detection of heterozygotes, along with a program to educate them to the undesirability of perpetuating the defective alleles.

Such an approach may seem inconsistent with the knowledge that every person probably carries between five and ten deleterious recessive alleles. If we were to suggest that all of those with defective alleles not produce offspring, would this not logically include every member of the human race? Not quite. Many of the defective alleles we carry are recessives that act early in the development of the homozygote. Since they cause no deaths of newborns and no stillbirths, and, in fact, may cause loss so early that it is undetected by the mother, they cause no social or personal trauma. At the other end of the spectrum are those with nonspecific effects, detectable as an overall loss of vitality only, when progeny of consanguineous marriage are compared to progeny at random in the population. These also are unmanageable in the sense that we cannot identify persons with these alleles. What are left are the genetic diseases with serious debilitating effects to varying degrees in the homozygotes. If the human could be manipulated genetically like animal or plant breeding stock, there would be no problem in removing the heterozygotes for the most serious medical diseases from the breeding population, without undue complications to its total reproductive capacity.

At the present time, concern with the impending disaster posed by an ever-expanding population has led many thoughtful persons to curb their reproductivity; many young people have decided simply not to have any children at all. From every point of view, these are precisely the groups on whom society should depend for the production of the next generation. If population size is to be limited, it would seem better that it be done on the basis of some randomly occurring genetic defect than by the voluntary self-sterilization of whole groups of sensitive and thoughtful youth.

Intervention in the Operation of the Somatic Cell

ENZYME THERAPY. In some cases the question arises as to whether or not a simple genetic defect caused by an enzyme deficiency might be ameliorated by injecting the missing enzyme. This possibility has great appeal and is very often used as an argument against the application of any measures that might be used to decrease the frequency of presumably deleterious genes in the population. The classic example in such discussions is diabetes and its treatment by insulin. How unnecessary, goes the argument, it would have been, many years ago, to have tried to eliminate any alleles predisposing a person to diabetes when subsequent work uncovered the cure in insulin. If this is the best example of *enzyme therapy*, it is not a particularly good one. Many cases of diabetes are not manageable by insulin, and, when they are, the treatment by insulin may not provide a complete cure. To be sure, the health of diabetics has been enormously improved by treatment with insulin, but the fact remains that on the average they are not as well as normal people, nor do they have as long life expectancies. Similarly, antihemophilic serum is of great help to hemophiliacs, but does not actually allow them to lead completely normal lives. Although such therapy is of inestimable value to the affected person, there is no doubt that, given a choice, that person would vastly prefer not to have the disease. Similarly, in considering a "cure" for sickle-cell anemia, it should be kept in mind that, if our experiences with other comparable diseases can serve as a guide, any cure will only relieve the condition to some extent, but not really cure it.

Furthermore, such therapy allows for increased reproductivity of the affected individuals, so that the defective allele may no longer be subject to the negative selection that is ordinarily thought to hold it at a low level in the population.

Figure 26-1 The anticipated consequences to the human morphology after introduction of the gene for wings from *Drosophila.*

This illustrates a dilemma that plagues medicine at the present time—that the effective treatment of the symptoms of individuals with genetic diseases may lead to their perpetuation in the population as a burden to future generations.

There are other problems associated with enzyme therapy. One is that in the vast majority of genetic disabilities an enzyme defect has not been found, or has been only tentatively identified, if at all. Purification of the enzyme, if possible, may be costly, and whether society will be willing to underwrite the expense involved may depend on its perception of the benefits to be gained in relation to the sums expended. Furthermore, enzymes very often perform their normal function in specific cellular environments and not in the bloodstream, where they would appear after injection. And finally in many cases the recipient would react to these foreign proteins by becoming sensitized to them. This would prevent their further use. So what seems like a simple straightforward way of handling a genetic problem in fact may be difficult, and in many cases impossible.

GENE REPAIR BY BASE MANIPULATION.   Gene repair has very often been suggested as a means of using the modern advances of molecular biology for the benefit of the human race. The vision of altering a defective gene in such a fashion as to enable it to carry on its normal function is one that has caught the attention of the public. However, this goal is very far from realization. At the present time, we do not even know how the mammalian chromosome is put together, to what extent the chromosomal proteins are involved, and how they are associated with the DNA. We do not know, even approximately, the location of the genes responsible for the most serious genetic diseases, and, even if we did, virtually all of the very important human characteristics, insofar as they are genetically determined, must have a polygenic and developmental nature with complicated interactions that will defy analysis for a great many years.

The Insertion of Foreign Genes

There is no reason that genes not normally found in human cells (or not found in some persons) cannot, in principle, be introduced from other organisms, with potentially beneficial results (Figure 26-1). As an example, humans lack a single enzyme needed to manufacture ascorbic acid (commonly referred to as vitamin C). Although the race manages to survive on the small amounts present in the diet, some workers have presented strong arguments that the optimal amounts for well-being and disease resistance are very much larger than those now commonly being taken by the average person. If this is the case, the insertion of the necessary gene into our genetic complement might have immediate beneficial effects. This insertion of genetic material might be accomplished in any one of a number of ways.

GENE ACTIVITY PROVIDED BY A MICROORGANISM.   One system currently under consideration involves the introduction of the genetic material from a virus into a defective human cell. According to this idea, if the virus is capable of carrying on a reaction that the cell cannot, its activity may substitute effectively for the missing one. There are some indications that blood enzyme levels may vary depending on the presence of virus. Activity of the enzyme thymidine kinase, found in the blood, has been increased by infection by herpes simplex, the common virus that causes cold sores. Another virus, Shope papilloma, has decreased the amino acid arginine in the blood of persons exposed, possibly by providing the enzyme that breaks it down.

There are several problems with this approach. One is that we do not yet understand how genes are controlled—turned on and off in some tissues and not in others—in higher organisms, and the deliberate injection of a virus amounts to the insertion of an unknown genome, uncontrolled in its activity and uncontrolled in its reproduction. Also, the recipient may develop antibodies against any such organisms, reducing the long-range effectiveness of any such treatment. But the most serious problem is the real possibility that such treatments might have a long-range carcinogenic effect—not detectable until many years after the treatment.

MICROORGANISMS AS CARRIERS OF NORMAL GENES. Normal genes may be introduced into cells lacking them by using viruses as the carrying agent. If the virus does not itself possess the desired gene, it is possible to add it to the virus from another organism. In principle, then, such a modified microorganism could be introduced into the affected person, with some hope that enough of the appropriate cells might incorporate the virus to restore normal enzymatic activity.

If we succeed in creating "new" microorganisms by inserting into them genetic material that has never been part of their genome in all of their previous history, there will always exist the possibility that something exceptionally virulent will be accidently produced. In the same way that we must rationally consider the problems that could arise if microorganisms of extraterrestrial origin were to suddenly appear on earth, we must also think of possible damage that would occur if uniquely new infective genomes, to which humans have never before been exposed, were to appear suddenly in the laboratory. In the tradition of the best science fiction, such a new genome could have disastrous consequences before isolation or some therapeutic measure could be devised.

An additional problem in such an approach is that one of the organisms commonly used in experiments involving gene transfer is the bacterium *Escherichia coli*, which commonly lives in the human intestinal tract. Clearly the opportunity for infection would be unlimited. The potential dangers in the introduction into common microorganisms of new characters such as resistance to antibiotics, production of poisonous substances (toxins), and induction of tumors have concerned many scientists in the field, who have voluntarily decided to limit future work in these areas except where adequate safeguards against any calamity can be maintained.

The Sanctity of the Gene Pool

Sometimes an argument is made against the elimination of defective alleles because to do so would alter the allele frequencies in the gene pool, variability upon which the species must rely for its ultimate survival. This argument is particularly appealing to those whose humanitarian instincts rebel against the suggestion of any limitation of human freedoms by curtailing reproductivity.

However, the extent to which our ultimate survival will depend on the composition of the gene pool is debatable. Even for those situations in which new capabilities might be created by mutational events technological skill will usually provide a swifter and more effective solution. How many mutations accumulated over how long a period of time would it take to change us so that we might fly, unaided, at the speed of sound at 50,000 ft? Or make a trip on our own to the moon? On the negative side, if our extinction becomes a possibility within the foreseeable future, we will probably be the instrument of our own destruction, and not the unwilling and unresponsive victim of some relatively minor environmental change.

In any case, it is very unlikely that the state of humanity will ever descend to the level at which our survival as a species might depend on low intelligence (as in homozygosity for PKU) or resistance to malaria (sickle-cell trait). Certainly one cannot justify the retention of alleles for Tay-Sachs disease, in which the homozygotes die miserable deaths at a few years of age, on the grounds that maybe, someday, this allele—perhaps in combination with others—will prove to be valuable. To condemn thousands of humans now and in the immediate future to a subnormal existence in favor of some future unknown, probably non-existent, advantage thousands of years hence seems a curious perversion of humanitarian ideals.

Summary

The preceding is a list of some of the "advances" that require public consideration—if not now, then within the next century. Geneticists sometimes allow their imaginations to run wild and in these moments of fancy make statements that might better come out of science fiction. For instance, there is the suggestion that genetic engineering might produce an organism with a large brain, so that it might philosophize, and with a patch of chlorophyll under its skin, so that it might photosynthesize. Such whimsical humor is appreciated by biologists, but when read by a nonbiologist it is apt either to be taken too seriously, as an immediate threat announcing the doom of humanity, or, at the other extreme, to be dismissed as the lunatic ravings of the mad scientist. In the meantime, the biologist is caught between two opposing forces, criticized on the one hand for progressing too rapidly, to the immediate threat of our existence, and on the other for not moving fast enough, as shown by the failure to discover "the cure for cancer." Perhaps some day the general public will understand that biological progress is certain even if its direction is unpredictable and, if applied wisely, must inevitably improve the lot of humanity.

References

ABELSON, P. 1971. Anxiety about genetic engineering (editorial). *Science,* **173:**285.

DAVIS, B. D. 1970. Prospects for genetic intervention in man. *Science,* **170:** 1279–83.

EDWARDS, R. G., and R. E. FOWLER. 1970. Human embryos in the laboratory. *Sci Am.,* **223**:44.

FOX, M. S., and J. W. LITTLEFIELD. 1971. Reservations concerning gene therapy *Science,* **173**:195.

FRIEDMAN, T., and R. ROBLIN. 1972. Gene therapy for human genetic disease? *Science,* **175**:949–55.

GURDON, J. B. 1968. Transplanted nuclei and cell differentiation. *Sci. Am.,* **210**:24–35.

LEDERBERG, J. 1970. Genetic engineering and the amelioration of genetic defect. *BioScience,* **20**:1307–10.

RABOVSKY, D. Molecular biology: gene insertion into mammalian cells. *Science,* **174**:933–34.

REVERS, C. 1972. Genetic engineers portend a grave new world. *Sat. Rev.,* **55**:23–27.

TUNNEY, J. V., and M. E. LEVINE. 1972. Genetic engineering. *Sat. Rev.,* August, 23–28.

ROGERS, S. 1970. Skills for genetic engineers. *New Sci.,* **45**:194–96.

WATSON, J. D. 1971. The future of asexual reproduction. *Intellectual Digest,* 69–74.

Questions

Useful terms: genetic engineering, eugenics, euphenics, cloning, carrier detection.

1. Can you give some simple examples of biological engineering, as it is broadly defined?
2. Would the possibility of predetermining the sex of offspring prior to birth represent a great advance in the control of human destiny?
3. Discuss the idea of sperm banks, with the pros and cons of their application.
4. In an apparently bona fide case of parthenogenesis in humans, what tests would you suggest to ascertain the likelihood of the phenomenon? List the tests in the order of their simplicity or feasibility.
5. Why has the prospect of cloning created considerable apprehension?
6. If amniocentesis leads to the detection of a defective fetus which is then not allowed to come to term, would the resulting decrease in the number of homozygotes for the recessive in question lead to a decrease in the frequency of the defective allele in the population?
7. Since every person carries somewhere between five and ten defective genes at least, is it reasonable to try to decrease the frequency of defective genes in the population? Would this logically require that no person produce offspring?
8. List a number of problems that may arise in the simplest application of genetic engineering, that of enzyme therapy.
9. How might new genes be introduced into the human germ plasm, and what beneficial effects might result from such an addition? Are there any problems attendant on such a procedure?
10. Would man's chances of survival in the long run be increased if his genetic variability were increased by exposing his germ cells to low doses of radiation to increase the mutation rate?

Index

When there are several pages per entry, **boldface** is used to indicate a definition or major discussion, if any. *Italics* signify illustrations.